Transboundary Environmental Impact Assessment in the European Union

Transboundary Environmental Impact Assessment in the European Union

The Espoo Convention and its Kiev Protocol on Strategic Environmental Assessment

Edited by Simon Marsden and Timo Koivurova

earthscan
publishing for a sustainable future
London • New York

First published 2011
by Earthscan
2 Park Square, Milton Park, Abingdon, Oxon OX14 4RN

Simultaneously published in the USA and Canada
by Earthscan
711 Third Avenue, New York, NY 10017

Earthscan is an imprint of the Taylor & Francis Group, an informa business

Earthscan publishes in association with the International Institute for Environment and Development

At Earthscan we strive to minimize our environmental impacts and carbon footprint through reducing waste, recycling and offsetting our CO_2 emissions, including those created through publication of this book. For more information on Earthscan publications, see www.earthscan.co.uk or write to earthinfo@earthscan.co.uk

British Library Cataloguing in Publication Data
A catalogue record for this book is available from the British Library

Library of Congress Cataloging in Publication Data
Transboundary environmental impact assessment in the European Union : the Espoo Convention and its Kiev Protocol on Strategic Environmental Assessment / edited by Simon Marsden and Timo Koivurova.
p. cm.
Includes bibliographical references and index.
ISBN 978-1-84971-348-1 (hardback)
1. Environmental impact analysis—Law and legislation—European Union countries. 2. Transboundary pollution—Law and legislation—European Union countries. 3. Convention on Environmental Impact Assessment in a Transboundary Context (1991) I. Marsden, Simon. II. Koivurova, Timo.
KJE6242.T728 2011
344.2404'6—dc22 2010047931

ISBN: 978-1-84971-348-1 (hbk)

Typeset in Garamond
by Domex e-Data Pvt Ltd
Cover design by Susanne Harris
Printed and bound in the UK by CPI, Antony Rowe.

The paper used is FSC certified.

MIX
Paper from responsible sources
FSC® C013604

Contents

List of Figures, Tables and Boxes

Figures

Tables

Box

List of Contributors

Simon Marsden is an Associate Professor at Flinders Law School in Adelaide, South Australia. He specializes in strategic and transboundary approaches to environmental impact assessment and the relationship between legal systems.

Timo Koivurova is a Research Professor and Director of the Northern Institute for Environmental and Minority Law, Arctic Centre, University of Lapland in Rovaniemi. He specializes in various aspects of international law applicable in the Arctic and Antarctic.

Jan de Mulder is an international policy advisor with the Public Governance Department of the Flemish government and also works at the Law Faculty and Maritime Institute at the University of Ghent in Belgium.

Ismo Pölönen is a Lecturer at the Department of Law, University of Eastern Finland. He specializes in legal issues relating to environmental impact assessment.

Gernot Stoeglehner is an Associate Professor at the Institute of Spatial Planning and Rural Development, University of Natural Resources and Applied Life Sciences Vienna, Austria. His research interests focus on environmental planning and assessment, including sustainable spatial planning and development, sustainability indicators as well as the energy and spatial planning link.

Gerald Schmid (formerly **Wegerer**) studied at the University of Natural Resources and Applied Life Sciences Vienna and the University of Fribourg, Switzerland in the field of Environmental Legislation in 2005. From 2005 to 2010, he worked at a civil engineer's office in the field of spatial planning and strategic environmental assessment.

Teresa Fidelis is an Assistant Professor in the Department of Environment and Planning at the University of Aveiro in Portugal. She conducts research into local and regional environmental planning and management.

Rita Albergaria currently manages a firm in the theme of environment and renewable energy, highlighting the work on solar energy. She has also done work in the promotion of integrated energy efficiency.

Acknowledgements

The genesis of this book was in discussions in 2010 between the two of us about ways of raising the profile of the Espoo Convention on the twentieth anniversary of its adoption in 2011; the coming into force of the Kiev Protocol in 2010 contributed an additional dimension to this. We decided to focus on application in the European Union primarily because of the largely comprehensive implementation by transposing legislation and extensive practice, and also reflecting our own interests. We approached the other contributing authors to add material not available to us and to diversify the legal focus of our own work. We would like to thank each of them for joining us in this project.

We are very grateful to our publisher for agreeing to take on the project in the first place and for seeing it through within the timeframes that we had originally agreed, publication by the 5th Session of the Meeting of the Parties to the Convention in June 2011, now also serving as the 1st Session of the Meeting of the Parties to the Protocol. We would also like to thank the three original reviewers of the proposal for the benefit of their wisdom in improving this for a wider audience, to interested colleagues for their support, encouragement and endorsement, and significantly to the Secretary of the Convention for preparing the Foreword.

Acknowledgement is due to the publishers of the previously published material for their agreement to reproduce this work in this edited book, the original source of which is indicated at the start of the notes to each chapter.

Simon Marsden, Law School, Flinders University of South Australia
Timo Koivurova, Northern Institute for Environmental and Minority Law, Arctic Centre, University of Lapland
April 2011

Foreword

The Convention on Environmental Impact Assessment in a Transboundary Context was adopted and signed 20 years ago in February 1991. The Convention was adopted by a meeting of the Senior Advisers to Governments on Environmental and Water Problems, held in the Finnish city of Espoo, and it became known as the Espoo Convention. It had been negotiated in the late 1980s under the United Nations Economic Commission for Europe (UNECE), a regional commission with members stretching from Canada and the US in the west to the Soviet Union – as it was then – in the east. It was signed by many states including the Soviet Union and the US.

Though the Russian Federation succeeded to signature of the Convention, it and the US have yet to ratify the Convention – ratification (or approval, acceptance or accession) is the essential step for a country to join the Convention and to be legally bound by it. However, the European core of the UNECE region has joined the Convention, with few exceptions. All 27 Member States of the European Union (EU), the EU as a 'regional economic integration organization', the countries of Eastern Europe (Belarus, Republic of Moldova, Ukraine) and south-eastern Europe (from Croatia to Albania), as well as Liechtenstein, Norway and Switzerland, have become Parties. Armenia, Azerbaijan, Kazakhstan and Kyrgyzstan, from the Caucasus and Central Asia, as well as Canada, have also joined so that by late 2010, the Convention had 45 Parties.

The Convention thus provides legal implementation for one of the principles of the 1992 Earth Summit in Rio de Janeiro: 'States shall provide prior and timely notification and relevant information to potentially affected States on activities that may have a significant adverse transboundary environmental effect and shall consult with those states at an early stage and in good faith'.

The Convention has been applied over 700 times to date, to activities planned in one Party likely to have an impact on the environment of another Party, and the vast majority of those cases have involved at least one EU member state, or a neighbouring state that has adopted EU environmental legislation.

This book, therefore, though focusing on transboundary environmental impact assessment in the EU, provides welcome insights into the legal and administrative systems that have framed the vast majority of practical applications of the Espoo Convention. It is important to recognize, though, that the book does not investigate the application of the Convention in an important but smaller group of countries: those countries that have inherited systems based on the Soviet Union's approach of state ecological expertise. However, this book does cover those important situations where one Party is applying EU legislation and another is not.

Nonetheless, this book should help developers, researchers, students and government officials understand how the Espoo Convention works in practice, and how governments have integrated it into their legal systems.

The Convention has now been supplemented by a Protocol on Strategic Environmental Assessment, which was adopted and signed in Kiev in 2003, entered into force in July 2010 and by later that year had 20 parties, again including the EU as a regional organization. The Protocol was negotiated once the similar EU Directive had been concluded and the Protocol text was heavily influenced by that of the Directive. The Protocol and the Directive apply similar procedures of environmental assessment to certain plans and programmes irrespective of whether they are likely to have significant transboundary environmental effects.

However, the Protocol has important provisions beyond the EU legislation with: a stronger emphasis on the role of health authorities (reflecting the input of the World Health Organization in the negotiations); a requirement that Parties 'endeavour to ensure that environmental, including health, concerns are considered and integrated to the extent appropriate in the preparation of its proposals for policies and legislation that are likely to have significant effects on the environment, including health'; and, not least, the opportunity for all Member States of the United Nations to join the Protocol, whether or not Parties to the Convention, and whether or not members of UNECE.

There is little literature dealing with the legal implementation and practical application of the Protocol on Strategic Environmental Assessment, so this book is a welcome initiative.

Nick Bonvoisin[1] Secretary to the Espoo Convention
Geneva, December 2010

1 The views expressed in this foreword are entirely those of the author and these views should not be taken out of context or cited without prior permission. They do not necessarily reflect the views of the UNECE, its Member States, or the parties to the Espoo Convention or to its Protocol.

Preface

Other than the Introduction and Part I, containing new chapters by the contributing editors and up to date as at the start of 2011, this book is based upon previously published work appearing in a number of legal and policy science journals. Summaries of these chapters from Parts II, III and IV follow, highlighting their continuing significance. Formatting has been changed and references to 'this article/paper' have been amended to 'this chapter' for consistency. The content of this work has however not been changed or updated, although most is very recent in origin, a key criterion in selecting it for inclusion.

Chapter 4, 'The Case of Vuotos: Interplay between International, Community and National Environmental Law', Timo Koivurova

This chapter studies the so-called Vuotos case, which has been one of the longest and best-known environmental struggles in Finland. The initial plans for the Vuotos Reservoir, or, as it was known then, 'the Kemihaara project', were completed in the 1960s, with the reservoir to be located in the municipalities in the northern Finland. The project was suspended for environmental reasons in 1974, and in 1982, but the project was resumed as the 'Vuotos project' in the mid-1980s, and the necessary preparatory work had been completed when Kemijoki Ltd applied to the Water Court of Northern Finland on 25 September 1992 for a permit to construct the Vuotos Reservoir and dam.

Transboundary Environmental Impact Assessment (TEIA) played an interesting and important role in the project, given that Swedish officials and the civil society were concerned about the possible effects of the construction of the reservoir on the Gulf of Bothnia, especially Swedish maritime areas. The Nordic Environmental Protection Convention (NEPC) – a treaty that was in force for the parties and contained an embryonic TEIA procedure – was intentionally brushed aside by the Finnish Ministry of the Environment, which notified Sweden of possible phosphorus and heavy metal effects if the artificial lake would be constructed in the Gulf of Bothnia on the basis of Article 3 of the Espoo Convention, even if the Espoo Convention had not entered into force at the time in 1995. This clearly showed the strength of the Espoo Convention, even before its entry into force, to influence state practice.

The case also highlighted the changed legal environment in the country, inasmuch as international and European law very much influenced how it was finally resolved. Although the Finnish Water Act was invoked by the courts as the main basis for their decision, Vuotos is a good example of how much the legal environment of Finland has changed, for the case hinged on not only the Finnish Water Act but also international and Community law. All these three levels of law were also taken into account by the three levels of administrative courts that needed to decide on the project's future, the Supreme Administrative Court ending up rejecting permit for the project.

The chapter argues that the Finnish administrative judiciary was able to apply both international law and Community law in a legally ideal manner, especially in view of the complexity of the case. Finland notified Sweden of the potential negative effects on the latter's maritime area even before the legally valid convention, the NEPC, required it to do so. Finland also allowed the Swedish private and public parties to the Vuotos case to participate in its national licensing procedure on the basis of the NEPC. Moreover, even if the Supreme Court made its decision on the basis of Article 2 (5) of the Finnish Water Act, the factual influence of the Wild Birds and Habitat Directives to the decision was, arguably, strong.

CHAPTER 5, 'MOX PLANT AND THE ESPOO CONVENTION: CAN MEMBER STATE DISPUTES CONCERNING MIXED ENVIRONMENTAL AGREEMENTS BE RESOLVED OUTSIDE EC LAW?', SIMON MARSDEN

The MOX Plant case was decided by the European Court of Justice (ECJ) in 2006. The dispute between Ireland and the UK that formed the basis for the case had earlier been considered by international tribunals under the OSPAR and UNCLOS Conventions, which contain provisions for protection of the marine environment of the North East Atlantic and the sea generally. The ECJ case concerned the relationship between international and European law, specifically whether EU Member States could take their disagreements to courts or tribunals outside the EU legal system when they had ratified agreements such as OSPAR and UNCLOS. Although the situation would have been different if the disagreement had been between an EU Member State and a non-EU Member State, for disputes between EU Member States, the ECJ decided this was a breach of the European Treaties and was not permitted.

The relevance of the case to the Espoo Convention is threefold: First, while a claim was not made by Ireland against the UK under the Espoo Convention (because Ireland was only a signatory to the treaty), the nature of the dispute concerned the likely significant transboundary environmental effects from a nuclear power facility in the Irish Sea. There was therefore potential for the Espoo Convention to determine this through the application of the Inquiry Procedure. This has only been applied once, in the context of the Danube Delta case (discussed in the chapter). Second, despite not being a party to it, Ireland cited the Espoo Convention as relevant contextual law in its arguments before the international tribunals to which it took its complaints. Third, because the Convention had been ratified by the EU, there was potential for the Procedure to have been directly applied by the EU. This would mean that it would affect the relations between Ireland and the UK regardless of whether it had been transposed fully by EU law, because it would bind them as EU Member States.

Together with the relevance of EU secondary law to the dispute, such as the access to information and Environmental Impact Assessment (EIA) Directives, which was also not argued in its dispute with the UK and which had transposed the majority of the provisions of the Espoo Convention, the application of the Inquiry Procedure is therefore considered in this chapter. Both at a substantive level (with reference to the Bystroe Canal Project, the 'Danube Delta' case), and from the procedural perspective of its incorporation into the EU legal system, (discussed in more detail in Chapter 3), it is argued that since the ECJ decided that Ireland could not take its dispute with the UK to an international tribunal, this may also have prevented Ireland taking its case to the Inquiry Procedure. This is because of the possibility that the Inquiry Procedure could also be considered to be a form of dispute resolution procedure, application to which could be held to be in conflict with the European Treaties.

CHAPTER 6, 'ENFORCING NON-DISCRIMINATION IN TRANSBOUNDARY ENVIRONMENTAL IMPACT ASSESSMENT: ADVANTAGES FOR EU CITIZENS FROM THE TRANSPOSITION OF THE ESPOO AND AARHUS CONVENTIONS?', SIMON MARSDEN

The relationship between the Espoo Convention and EU law has been affected by the operation of the Aarhus Convention on Access to Information, Public Participation in Decision-making and Access to Justice in Environmental Matters. This chapter examines the role of citizens in affected states to input the procedure in origin states, specifically whether there is a difference between the opportunities available to those in EU affected states and those in non-EU affected states. The principle of non-discrimination provides these opportunities in relation to the procedure of the state of origin regardless of the state of residence. The Aarhus Convention has also affected the provisions for strategic environmental assessment (SEA) in EU law, in particular provisions for public participation in relation to plans and programmes. SEA is examined in the final part of this book.

The non-discrimination principle is recognized by the Espoo and Aarhus Conventions and the Treaty on European Union. It therefore permits individuals and non-governmental organizations (NGOs) to exercise rights of access to information, public participation in decision-making and access to justice in another state in the context of a TEIA procedure operated by the other state. The chapter challenges the argument that an absence of an enforceable context for TEIA makes the procedure less effective than a domestic EIA procedure by also analysing the enforcement opportunities in the EU legal system, which relate to the third pillar of the Aarhus Convention, access to justice. These include the role of the European Commission as 'guardian of the Treaties' and the ECJ in respect of infringement proceedings against Member States allegedly in breach.

The chapter examines the objective and application of the Espoo and Aarhus Conventions, the non-discriminatory procedures within them and treaty reviews of implementation and compliance that have analysed their effectiveness. It also evaluates the implementation of both treaties by EU law, specifically the EIA and Public Participation Directives, reviews of implementation and compliance by the European Commission, and the role of the Commission and ECJ in securing access to justice. While direct access to the ECJ remains problematic for individuals and NGOs given the restricted standing provision in the Treaty on the Functioning of the European Union, subject to Member State access restrictions, references for preliminary rulings may be possible via national courts. Complaints can also be made directly to the European Commission, and petitions can be issued to the European Parliament.

CHAPTER 7, 'TRANSBOUNDARY EIA: IBERIAN EXPERIENCES', RITA ALBERGARIA AND TERESA FIDELIS

Experiences with TEIA have been discussed progressively in the literature following the coming into force of the Espoo Convention and its implementation by EU law. This chapter considers application of the Convention and EIA Directive in respect of the Sela and Alqueva Dams, which were water management projects proposed in the late 1990s. They continue to highlight the significance of transboundary watercourses in international law, the diversion, damming or pollution of which has been examined by international courts on many notable occasions including the *Lac Lanoux Arbitration* (France v. Spain) (1957), the *Gabcikovo-Nagymaros case* (Hungary v. Slovakia) (1997) and most recently the *Pulp Mills on the River Uruguay* (2010). The actions taken by Ukraine in the *Bystroe Canal case* which were challenged by Romania under the Espoo Convention and several other international agreements, also demonstrate the potential conflicts in this area.

Both of the dams on the Iberian Peninsula were subject to the Luso-Spanish International Convention 1968, which contained provisions for hydraulic development in the international basin of the Guadiana River, where the dams were due to be built. The chapter summarizes provisions of the Convention and Directive before examining the case studies presented, which are analysed in accordance with them. Procedural strengths and weaknesses are identified, and a good practice model for TEIA to apply in the relationship between the two states is considered, which aims to comply with the requirements of both the Convention and Directive.

Strengths identified include the evaluation of the Environmental Impact Statement (EIS) and promotion of public participation by a bilateral team who also made the final decision (Sela Dam). Impact identification was also carried out by joint working groups before the EIS was approved. In relation to the Alqueva Dam, inter-ministerial documentation was exchanged allowing the affected state to evaluate the significance of any impacts. The EIS was also carried out by a team independent of the proponent. Weaknesses identified in the Sela Dam process included the failure to involve regional and local entities at an early stage, and the need for the Spanish public to go to Portugal to take part in hearings; other problems identified were late public consultation and the conduct of the EIS falling to the proponent. In relation to the Alqueva Dam, bilateral cooperation was lacking in the evaluation of transboundary effects, and alternatives and mitigation measures were not considered. Working groups were also not involved at a regional level and local bodies and the affected public was not involved at an early stage, with opinions not incorporated into the EIS.

Based on these findings, a good practice model is developed to improve future application. This integrates in a single procedure the bilateral consultation process for EIA between the two states, aiming to comply with all requirements of the Convention and EU Directive(s). A permanent bilateral body is proposed to facilitate relations and ensure greater efficiency. The position of the national and regional bodies is also clarified.

Chapter 8, 'Transboundary Environmental Assessment in the Arctic', Timo Koivurova

This chapter examines the extent of applicable international treaties and other regulations that oblige the concerned states to perform TEIA in the Arctic, and whether these have induced any state practice. It will be shown that there are few international treaties providing for TEIA in the Arctic, most prominently that of the Espoo Convention. Even if the Arctic-wide cooperation between the eight Arctic states (the Russian Federation, the United States, Canada and the five Nordic States) – starting with the 1991 Arctic Environmental Protection Strategy, which was merged into the 1996 Arctic Council – promoted the swift ratification of the Espoo Convention, three Arctic states are still to become parties to the Convention, the USA, Russia and Iceland. Iceland is candidate country of the EU and if accepted as a Member State it will be bound by the Convention regardless of whether it becomes a party in its own right.

More importantly, the cooperation process between the eight Arctic states has adopted an instrument aiming to influence how EIA and TEIA should be undertaken in the Arctic conditions, Guidelines for Environmental Impact Assessment in the Arctic, which was explicitly based on the Espoo Convention. Unfortunately, as will be shown in the chapter, these guidelines have not been made use of by the Arctic states, even if in the 1997 Alta ministerial meeting they agreed to apply these Guidelines in practice. The chapter will contemplate some of the reasons for this, and illustrate the problems with a case-study in the Finnish Arctic.

The Arctic is changing rapidly. As the 2004 Arctic Climate Impact Assessment and more recent studies and assessments have shown, the Arctic can be seen as the barometer of climate change. We can already witness the consequences of climate change in the region simply because snow and ice react very swiftly to climate change and these consequences will be more intense in the Arctic. Together with accelerating economic globalization, this will mean vast transformation for the Arctic region. The previously inaccessible region, and its vast natural resources, will be open for exploitation. This type of change calls also for quality EIA and TEIA procedures, and the argument will be made that it is of much importance to have the three Arctic states still non-parties to the Espoo Convention to ratify the Convention and commence the work in the Arctic Council to revise and further develop the Guidelines into a living instrument.

Chapter 9, 'Transboundary Environmental Impact Assessment in the Case of the Baltic Sea Gas Pipeline', Timo Koivurova and Ismo Pölönen

The Baltic Sea Gas Pipeline (BSGP) was proposed by the Nord Stream Company in 2006 to run from Portovaya Bay near Vyborg on the Russian coast of the Gulf of Finland to Greifswalder Bodden in Germany. The BSGP is the longest offshore pipeline ever planned and the TEIA procedure applied to it proved to be the most challenging TEIA procedure ever conducted by the international community; the procedure involved five origin and nine affected states. The project was of major importance to the EU, given that out of the nine littoral states to the Baltic Sea, only one, Russia, is not a Member State. Moreover, Russia is not yet a party to the Espoo Convention, but agreed to apply the Espoo Convention to the project as far as permitted by its national legislation.

As the BSGP passes through the jurisdiction of four contracting parties (Finland, Sweden, Denmark and Germany) of the Espoo Convention, it was clear that the Convention had to be applied to the project. The scope of the Convention covers large-diameter gas pipelines that are likely to cause significant adverse transboundary impacts. However, the BSGP is also to be located in marine areas under the jurisdiction of Russia, causing an obvious problem for the smooth application of the Convention: Russia has signed the Convention but has not ratified it, meaning that it is not legally bound by the instrument.

The chapter proceeds as follows. It first takes up the important question why the BSGP presents a challenge to the Espoo Convention. This will show what the points of contact of the Espoo Convention had to confront when they started to design how to organize the TEIA. This is followed by an examination of how the

TEIA in effect proceeded, which is needed in order to understand how the system ultimately performed in practice. Since the TEIA had to be linked to the five national EIA procedures of the four origin states and Russia, the next section studies how this linking was accomplished in the case of one of these systems, the Finnish EIA procedure. It is useful to take the Finnish case as an illustrative example, given that it has been the Finnish EIA authorities – both those supervising the national EIA and those overseeing the international EIA – that have most challenges in organizing the EIA procedures, as shown in Chapter 3.

Even though an argument will be made that the TEIA was organized in a responsible manner by the concerned states, it is also clear that many difficulties were encountered; these manifested themselves in the very critical comments by the affected states (and others) on the impact assessment studies made by the project proponent, Nord Stream AG. There was a difference of view as to how the BSGP TEIA should have been organized, many parties and the secretariat of the Espoo Convention perceiving that an SEA kind of procedure would have suited better the BSGP type of project. Yet, it will be argued that the points of contact – by focusing on the environmental dimensions of the project and by creating a community approach to organizing such a complex TEIA – provided an important and valuable precedent for the future multi-jurisdictional large-scale projects.

Chapter 10, 'Strategic Environmental Assessment and Sustainability in Europe: How Bright Is the Future?', Simon Marsden and Jan De Mulder

The relationship between EIA and SEA in a transboundary context is a developing one. While the Protocol focuses on SEA in a domestic context, provision is made for application across borders, the focus of EIA under the Convention; as indicated in Chapter 9, recent guidance under the Convention in relation to complex projects has also suggested strategic matters be considered in the EIA. This chapter provides an introduction to SEA and the two legal instruments which have been responsible for its greatest development, the SEA Directive and SEA (Kiev) Protocol. It explains objectives, traces historical development, examines the relationship with environmental policy integration and sustainable development, and analyses the substantive provisions of both laws. The context of sustainable development and the relationship of SEA with sustainability impact assessment, which operates at the level of the EU institutions and in many Member States, is also analysed.

While published at a time before experience with the SEA Directive was advanced, and before the European Commission implementation review had been completed (originally due 2006, delayed until 2009), similarities and differences between the two instruments are examined, pointing to the need for future changes to the SEA Directive, which, in preceding the SEA Protocol, was not fully compliant with it. The Protocol, for example, requires the evaluation of health effects, and includes regional development and mining as additional sectors. Importantly, it also contains a discretionary provision enabling the public to comment on the content of the environmental report, and encourages application to policies and legislation, requiring Parties to report on their efforts in this respect.

The Directive is however more advanced in some respects than the Protocol, with the latter not containing a provision for information on the decision, and the monitoring requirement considered weaker than the Directive's. Both legal instruments were at the time not in compliance with the Aarhus Convention, notably failing to include definitions of the 'public concerned'. As the Convention was not in force until October 2001 and the Directive was adopted in May of the same year, this became a problem later that year, and was hence addressed as a result of amendments made via the Public Participation Directive. Yet the adoption of the Protocol in May 2003 failed to take into account the wording of the Aarhus Convention, and since it entered into force in July 2010, the matter remains to be resolved.

The chapter discusses the different policy contexts between the SEA Directive and Protocol, which have arguably resulted in a weaker framework for sustainable development under the latter, reflecting different approaches by international and European law at the time of the adoption of the instruments. Whether this is reflected through the Protocol's implementation remains to be seen, but in regard to the EU Member States, the European approach would undoubtedly apply, indicating potential differences of approach between EU and non-EU Member States as a result of transposition.

CHAPTER 11, 'THE SEA DIRECTIVE AND SEA PROTOCOL ADOPTED TO SPATIAL PLANNING: SIMILARITIES AND DIFFERENCES', GERNOT STOEGLEHNER AND GERALD WEGERER

The Kiev (SEA) Protocol came into force on 11 July 2010 and changes to the SEA Directive are expected to incorporate additional provisions found in the Protocol that are currently not found in the Directive. As indicated in Chapter 10, these include additional sectors of application (mining and regional development), emphasizing the importance of health and public participation (reflecting the role of the Aarhus Convention), and voluntary application to policies and legislation. The chapter aims to clarify whether SEA practice in the EU will differ greatly if limited to these matters and what the costs and benefits will be. The focus is on spatial planning in the EU Member State of Austria. Differences between the two legal instruments are identified in detail and amendments suggested for implementing the Kiev Protocol in full.

In describing the obligations of the SEA Directive, the authors distinguish between its obligatory and non-obligatory scope, explaining that with respect to the latter there remain limitations on the exercise of discretion. They further highlight the choices available to the Member States as to the approach they can take to transposing obligations into their domestic law. With respect to Austrian spatial planning, SEA is implemented in existing processes, with amendments made to the provincial spatial planning laws. The SEA Directive emphasizes that significant additional effort is to be avoided where possible by applying best practice planning methods.

In pointing out differences in the SEA Protocol, transboundary consultations are highlighted as being more detailed than in the SEA Directive. Hence the state that is adopting the plan has to inform the neighbouring states about the draft plan and environmental report. The affected state can examine the documents and decide whether it wishes to be consulted. The origin state must indicate a reasonable timeframe for the affected state to do this and measures to prevent, reduce or mitigate adverse effects must be properly addressed during consultation. In decision-making, special emphasis is laid on these measures as identified in the environmental report.

The costs of introducing amendments to the existing processes in Austria as a result of the SEA Protocol are analysed in the final section. The main costs of introducing the SEA Directive are first identified as the appraisal of alternatives, the environmental report, transboundary consultations and monitoring. Added to these costs from the Protocol are those resulting from enlarged public participation and the increased integration of health aspects in the planning process. With respect to participation, there is a need for it to feature in screening, scoping and information about monitoring. The authors calculate that delays of three to five months may result from the addition of the new procedures, which may cause serious problems for amendments of plans. Making information available on monitoring, which will need to be adapted to the audience concerned, may also cause delays.

Concerns in relation to health are cited as having potential to result in major problems, as the Protocol intends that health professionals should undertake the appraisal of health effects, not environmental or social scientists. Issues of capacity-building for these professionals are identified as needing to be addressed, and significantly, the concern is raised that including health effects into an SEA report 'might bear the risk that environment will mainly be seen from an anthropocentric point of view that thinks little of other issues like flora and fauna'. The authors therefore conclude with some scepticism about the additional benefits arising from the Protocol in relation to these costs, and only time will tell whether this is correct and whether this is what was intended. If not, changes or guidance may be needed in the future.

Chapter 12, 'The Institutional Context for Transboundary Environmental Impact Assessment in Belgium: Multi-level Setting – A Matter of Smooth Governance?', Jan De Mulder

Belgium is unusual in having arrangements for TEIA between its three territorial regions of Flanders, Wallonia and Brussels, which together with its three language-related communities make up the federal state. The regions and communities are able to conclude international agreements in areas of allocated constitutional responsibility, which include most environmental issues. This chapter examines the role of the cooperation agreement which provides for intra-Belgium TEIA and its effect upon TEIA as it operates traditionally on a bilateral basis between Belgium and other states, here notably in relation to the Netherlands. The relationship between TEIA externally and the intra-Belgium system is also analysed, in particular its role in inspiring internal policy development. Furthermore, the relationship between national and TEIA is recognized generally, as TEIA will reflect the strengths and weaknesses of these systems.

The chapter begins by discussing the constitutional situation in Belgium and the transposition of EU law. It then gives a brief overview of major relevant legal frameworks at the federal and Flemish regional level, before considering some examples of approaches in relation to the Netherlands, which shares a border with the Flemish region of Belgium. These approaches importantly include SEA, which was applied to the 2010 Development Outline for the Scheldt River Estuary, hence explaining the significance of this chapter for Part IV of the book, notably the potential of transboundary SEA under the Kiev Protocol and SEA Directive.

In relation to the Scheldt River, management has been sensitive for some time, based on the historical relationship between Belgium and the Netherlands, as the former was originally part of the latter. Access to ports such as Antwerp is vital to the Belgium economy, flood safety is a concern to both states, and the ecological value of the estuary is considerable and is again emphasized by both. In 1998 therefore, Belgium and the Netherlands decided to develop a joint long-term vision for the Estuary to accommodate such concerns in the context of the Technical Scheldt Commission (which also includes Luxembourg). The 2010 Development Outline was to be the means of defining projects and measures which must be begun no later than 2010 to ensure the long-term vision by 2030. Treaties were finalized between the states to take matters forward, including the establishment of a Flemish-Dutch Scheldt Commission.

A joint SEA was decided upon by the parties, which required a procedure acceptable to both. A number of TEIAs and SEAs followed, and the experience of the consultants who steered the SEA of the 2010 Development Outline is examined in the chapter. Conclusions identified include the weakness that a lack of a hierarchy can bring, with there being no central authority responsible for impact assessment able to decide on disputes other than the Constitutional Court. The diversity of approaches taken to EIA and SEA would also appear to have weakened rather than strengthened practice. The author suggests that the creation of a joint body in relation to TEIA (as suggested by Espoo Convention guidance) is an option that should be considered, which could itself be the subject of a new cooperation agreement on impact assessment.

List of Acronyms and Abbreviations

ABNJ	Areas Beyond National Jurisdiction
ACIA	Arctic Climate Impact Assessment
AEPS	Arctic Environmental Protection Strategy
AHDR	Arctic Human Development Report
AIA	Autonomous Institutional Arrangement
AMAP	Arctic Monitoring and Assessment Programme
ATS	Antarctic Treaty System
BNFL	British Nuclear Fuels Ltd
BSGP	Baltic Sea Gas Pipeline
CBA	Cost-Benefit Analysis
CBD	Convention on Biological Diversity
CC	Compliance Committee
CIA	Cumulative Impact Assessment
COP	Conference of the Parties
CSCE	Conference on Security and Co-operation in Europe
CSR	Corporate Social Responsibility
DO	Development Outline
EA	Environmental Assessment
EAP	Environment Action Programme
EBRD	European Bank for Reconstruction and Development
EC	European Community
ECJ	European Court of Justice
EEA	European Economic Area
EEC	European Economic Community
EEZ	Exclusive Economic Zone
EFTA	European Free Trade Association
EIA	Environmental Impact Assessment
EIS	Environmental Impact Statement
EU	European Union
FANC	Federal Agency for Nuclear Control
FME	Finnish Ministry of the Environment
HELCOM	Helsinki Commission
HIA	Health Impact Assessment
IC	Implementation Committee
ICJ	International Court of Justice
IEEP	Institute for European Environmental Policy
IGO	Inter-Governmental Organization

ILA	International Law Association
ILC	International Law Commission
ITLOS	International Tribunal on the Law of the Seas
LREC	Lapland Regional Environmental Centre
MoE	Ministry of the Environment
MOP	Meeting of the Parties
MOX	Mixed Oxide Fuel
MP	Member of Parliament
MUMM	Management Unit of the North Sea Mathematical Models and the Scheldt Estuary
NAAEC	North American Agreement on Environmental Cooperation
NACEC	North American Commission for Environmental Cooperation
NAFTA	North American Free Trade Agreement
NCP	Non-Compliance Procedures
NEPA	National Environmental Policy Act
NEPC	Nordic Environmental Protection Convention
NGO	Non-Governmental Organization
OECD	Organisation for Economic Cooperation and Development
OSPAR	Convention for the Protection of the Marine Environment of the North-East Atlantic
PPP	Plans, Programmes or Policies
SCI	Sites of Community Importance
SEA	Strategic Environmental Assessment
SEPA	Swedish Environmental Protection Agency
SIA	Sustainability Impact Assessment
SPA	Special Protection Area
SRI	Socially Responsible Investment
TEA	Transboundary Environmental Assessment
TEC	Treaty of the European Community
TEIA	Transboundary Environmental Impact Assessment
TERM	Transport and Environment Reporting Mechanism
TEU	Treaty on European Union
TFEU	Treaty on the Functioning of the European Union
TIA	Transboundary Impact Assessment
TSC	Technical Scheldt Commission
TSEA	Transboundary Strategic Environmental Assessment
UN	United Nations
UNCED	United Nations Conference on Environment and Development
UNCLOS	UN Convention on the Law of the Sea
UNDP	United Nations Development Programme
UNECE	United Nations Economic Commission for Europe
UNEP	United Nations Environment Programme
UREC	Uusimaa Regional Environmental Centre
WHO	World Health Organization
WSSD	World Summit on Sustainable Development

1
Introduction

Simon Marsden and Timo Koivurova

The evolvement of the idea of transboundary environmental impact assessment (hereafter transboundary EIA or TEIA) is connected with the development of EIA in general, because a TEIA procedure is simply a means of extending the national procedure to foreign impacts and foreign actors. The first national EIA procedures appeared in the late 1960s and early 1970s, and they have since spread worldwide. However, the first TEIA procedure of international importance was the 1985 EIA Directive of the then European Community (EC).[1] Hence from the beginning, the EC, eventually replaced by the European Union (EU),[2] has played a key role in expanding EIA to transboundary impacts. Currently, most of the increasing TEIA practice takes place between Member States of the EU, or between Member States and non-Member States of the EU.

Chapter outline

This introduction first explains objectives and reasons for this edited book of materials on TEIA in the EU, and its intended readership. Second, the structure and content of each chapter is outlined. Third, an overview of TEIA procedure and practice is given. Fourth, an outline of the applicable law in the EU is presented, summarizing briefly the relevant international law and implementing EU secondary law, and the relationship between them. Fifth, the role of applicable institutions is examined in relation to review, compliance and enforcement. A brief conclusion follows.

Objectives, reasons and readership

In 2011 the Espoo Convention on Transboundary Environmental Impact Assessment (TEIA)[3] celebrates the 20th anniversary of its adoption, and 2010 marked the coming into force of its Kiev Protocol on strategic environmental assessment (SEA), which applies both in a domestic and transboundary context.[4] In June 2011, the fifth session of the Meeting of the Parties (MOP) to the Espoo Convention will be held, with the first session of the MOP to the Convention serving as the MOP to the Protocol, also to meet. While both agreements are applicable to Parties beyond the EU, EU Member States are in the majority of the Parties to the Convention[5] and Protocol,[6] and reflecting this, most of the commentary on the application of the Convention and Protocol has been in relation to the EU to date. Other than individual papers, however, published research in relation to TEIA has not focused on the Espoo Convention, discussed it specifically from an EU perspective, or combined it with consideration of the Kiev Protocol as applied in a transboundary context.[7] This book combines and addresses each of these matters, as well as the contribution of the Aarhus Convention[8] in the context of TEIA.

The primary objective of this book is hence to recognize and celebrate the events of 2010 and 2011 by collating a selection of key research materials by the editors and a few others which illustrate the potential of and practice with these treaties in the context of the EU,[9] and as a supplement to the guidance on the practical application of the Convention that is available in relation to all Parties.[10] Given the contributing editors' backgrounds and interests, most of these papers are written from a legal perspective and consider substantive and procedural law and underlying theory, although a few, including the work of other contributors, concern policy and practical implementation matters. They were chosen to reflect recent and current topical issues of interest and significance to EU Member States in particular, together with non-EU Parties to the Convention and Protocol, and those interested in joining. The book is therefore addressed principally to those involved in implementing the requirements of the Convention and Protocol under EU law, and is focused on legal aspects in particular. Environmental lawyers, policy-makers, planners and managers involved at the intersection of international, European and domestic law, as well as academics and graduate students teaching and studying degrees in these areas should also find the materials of interest and assistance.

Structure and content

The book has 12 chapters arranged into four parts. Following this Introduction, the first part, containing new work by the editors analysing recent international and European legal developments, considers TEIA in international and European law. Chapter 2 focuses on development within public international law, from the Nordic Environment Protection Convention (NEPC)[11] to the Espoo Convention, and other agreements such as the North American Agreement on Environmental Cooperation (NAAEC);[12] customary international law in relation to transboundary EIA is also examined, in particular decisions of the International Court of Justice (ICJ), notably the Nuclear Tests,[13] Danube Dam[14] and recent Pulp Mills decision.[15] Chapter 3 examines the transposition of the Espoo Convention and Kiev Protocol into EU law, together with related provisions in the Aarhus Convention. It focuses on the EIA, Public Participation and SEA directives, analysing substantive provisions of relevance to transboundary EIA, recent implementation reviews conducted by the European Commission, and rulings by the European Court of Justice (ECJ) in relation to these matters. The importance of the Aarhus Convention is recognized in this Introduction, and is included in the overview of applicable law, especially its provisions in relation to public participation and access to justice.[16]

The second, third and fourth parts, respectively entitled The Relationship between International and European Law, Convention Law and Practice, and Protocol Law and Potential, contain previously published materials in a range of legal and environmental journals, which are collated in the book. There are three chapters in each part, and for each chapter the editors have summarized their content and significance. Chapters 4, 5 and 6 concern the relationship between international and European law. In Chapters 4 and 5, which concern the transboundary relationships between Finland and Sweden on one hand and the UK and Ireland on the other, the Convention was not applied because it was not in force, however arguments are made or demonstrated for its application, and overlapping international and European law considered. Chapter 6 compares and contrasts opportunities to enforce transboundary EIA provisions for public participation in EU and non-EU Member States. Chapters 7, 8 and 9 review Convention law and practice in three very different regions: Iberia, the Arctic and the Baltic Sea. Chapters 10, 11 and 12 examine Protocol law and potential, making comparisons between the SEA Protocol and Directive in the context of spatial planning in Austria, and the transboundary experiences of Belgium and its relationship with the Netherlands. Appendices include the text of the Convention and Protocol.

TRANSBOUNDARY EIA/SEA PROCEDURE AND PRACTICE

TEIA applies EIA procedures to proposals in a state which have potentially significant environmental effects upon another state. For the most part, these proposals relate to individual projects, although they may also be plans or programmes, or policies or legislation, in which case the procedure is termed transboundary SEA (TSEA).[17] Proposals may either relate to the territory of the origin state, or be activities that physically cross the boundary between origin and affected states. Complex proposals can involve more than one origin and affected state and may raise important strategic issues despite being essentially of a project nature.[18] Examples of proposals were given by Parties to the Convention and others in the Second Review of Implementation. Activities most commonly cited, and which illustrate both proposals located within a state and crossing the boundaries of states, were thermal and nuclear power stations, and motorways, express roads and railways.[19] Other types of proposals cited, which illustrate the project specific (EIA) and plan and programming type proposals (SEA) were flood protection, hydropower and marine dredging projects on one hand, and spatial planning, land use and other plans on the other.[20]

Transboundary EIA/SEA have therefore developed legal structures, policies and practices to meet the development challenges of the 21st century. Addressed initially to specific project proposals and increasingly to planning and programming proposals, they are designed, like domestic EIA and SEA, to provide information to decision-makers about significant potential environmental effects, views of affected publics and states concerned, and alternatives, in order to either approve proposals subject to conditions, or reject them. Participation is based on the principle of non-discrimination, enabling those in an affected state to participate in the procedure of an origin state, which is examined in Chapter 6 in detail.

Experience with the Convention reported in the Second Review of Implementation supports the prevention, reduction and control of possible significant transboundary effects. TEIA is an important component of the no-harm or harm prevention principle, that states have the right to develop their own resources provided they do not cause harm to other states,[21] which has resulted in numerous requirements in a broad range of international treaties, as outlined in Chapter 2. The Second Review gives as examples: mitigation of possible impacts, including the addition of stricter environmental protection measures in relation to interim storage facilities for spent nuclear fuel, a flood protection project, the design of a goldmine and either the halting of dredging or reducing the extent of a dredging area. Evidence of a project being refused as a result of the Convention's application was given by one state, and greater awareness of environmental consequences of projects leading to improved public participation was reported by another.[22]

Transboundary EIA/SEA Law

The EU confirms in Article 3(5) of the Treaty on European Union (TEU) that it adheres to 'the strict observance and the development of international law'.[23] It also confirms in Article 8(2) that relations with non-EU states are based on the principles of cooperation and good neighbourliness, also key principles underlying TEIA.[24] The non-discrimination principle is found in Article 2 TEU, and in Article 18 of the Treaty on the Functioning of the European Union (TFEU).[25] The EU has responded to the no-harm principle with principles of its own, including the preventive principle found in Article 191(2) TFEU. In accordance with competences granted for the environment in this Article, and with external affairs powers,[26] it has also signed and ratified the UNECE agreements and incorporated them into EU law by adopting and approving legislation of its own to transpose them. The EIA,[27] Public Participation[28] and SEA[29] Directives contain the primary provisions, in the case of the latter preceding and influencing the development of the Protocol. These provisions are cited here briefly and discussed in detail in Chapter 3 of this book.

Espoo Convention and EIA Directive

The Espoo Convention came into force in 1997 after adoption in 1991, the first major treaty to detail provisions for transboundary EIA. It remains the most significant as a result of the large number of Parties. The historical development of the treaty has been analysed elsewhere,[30] and its procedural provisions, relationship with the Aarhus Convention, implementation experience (to 2007) and SEA Protocol are outlined by Schrage, until recently the Secretary to the Convention, who notes that the definition of transboundary impact 'explicitly excludes impacts of a global nature and therefore concentrates on transboundary impacts of a local or sub-regional character in the ECE region'.[31] The relationship of the Convention with the EIA Directive is discussed in many chapters of this book, including Chapters 3, 6 and 7. Article 7 of the amended EIA Directive transposes most of the obligations in the Espoo Convention into EU law.

Kiev Protocol and SEA Directive

The Kiev Protocol came into force in 2010 after adoption in 2003, also the first major international instrument to detail provisions for SEA, albeit primarily in a domestic context. These provisions, however, also apply in a transboundary context, and application of them is likely to grow as development proposals become more complex and involve more states. Its historical development has also been analysed elsewhere, notably by de Mulder, co-author and author of Chapters 10 and 12 and first chair of the Meeting of the Signatories;[32] procedural provisions of the Protocol are further outlined in this book in Chapter 11. Articles 7 and 8 of the SEA Directive transpose the transboundary obligations of the Kiev Protocol, which are found in Articles 10 and 11 of the legal instrument.

Aarhus Convention and Public Participation Directive

The Aarhus Convention came into force in 2005 after its adoption in 1998, and relates access to information, public participation and access to justice to environmental protection. It has served to expand existing provisions in the EIA and SEA Directives in relation to public participation and access to justice in particular, and is hence an important part of the practical workings of EIA and SEA. The relationship between EIA, SEA and the Aarhus Convention has been explored by Ryall in relation to the principle of effective judicial protection in the EU.[33] Article 3 of the Public Participation Directive amends the EIA Directive, with Article 3(5) specifically concerned with participation in TEIA; Article 3(7) implements the access to justice provisions in EIA; Article 2(2) provides for public participation in relation to the preparation of plans and programmes not covered by the SEA or Water Framework[34] Directives.

Institutions: Reviews, compliance and enforcement

At the level of both international and European law, institutions enable review, compliance and enforcement. Under the Espoo Convention, there is a permanent Secretariat, Working Group on EIA and Implementation Committee, and potential for an Inquiry Commission to be established when needed.[35] The same institutions also apply to the SEA Protocol,[36] and under the Aarhus Convention there are similar institutions, one of particular significance being the Compliance Committee. In relation to EU law, the European Commission[37] and ECJ[38] have important roles in review, compliance and enforcement. This section explains the role of these institutions in reviews of implementation in international and EU law, and dispute settlement and avoidance in international and EU law.

Reviews of implementation in international law

The Espoo Implementation Committee receives reports from Parties to the Convention in relation to compliance. The Second Review of Implementation, for the period 2003–2005 is discussed above in relation to examples of proposals and effects of the procedure on decision-making. As a Party to the Convention, the EU reported by letter to explain its status and competence, making reference to transposing law, research projects in progress and the status of ratification of the amendments to the Convention and Protocol.[39] The Third Review of Implementation is currently in progress under a revised questionnaire for the period 2006–2009, and is expected to be reported on to the MOP in 2011.[40] With respect to this, the EU again reported by way of a letter to the Convention Secretariat advising on these matters and on the recent reviews of the EIA and SEA Directives.[41] The Implementation Committee has noted the review examining the implementation of the provisions concerned with public participation, finding that concerned Parties have a common responsibility for providing equivalent opportunities for public participation in the affected Party, and that 'synergies should be sought with national reporting on implementation of Article 6 of the Aarhus Convention (on public participation)…'.[42] In 2008, the EU as a Party to the Aarhus Convention submitted a report via the Commission in respect of its obligations under this agreement.[43]

Reviews of implementation in EU law

The European Commission is designated the 'Guardian of the [EU] Treaties', and as such has a vital role in relation to Member State compliance with EU law further to the TEU as noted. The role of periodically reviewing specific secondary EU law is examined in detail in Chapter 3, together with preparing annual reports in the application of EU law generally. Each directive contains specific provisions for regular reviews in order for implementation experience to be gathered and suggestions made for future reform. This has included consideration being given to consolidating them, as seen in Chapter 3. The EIA Directive has been reviewed four times (1993, 1997, 2003 and 2009), the SEA Directive once (2009) and the Public Participation Directive once (2010).

Dispute settlement and avoidance in international law

Dispute settlement in international law is often provided for by treaties stipulating options to take court proceedings when Parties cannot resolve their differences amicably. None of the provisions in the Espoo Convention,[44] Kiev Protocol[45] and Aarhus Convention[46] have been used in this way. As an alternative, the Espoo Convention Implementation Committee[47] and the Aarhus Convention Compliance Committee[48] have been called on to look at specific instances of non-compliance with obligations under the Conventions,[49] the latter of which may relate to TEIA. Importantly, these procedures are not intended to be dispute settlement procedures, but rather to assist Parties with compliance, hence to avoid disputes becoming litigious. The Aarhus Compliance Committee, while it does not receive reports from the Parties, has a very important role in relation to non-compliance, based not only upon issues being raised by other Parties or the Secretariat, but also by members of the public. Many of these 'public communications' have concerned EIA-related matters in the EU. Chapter 6 (Marsden) discusses the role of these bodies and some of the cases to date.[50]

Dispute settlement and avoidance in EU Law

The European Commission has significant roles in ensuring compliance and initiating enforcement action under EU law. Where a Member State fails to comply with EU law, either by action or omission, the Commission is able to take infringement proceedings against it to try and ensure compliance, with cases referred to the ECJ where necessary. Infringement proceedings are the pre-litigation stage, allowing the Member State to voluntarily conform to EU law. A letter of formal notice will first be issued by the Commission for the Member State to submit its observations on the problem within a time limit, and a reasoned opinion will set out the Commission's position on the infringement. Failure to provide a satisfactory response or none at all may result in the Commission referring a case to the ECJ. The ECJ enforces EU law, and if the Commission succeeds in a case against a Member State, there may be monetary and other consequences for failure to comply with an order.

Conclusion

TEIA has a significant presence in the European landscape, and by weight of numbers and environmental advocacy, the EU has had a profound influence on the development and expanding role of the Espoo Convention, notwithstanding some differences of approach and difficulties of implementation. For similar reasons, the EU has had a fundamental role in the establishment of the Aarhus Convention and its position in broadening and deepening provisions for participation and access to justice in respect of TEIA. The procedures and scope of the EU SEA Directive was also pivotal to the establishment of the Kiev Protocol, whatever the current differences between these laws which will need to be addressed now the Protocol is in force. The chapters that follow analyse these provisions and challenges from theoretical and practical legal perspectives, emphasizing potential where experience is yet to be gained. Overall, they present an extremely positive report card on the 20th anniversary of the Convention, the geographical scope of which is likely to extend far beyond the EU and the Economic Commission for Europe (ECE) region in the future.

Notes

1 Council Directive of 27 June 1985 on the assessment of the effects of certain public and private projects on the environment (85/337/EEC) (OJ L175, 5.7.1985, p40).
2 Following the coming into force of the Lisbon Treaty on 1 December 2009, EU law has replaced EC law in full, and reference is to be made to the Consolidated Version of the Treaty on European Union (TEU), OJ C115/13, and the Treaty on the Functioning of the European Union (TFEU), OJ C115/47. For implications of the changes see De Witte, B. (2009) 'International Law as a Tool for the European Union' 5 *European Constitutional Law Review* vol 5, p265; and Lee, M. (2008) 'The Environmental Implications of the Lisbon Treaty' 10 *Environmental Law Review*, vol 10, p131.
3 Convention on Environmental Impact Assessment in a Transboundary Context (Espoo) *ILM* (1991), vol 30, p802.
4 Protocol on Strategic Environmental Assessment to the Convention on Environmental Impact Assessment in a Transboundary Context 2003 (Kiev).
5 As at October 2010, there are 27 EU Member States, and all are now Parties to the Espoo Convention in their own right; the EU is also a Party, so the agreement is mixed. The Convention has 30 Signatories and 45 Parties, and in addition to the 27 EU Member States, two of the EU Candidate Countries are Parties to the Convention, Croatia and the former Yugoslav Republic of Macedonia, and one – Iceland – is a Signatory. Four of the Potential EU Candidate Countries are also Parties

to the Espoo Convention, Albania, Bosnia and Herzegovina, Montenegro and Serbia. Iceland and Norway (a Party to the Convention) are bound by EU law implementing the Convention and Protocol as a result of their membership of the European Free Trade Association (EFTA). If these Candidate and Potential Candidate Countries are accepted into the EU, this would increase the number of EU Member States to 35 out of the current 45 Parties, although it is noted that if the First Amendment to the Convention comes into force, states beyond the ECE Region (56 countries in total, not all of whom are Parties/Signatories), may also be permitted to join the Convention which would dilute EU dominance. Of the Signatories to the Espoo Convention who have not ratified – Russia and the US – Russia has agreed to comply with the Convention in the Nord Stream Project to the extent permitted by its national legislation.

6 The Protocol has 38 Signatories and 19 Parties, including the EU. 13 Parties are EU Member States and the others are Albania, Croatia, Montenegro, Norway and Serbia. The Protocol already permits non-UNECE states to join, see Article 23(3), upon approval of the MOP.

7 See (2008) 26(4) *Impact Assessment and Project Appraisal*, special issue on environmental assessment and transboundary impact assessment.

8 Convention on Access to Information, Public Participation in Decision-Making and Access to Justice in Environmental Matters (Aarhus), *ILM* (1998), vol 38, p517.

9 For related recent works by the editors, note in particular, Marsden, S. (2008) *Strategic Environmental Assessment in International and European Law*, Earthscan, London, and Bastmeijer, K. and Koivurova, T. (eds) (2008) *Theory and Practice of Transboundary Environmental Impact Assessment*, Martinus Nijhoff.

10 For recent publications, see ECE, *Guidance on Notification According to the Espoo Convention*, ECE/MP.EIA/12, UN New York and Geneva, 2009; ECE, *Guidance on the Practical Application of the Espoo Convention*, ECE/MP.EIA/8, UN New York and Geneva, 2006; ECE, *Guidance on Public Participation on EIA in a Transboundary Context*, ECE/MP.EIA/7, UN New York and Geneva, 2006. In respect of SEA, see UNECE and Regional Environmental Center for Central and Eastern Europe, *Protocol on SEA: Resource Manual to Support Application of the Protocol on SEA*, Draft Final, 2007, A4.5 concerns transboundary consultations.

11 Convention on the Protection of the Environment between Denmark, Finland, Norway and Sweden (Stockholm) *ILM* (1974), vol 8, p591.

12 North American Agreement on Environmental Cooperation Between the Government of Canada, the Government of the United Mexican States and the Government of the United States of America, 1993.

13 Dissenting Opinion of Judge Weeramantry in the International Court of Justice Nuclear Tests Case (NZ v. France), available at www.icj-cij.org/docket/files/97/7567.pdf

14 Separate Opinion of Vice-President Weeramantry in the International Court of Justice, in the Case Concerning the Gabcíkovo-Nagymaros Project (Hungary-Slovakia), available at www.icj-cij.org/docket/files/92/7383.pdf?PHPSESSID=1917 14d456bec3b9d512e041a16f0c7a

15 Case Concerning Pulp Mills on the River Uruguay (Argentina v. Uruguay), 20 April 2010, judgment available at www.icj-cij.org/docket/files/135/15877.pdf

16 See generally, Marsden, S. (2011)[Still in press] Public Participation in Transboundary Environmental Impact Assessment – Closing the Gap between International and Public Law?, in *Environmental Discourses in Public and International Law*, Cambridge University Press, Cambridge.

17 See Bonvoisin, N., Transboundary issues in SEA, in Sadler, B., Aschemann, R., Dusik, J., Fischer, T., Partidário, M., and Verheem, R. (eds) (2010) *Handbook of Strategic Environmental Assessment*, Earthscan, London, Chapter 25; Albrecht, E., (2008) Transboundary consultations in strategic environmental assessment, *Impact Assessment and Project Appraisal*, vol 26, no 4, pp289–298.

18 See United Nations, Economic and Social Council, ECE, MOP to the Espoo Convention, Working Group on Environmental Impact Assessment, Twelfth Meeting, Geneva, 11–13 May 2009, Item 5(a) of the provisional agenda, Exchange of Good Practices, Large-Scale Transboundary Projects, Application of the Convention to complex activities, Note by the Secretariat, ECE/MP.EIA/WG.1/2009/4, 2 March 2009. This is critiqued by Koivurova and Pölönen in Chapter 9, pp175–179 and conclusions, where the difficulty that Parties cannot legally oblige the proponent of the Nord Stream Project to assess strategic issues (particularly geopolitical, energy and climate) is highlighted. Note also, Hernández, F., *Analysis of the Espoo Convention as Applied to Mega Projects: The Case of Nord Stream,* thesis submitted to Lunds University for the degree of Master of Science, pp23–25, 40–42.

19 Economic Commission for Europe, *Review of Implementation of the Espoo Convention*, United Nations, New York and Geneva, 2008, ECE/MP.EIA/11, pp31 and 40. Cross border projects also cited include oil and gas pipelines, power lines, and tunnels and bridges. Nuclear power plants are currently of significant interest to the Convention because of the plans for construction and decommissioning in many states. See Economic Commission for Europe, MOP to the Convention on Environmental Impact Assessment in a Transboundary Context, Working Group on Environmental Impact Assessment, 14th Meeting Geneva, 24–26 November 2010, Item 4(d) of the provisional agenda, Exchange of good practices, Application of the Convention to nuclear energy-related activities, background note draft, ECE/MP.EIA/WG.1/2010/INF.7.

20 Ibid., p.32. For example, note Kment, M. (2009) Cross-border Development Plans – An Outline of a Useful Planning Instrument for European Border Regions *Journal for European Environmental and Planning Law*, vol 6, no. 2, pp233–253.

21 See *Trail Smelter Arbitration (United States of America v. Canada)* (1938) RIAA vol 3, 1905; 9 ILR 315*; Corfu Channel Case (United Kingdom v. Albania) (Merits)* (1949) ICJ Reports 4; *Lac Lanoux Arbitration (France v. Spain)* 24 ILR 101.

22 Ibid. see note 19 above, p.33.

23 Consolidated Version of the Treaty on European Union, OJ C115/13, 9.5.2008. See also Article 21(1) and Article 21(2)(f), which states: 'The Union shall define and pursue common policies and actions, and shall work for a high degree of cooperation in all fields of international relations, in order to help develop international measures to preserve and improve the quality of the environment and the sustainable management of global natural resources, in order to ensure sustainable development.'

24 The EU has supported the Convention in many ways, recently organizing with the support of the Secretariat a one day seminar on large-scale projects involving several countries, held in May 2009. The European Commission has also led a project to assist Ukraine in the implementation of the Espoo and Aarhus Conventions.

25 Consolidated Version of the Treaty on the Functioning of the European Union, OJ C115/47, 9.5.2008.

26 See Article 21 TEU, and Article 216 TFEU.

27 Council Directive 97/11/EC of 3 March 1997 (97/11/EC) (OJ L73, 14.3.1997, p5).

28 Directive 2003/35/EC of the European Parliament and of the Council of 26 May 2003 providing for public participation in respect of the drawing up of certain plans and programmes relating to the environment (2003/35/EC) (OJ L156, 25.6.2003, p17).

29 Directive 2001/42/EC of the European Parliament and of the Council of 27 June 2001 on the assessment of the effects of certain plans and programmes on the environment (Directive 2001/42/EC) (OJ L197, 21.7.01, p30).

30 Connelly, R., (1990) The UN Convention on EIA in a Transboundary Context: A Historical Perspective *Environmental Impact Assessment Review*, vol 19, no 1, pp37–46.

31 Schrage, W., The Convention on Environmental Impact Assessment in a Transboundary Context, in Bastmeijer, K. and Koivurova, T. (eds) (2008) *Theory and Practice of Transboundary Environmental Impact Assessment*, Martinus Nijhoff, 32. Opening the Convention to non-ECE states may of course change this in the future.

32 See also De Mulder, J., The Expansion of Environmental Assessment in International Law: The Protocol on Strategic Environmental Assessment to the Espoo Convention (2006) *Environmental Law and Management*, vol 18, pp269–281.

33 Ryall, A. (2009) *Effective Judicial Protection and the Environmental Impact Assessment Directive in Ireland*, Hart Publishing, Oxford.

34 2000/60/EC of the European Parliament and of the Council of 23 October 2000 establishing a framework for Community action in the field of water policy.

35 See Koivurova, T. The Convention on Environmental Impact Assessment in a Transboundary Context, in Ulfstein, G., Marauhn, T. and Zimmermann, A. (eds) (2007) *Making Treaties Work: Human Rights, Environment and Arms Control*, Cambridge University Press, Cambridge, pp218–241.

36 In 2007, the Working Group on EIA examined the structure and functions of the Implementation Committee when considering matters under the Kiev Protocol, see United Nations, Economic and Social Council, ECE, Meeting of the Parties to the Convention Espoo in a Transboundary Context, Working Group on EIA, Tenth Meeting Geneva, 21–23 May 2007, Item 4(a) of the provisional agenda, Activities Relating to the Convention Listed in the Work Plan (Decision III/9) Compliance with and implementation of the Convention, Report of the 11th Meeting of the Implementation Committee, ECE/MP.EIA/WG.1/2007/4, paragraph 14, 12 March 2007.

37 Article 17(1) TEU states: 'It shall ensure the application of the Treaties, and of measures adopted by the institutions pursuant to them. It shall oversee the application of Union law under the control of the Court of Justice of the European Union.'

38 Article 19(1) TEU states: 'It shall ensure that in the interpretation and application of the Treaties the law is observed.' The functions of the ECJ are set out in Article 19(3), which include (a) ruling on actions brought by the Commission.

39 Espoo Convention Review of Implementation 2nd Review, Report of the EC for 2003–2005 on the Implementation of the Espoo Convention on Environmental Impact Assessment in a Transboundary Context, 8 February 2006, available at www.unece.org/env/eia/documents/Review%202003-2005/Questionnaire %20-%20 2003%20-%20EC.pdf

40 As of 31 August 2010, only 30 of the 45 Parties had completed and returned the questionnaire; the Implementation Committee has advised the Secretariat not to process reports after 31 December 2010, and that failure to report may be considered a matter of non-compliance. See United Nations, Economic and Social Council, ECE, MOP to the Convention on EIA in a Transboundary Context, Implementation Committee, Nineteenth Session, 31 August–2 September 2010, Geneva, Report of the Implementation Committee on its Nineteenth Session, ECE/MP.EIA/IC/2010/4, 23 September 2010, paragraphs 40–41.

41 Espoo Convention Review of Implementation 3rd Review, Report of the European Union for 2006–2009 on the implementation of the Convention on Environmental Impact Assessment in a Transboundary Context (1991 Espoo Convention), 22 June 2010, available at www.unece.org/env/eia/documents/Review_2006_2009/Questionnaire2006_09_EU_reply_en.pdf

42 Ibid., paragraph 19(b). Bilateral agreements were regarded as having potential to resolve many of these issues, paragraph 19(f).

43 Commission of the European Communities, Aarhus Convention Implementation Report, EC, Brussels, 7 May 2008 SEC(2008) 556.

44 Article 15 Settlement of Disputes, Appendix VII Arbitration.

45 Article 16 Settlement of Disputes, Annex II Arbitration.

46 Article 20 Settlement of Disputes (same as for Convention).

47 Decision II/4 Review of Compliance, 2nd Meeting of the Parties, Annex IV, pp72–76, ECE/MP.EIA/4; Article 14 bis Review of Compliance (second amendment), to be available to Protocol when in force.

48 Article 15 Review of Compliance.

49 For recent literature, see Fasoli, E. Procedures and Mechanisms for Review of Compliance under the 1991 Espoo Convention on Environmental Impact Assessment in a Transboundary Context and its 2003 Protocol on Strategic Environmental Assessment, 181–203; Jendroska, J. Practice and Relevant Cases that Emerged in the Context of the Espoo Convention Implementation Committee, pp319–335, both in Treves T. et al (eds) (2009) *Non-Compliance Procedures and the Effectiveness of International Environmental Agreements*, TMC Asser Press, The Hague.

50 These include a case against Finland which was dismissed, and the ongoing dispute between Romania and the Ukraine, which in addition to the major claims against the Ukraine, also included claims (dismissed) against Romania.

Part I

International and European Law

2

Transboundary Environmental Impact Assessment in International Law

Timo Koivurova

Chapter outline

This chapter examines the development of transboundary environmental impact assessment (TEIA) in international law. After an introduction, the chapter proceeds to consider the development of TEIA through treaty and soft law instruments and discusses its position in customary or 'general' international law. Development of transboundary norms to address potential significant environmental effects in shared areas, and in areas beyond national jurisdiction (ABNJ), are next considered, together with the role of TEIA in relation to international financial institutions. The Espoo Convention and Kiev Protocol are also introduced and their provisions are analysed in this chapter, and the relationship between the Espoo Convention in particular and other aspects of international law is explained and evaluated in the light of recent jurisprudence of the International Court of Justice (ICJ).

Introduction

Environmental Impact Assessment (EIA) can be defined – at its minimum – as a government-controlled procedure involving the public within which scientific assessments are conducted of the potential environmental impacts of a proposed activity that may be environmentally harmful. Its goals include improving the quality of information so decision-makers can make better decisions from the viewpoint of the environment, raising the level of public participation in relation to environmental protection. TEIA is often perceived only as an extended EIA procedure between territorial states that includes transboundary impacts and foreign actors into the domestic EIA of the origin state. Yet today the concept has been enlarged to cover EIA procedures designed to evaluate possible impacts by human activities on the environment in ABNJ of states.

It was the spread of national EIA procedures that made it possible to conclude legally relevant instruments and treaties on TEIA. Since TEIA tries to integrate foreign impacts and foreign actors into the operation of the national EIA procedure of the origin state in cases where transboundary impacts are likely, this is best done by an international treaty, given that states' jurisdiction is confined to certain pieces of territory.

Treaties or other instruments that try to organize a TEIA procedure must take into account the basic fact that states have differing national EIA procedures. From the viewpoint of the functioning of a TEIA procedure, the ideal situation would obviously be one in which all the contracting states had an identical EIA procedure, for the potentially affected state and its public would know when and how to participate in the EIA procedure taking place in accordance with the EIA procedure of the origin state. While this has to a significant extent been achieved by the implementation of the Espoo Convention and Kiev Protocol in relation to European Union (EU) Member States, domestic EIA procedures still differ to some extent, making the organization of TEIA complicated. With respect to non-EU Member States, the situation is of course even more complicated.

DEVELOPMENT OF TRANSBOUNDARY EIA IN TREATY AND SOFT LAW

The first national EIA procedures appeared at the end of the 1960s and beginning of the 1970s. The then European Economic Community (EEC) was at the forefront of also introducing a legally binding TEIA procedure with its 1985 EIA Directive.[1] This Directive, which will be studied more closely in Chapter 3, was however preceded firstly by a type of embryonic TEIA procedure fleshed out in the 1974 Nordic Environmental Protection Convention (NEPC), and secondly by a resolution of the US Senate in 1978 containing the outline of a general convention for TEIA which urged other nations to participate in the negotiations for such a convention.[2] It is of interest to briefly study these early procedures before considering other provisions for TEIA in treaty and soft law (including the Espoo Convention and Kiev Protocol), which followed after the EIA Directive.

Nordic Environmental Protection Convention

The 1974 NEPC[3], which is still in force, contains two fairly rudimentary TEIA procedures, the choice of which depends on which governmental authorities of the origin state are in charge of the licencing decision on the environmentally harmful activity. The scope of both procedures relies on the concept of 'environmentally harmful activities', which is defined in detail in Article 1(1); the definition only includes stationary activities.

The environmental effects of these stationary activities are not defined exhaustively since the Article only enumerates certain pollution effects and provides that other pollution effects are also covered.[4] Although both of the TEIA procedures of the NEPC are based on the concept of environmentally harmful activities, they have a somewhat different scope in that the procedure involving the examining authority and the supervisory authority of the concerned states is triggered when a 'nuisance of significance'[5] takes place whereas that involving the central governments requires that 'considerable nuisance'[6] is caused. Hence, the threshold at which transboundary pollution is deemed to occur is lower in the former procedure.

If the licencing decision is made by a 'Court or the Administrative Authority'[7], the procedure is carried out by this examining authority and the supervisory authority[8] of the potentially affected state. All environmentally harmful activities that 'may entail nuisance of significance in another Contracting State'[9] are covered by a procedure carried out by the licencing authority of the origin state and the supervisory authority of the affected state. The origin state's licencing authority is obligated to 'send as soon as possible a copy of the documents of the case to the supervisory authority of the other State'.[10] The licencing authority must also allow the supervisory authority to give its opinion on the matter.

If the supervisory authority so requests, the permit authority must require from the applicant, 'such additional particulars, drawings and technical specifications as the examining authority deems necessary for evaluating the effects in the other State'.[11] Hence, the origin state is obligated to conduct a TEIA. In addition, the supervisory authority is entitled to conduct its own investigations on the potential transboundary impacts 'as it deems necessary'[12], which means that the affected state also has a right to have such information taken into account in the licensing procedure of the origin state. Interestingly, 'if necessary for determining the damage caused in another State'[13] the supervisory authority of the potentially affected state is obligated to organize an on-site inspection upon request by the examining authority. In the estimate by one author, 'this provision is intended to enable the examining authority to gather all possible evidence as to the damage caused before it reaches a decision in the matter'.[14]

Environmentally harmful activities 'which may entail considerable nuisance in another Contracting State',[15] and which are controlled by the central government of the origin state, are governed by a procedure that is carried out by central governments of the concerned states. If the potentially affected state so requests, the central governments must arrange consultations on the matter. In these cases, both of the concerned states are also entitled to demand that an advisory opinion be given by an inquiry commission.[16] When such a commission is appointed, the dispute cannot be decided until the commission has given its opinion.[17]

Proposed Convention of the US Senate

The scope of the above-mentioned 1978 proposed convention by the US Senate[18] was broad since all major projects that might reasonably be expected to have adverse effects on the environment of other states or ABNJ were to be assessed, and the results of such 'international environmental assessments' were to be communicated to the potentially affected state as well as the United Nations Environment Program (UNEP), especially in cases of potential damage to the environment of ABNJ.[19] In case the potentially affected state or, in the case of global commons (ABNJ), if UNEP so requested, the origin state was to consult these affected subjects 'with a view towards preventing or minimizing any potential adverse environmental consequences beyond its territory'.[20] If the potentially affected state or UNEP so requested, the origin state was to 'refrain from initiating the project or activity for a period of not more than 90 days after submission of an international environmental assessment'.[21] However, if the postponement of the project '[involved] serious risks to the safety of life or property or would otherwise be clearly infeasible', the origin state could go ahead with the project.[22] The Senate's call for negotiations of an international convention on this basis never took off, but internally the US President adopted in the same year an Executive Order, which placed federal agencies under an obligation to assess the environmental effects of major federal actions with respect to the 'environment outside the United States, its territories and possessions'.[23]

UNEP Goals and Principles of EIA

An important legally relevant instrument to the development of TEIA was the 1987 Goals and Principles of Environmental Impact Assessment adopted by UNEP.[24] Under this soft-law instrument, states are generally encouraged to conclude arrangements for cooperation not only with regard to potential transboundary impacts of an inter-state character but also, interestingly, with regard to potential pollution affecting ABNJ.[25] Principle 12 of the Goals and Principles requires that if the national EIA procedure indicates that the environment of another state is 'likely to be significantly affected by a proposed activity',[26] the origin state should notify the potentially affected state and 'transmit to the potentially affected state any relevant information from the EIA'.[27] If an agreement is reached between the states, timely consultations should take place between them. The principal importance of the document is that it was used as a model in the negotiations for the Espoo Convention.[28]

Espoo Convention

The first full-fledged TEIA procedure was not adopted until 1991, with the conclusion of the Espoo Convention on Environmental Impact Assessment in a

Transboundary Context.[29] The Convention regulates situations where a significant adverse transboundary impact is likely to be caused to a state's environment by a proposed activity in another contracting state (the origin state). It aims to manage these situations by requiring the parties to cooperate with each other before the activity is undertaken. In order for this procedure to function effectively, the Espoo Convention requires the states parties to establish national and/or sub-national EIA procedures that allow for the integration of foreign impacts and foreign actors.[30]

The TEIA procedure prescribed in the Convention differs in respect of activities listed in Appendix I and those that are not listed. The Convention contains a presumption that a TEIA procedure must be initiated in regard to the activities listed in Appendix I of the Convention. These activities are not, however, automatically included in the scope of the Convention; rather, it is presumed that in normal circumstances they are 'likely to cause a significant adverse transboundary impact'. In order to enforce this presumption with regard to the listed activities, the Convention contains a special mechanism for evaluating whether the criteria provided in Article 3(1) are fulfilled. If the origin state decides not to initiate a procedure with regard to an activity listed in Appendix I, the concerned states are, according to Article 3(7), obligated to discuss whether the TEIA procedure should be applied. If discussions between the states fail, either party has the right to initiate an inquiry commission procedure in which an expert panel gives its opinion on whether the activity is likely to cause a significant adverse transboundary impact. The Convention contains a reverse presumption with respect to those activities not listed in Appendix I: it is presumed that they do not fulfil the criteria laid down in Article 3(1) in normal circumstances. This presumption can be rebutted by showing that exceptional circumstances exist; for instance, the activity may be located very near an international border. The criteria for these exceptional circumstances are enumerated in Appendix III of the Convention and the procedure for states to negotiate on these proposed activities in Article 2(5).

The transboundary EIA procedure can thus commence in respect of both activities listed in Appendix I and other activities. When the origin state first notifies the potentially affected state on the basis of Article 3(1), the affected state must next confirm that it wants to participate in the procedure.[31] The origin state is then obligated to study the transboundary impacts together with the affected state and allow the public[32] of that state to participate in the process on the same terms as its own public would be entitled to.[33] After the impact assessment, the affected state has an opportunity through consultations with the origin state to comment on the proposed activity and its likely impacts; the public of the affected state is entitled to provide its comments on the proposed activity on the same terms as apply to the public of the origin state.[34] The final decision taken on the proposed activity in the origin state must take due account of the comments from the potentially affected state and its public and must be delivered to the affected state.[35] The states' parties are not required to determine whether the impacts studied ultimately materialize, as post-project analysis is optional.[36]

The institutional development within the Espoo Convention has progressed over time. The MOP, representing the voice of states' parties, has taken place four times and has adopted many important decisions pertaining to the institutional structure of the Espoo Convention. The Bureau of the Convention and the Working Group on EIA were established as organs coordinating the work pertaining to the development of the system of the Convention between the MOP. Secretariat tasks are handled by the United Nations Economic Commission for Europe (UNECE). The Implementation Committee studies potential non-compliance with the obligations of the Convention and reviews the Convention's implementation. The system of the Espoo Convention can most appropriately be qualified as an example of a recent phenomenon in international law that has developed in the field of environmental protection in particular and which Churchill and Ulfstein refer to as an autonomous institutional arrangement (AIA).[37]

Kiev Protocol

The SEA Protocol to the Espoo Convention was signed in an extraordinary meeting of the parties in Kiev, 21–23 May 2003, by 35 countries, together with the EC. The Protocol primarily harmonizes the SEA procedures for the contracting states but it also includes provisions on transboundary SEA. Where the proposed plan or programme is likely to have significant transboundary effects, including health impacts,[38] Article 10 of the Protocol is applicable. If the potentially affected state opines that it is likely to be significantly affected by the proposed plan or programme, it has a right to request that the origin state commences the transboundary procedure. However, should the origin state consider that no such effect is likely, there exists no procedure, such as the inquiry commission procedure of the Espoo Convention, for the potential affected state to dispute the finding of the origin state.

The origin state is obligated to notify the potentially affected state before the adoption of the plan or programme 'as early as possible'. The notification must contain the environmental report and information regarding the decision-making procedure. The potentially affected state is obligated to respond within the time specified in the notification as to whether it wishes to participate in the procedure. If it so indicates, the two parties are to initiate consultations about the likely transboundary effects of the plan or programme and the measures to mitigate its environmental effects. Significantly, the parties are also obligated to organize the participation of the public and the environmental and health authorities of the potentially affected state on the same terms as apply to the public in the state of origin.

According to Article 11, all states' parties are obligated to ensure that the final decision on the plan or a programme takes due account of the environmental report and the comments received from the public, the expert authorities and, in the case of potential transboundary impact, the authorities and public in the potentially affected state. All the participants of the SEA procedure must also be sent the final decision, including a summary statement of how the environmental concerns were taken into account in the decision, how the comments by the public and expert authorities were considered in the decision-making and why a certain decision was taken vis-à-vis other reasonable alternatives. Importantly, as stipulated in Article 12, all the parties are also required to monitor the environmental effects of the implementation of the plans and programmes.

Draft North American TEIA Agreement

The draft TEIA, was negotiated under the umbrella of the North American Commission for Environmental Cooperation (NACEC), even though two of the parties are also signatories to the Espoo Convention. The draft TEIA is different in structure from the TEIA procedure of the Espoo Convention, as the national EIA procedures of both Canada and the US differ generally from those in Europe.[39] Although the negotiations between Canada, the US and Mexico produced a draft agreement, it is yet to be negotiated into an international treaty between the concerned states and currently there is no sign that such a process would be set in motion.

The draft TEIA contains two separate procedures: a notification procedure and a TEIA procedure. The origin state is obligated to notify the potentially affected state if certain objective criteria are fulfilled. If the proposed project is closer than 100 kilometres to the international border of the states, or if it belongs to the classes of activities defined in Appendix I of the TEIA, notification must be carried out by the origin state. This is a very objective criterion for initiation of the notification procedure, even more objective than that of the Espoo Convention, in which the notification duty is based on two criteria, i.e., that the proposed activity must be among the categories of activities listed in Appendix I of the Convention and that it must be likely to cause significant adverse transboundary impact on the environment of other contracting states. However, the notification duty of the Espoo Convention triggers the full TEIA procedure, whereas in the draft TEIA it is only a duty to notify other contracting states about the proposed activity, and nothing else. In the draft TEIA, the duty to conduct a TEIA is separated from the duty of notification. At this stage, the origin state is left with much larger discretion as to whether to initiate a TEIA: the only criterion is whether significant adverse transboundary environmental impact is likely from the proposed activity. The determination of whether such effects will ensue from the proposed project is made by the origin state alone, guided by the criteria in Appendix III of the instrument. The potentially affected state has no right to dispute the findings of the origin state other than by appeal to the general dispute settlement provisions of the instrument.[40]

ILC Draft Articles on Prevention

The International Law Commission (ILC) has provisionally adopted draft articles on 'prevention of transboundary damage from hazardous activities'. It seems clear that these articles will remain drafts even if there were plans in place to aim for an international treaty when these were adopted. In any case, the draft articles may well have consequences for the development of customary law on the subject because they have already been adopted by the ILC and are general in scope.

The draft articles apply to a broad set of situations. They include all activities which 'involve a risk of causing significant transboundary harm through their physical consequences'[41] and are not prohibited by international law. The required physical consequences are not confined to likely impacts; only the existence of risk is required. According to Article 2(a), this includes both 'a low probability of causing disastrous harm and a high probability of causing other significant harm'. 'Harm' is defined to include harm to 'persons, property or the environment'.[42] The geographical scope of the draft articles is broad since the link between the activity and the origin state is established through the criterion of 'jurisdiction or control', which includes the areas of jurisdictional competence of states.[43] A potentially affected state is also defined inclusively as any state 'which has jurisdiction or control over any other place where such harm is likely to occur'.[44]

The states are required to establish prior procedures – both impact assessment[45] and licensing procedures – for the environmentally harmful activities.[46] The origin state must notify the potentially affected state if the assessment indicates a risk of significant transboundary harm.[47] This notification must be done in a timely manner, and the origin state must transmit to the potentially affected states 'the available technical and other relevant information on which the assessment is based'.[48] The potentially affected state must respond to this notification within a reasonable time.[49]

If the origin state does not notify the potentially affected state, and the affected state has reasonable grounds to believe that the draft articles apply, the latter can make a request for the origin state to apply Article 10.[50] If the origin state in turn finds that the draft articles are not applicable, it must 'so inform the other state within a reasonable time, providing a documented explanation'.[51] In such an eventuality, the potentially affected state has a right both to enter into consultations[52] and to request that the origin state takes measures to minimize the risk or 'suspend the activity in question for a period of six months unless otherwise agreed'.[53] If the origin state refuses to apply the draft articles, Article 17(2) provides that if the concerned states have not reached an agreement within six months, the potentially affected state can initiate a fact-finding procedure. The report of the fact-finding commission has only an advisory character.[54] The origin state must ensure that the public likely to be affected is provided with information relating to the activity, the risk involved and the harm that may ensue.[55] In addition, the public is guaranteed an undefined right to make its views known.[56]

TRANSBOUNDARY EIA AND CUSTOMARY INTERNATIONAL LAW

There is a clear connection between TEIA and the no-harm Principle (sometimes referred to as no-significant harm principle), given that TEIA is often seen as the procedural component of the no-harm principle, which is articulated in principle 2 of the Rio Declaration and has been endorsed by the ICJ as being part of general international law.[57] Principle 2 reads:

> *States have, in accordance with the Charter of the United Nations and the principles of international law, the sovereign right to exploit their own resources pursuant to their own environmental and developmental policies, and the responsibility to ensure that activities within their jurisdiction or control do not cause damage to the environment of other States or of areas beyond the limits of national jurisdiction.*

The primary rule here is that of no-harm, namely that the states have responsibility to ensure that activities within their jurisdiction or control do not cause damage to the environment of other states or to ABNJ. This primary rule is not an absolute one, that is, an obligation of result, requiring the origin state to prevent damage to the environment of other states or in ABNJ. It is a due diligence principle, which requires states to take all appropriate measures to prevent or minimize the risk of significant transboundary harm, which includes domestic policies and legislation on private and public conduct capable of preventing or at least minimizing the risk of significant transboundary harm and to take account of evolving standards of technology.[58] Given that the no-harm rule is an obligation of diligent prevention and control of foreseeable risks, it is equally clear that in most cases of potential transboundary harm, TEIA must be undertaken.

Pulp Mills on the River Uruguay case

This issue was authoritatively dealt with by the ICJ in the recent *Pulp Mills* case between Uruguay and Argentina.[59] In this dispute over the potential and actual environmental harm caused by the construction and operation of pulp mills located close to the mutual border river between the disputing states, the ICJ, in 2010, made some interesting statements over the status of TEIA in general international law as part of the overall no-harm and due diligence principles (even if it resolved the dispute mainly by applying the 1975 Statute of the River Uruguay).

The Court confirmed that the no-harm principle is a customary rule, and as such has its origins in the due diligence that is required of a state in its territory. For this reason, the Court argued, a state is obligated to use all the means at its disposal in order to avoid activities that take place in its territory, or in any area under its jurisdiction, causing significant damage to the environment of another state.[60]

The Court, importantly, elaborated also on the procedural requirements under general international law in cases of breach of the no-harm and due diligence principles. It stated that due diligence cannot be considered to have been exercised, if the party planning the works liable to affect the quality of the border river did not undertake an EIA on the potential effects of such undertakings and did confirm that customary or general international law requires the origin state to undertake an EIA in cases where there is a risk of the proposed industrial activity having a significant adverse transboundary impact. In the words of the Court:

> *In this sense, the obligation to protect and preserve, under Article 41(a) of the Statute, has to be interpreted in accordance with a practice, which in recent years has gained so much acceptance among States that it may now be considered a requirement under general international law to undertake an environmental impact assessment where there is a risk that the proposed industrial activity may have a significant adverse impact in a transboundary context, in particular, on a shared resource.*[61]

Yet the Court was careful in navigating between views that argue on the one hand that there already exists a fairly specific scope and content for TEIA based e.g., on the Espoo Convention, and on the other, that it is enough that some sort of TEIA is done.[62] According to the Court, the starting-point is that general international law does not specify any clear scope and content for EIA. Yet, it did also state:

> *... it is the view of the Court that it is for each State to determine in its domestic legislation or in the authorization process for the project, the specific content of the environmental impact assessment required in each case, having regard to the nature and magnitude of the proposed development and its likely adverse impact on the environment as well as to the need to exercise due diligence in conducting such an assessment. The Court also considers that an environmental impact assessment must be conducted prior to the implementation of a project. Moreover, once operations have started and, where necessary, throughout the life of the project, continuous monitoring of its effects on the environment shall be undertaken.*[63]

The Court clearly perceives that EIA and other procedural duties form part and parcel of the principles of no-harm and due diligence. What is interesting is how much emphasis the Court places on the exact requirement to carry out an EIA as a mandatory requirement of no-harm and due diligence in transboundary pollution cases. Another interesting feature is that the ICJ emphasizes one aspect that is left optional, for instance in the TEIA prescribed by the Espoo Convention, namely the continuous monitoring of the effects of the project.[64] Here the Court seems to have followed its reasoning in its earlier judgment in the *Gabcíkovo-Nagymaros* case, where it essentially took the view that evaluation of the environmental risks of proposed activities and monitoring of the materialized impacts should be seen as part of the concept of sustainable development[65], a view that was even further clarified by the then Vice-President of the ICJ, Christopher Weeramantry, in his separate opinion.[66]

Transboundary EIA in shared areas

Shared areas contain resources that are not fully subject to the exclusive jurisdiction or control of a state but cannot be deemed to be common property of all states either. Usually, there is a certain measure of common interest in exercising shared rights over the resource in question, which is most often a geographically confined one. Well-known examples are enclosed or semi-enclosed seas (such as the Baltic Sea, where the TEIA can become quite a complex procedure to organize, as will be shown in Chapter 9) and international rivers. The role of TEIA is different in the management of shared resources than it is in situations of potential transboundary pollution between states. The clearest difference is that TEIA is normally only one of the policy tools to protect the environment and regulate the use of a shared resource.

For a long time, the law relating to transboundary watercourses has been considered *lex specialis* from the operation of the no-harm principle. The 1966 'Helsinki Rules on the Uses of the Waters of International Rivers' (hereinafter 'the Helsinki Rules')[67] are widely taken as representing the basic principles governing the use of international rivers.[68] Chapter 2 in particular of these rules, which relates to the principle of equitable utilization, has received the status of *lex lata* and was endorsed in the 1997 Convention on Non-Navigational Uses of International Watercourses prepared by the ILC.[69] Yet, the recent judgment by the ICJ in the *Pulp Mills* case provides nuance to this view. According to paragraph 204 of the judgment:

> *... a practice, which in recent years has gained so much acceptance among states that it may now be considered a requirement under general international law to undertake an environmental impact assessment where there is a risk that the proposed industrial activity*

> *may have a significant adverse impact in a transboundary context, in* **particular, on a shared resource.** *(emphasis mine)*

Hence, it seems that the principles of no-harm and due diligence require TEIA in particular on a shared resource such as a border river, an argument that consolidates the view that no-harm principle – together with associated TEIA – complements the considerations of equitable use in international rivers and drainage basins.

TRANSBOUNDARY EIA IN AREAS BEYOND NATIONAL JURISDICTION

In a typical case, TEIA between states takes place between the sovereign areas of states, where both the origin and the affected state have exclusive jurisdiction. Hence, the treaties regulating such TEIA need to be built on those premises, as a reciprocal affair between two or more states regarding the transboundary impact. International areas and parts of shared areas are not the subject of exclusive territorial jurisdiction of one state, which creates differences in respect of TEIA issues. The terms 'origin state' and 'affected state' that are normally used with regard to territorial TEIA between states are not well suited to TEIA in shared and international areas, where activities as well as their impacts may take place outside the territory of a state; for example, if the environmental impacts of an activity take place in international areas, the term 'affected state' does not apply. This may have consequences for the various procedural and substantial components of a TEIA procedure, for instance, for questions of who should be notified and consulted.

Areas beyond national jurisdiction (ABNJ) are those areas or resources that lie outside of the jurisdiction of any nation or group of nations. Space, the deep seabed, the high seas and, to a different degree, Antarctica, are all ABNJ, areas which face challenges from the viewpoint of environmental protection, given that they and their resources are 'common things', available for consumption by everyone. From the legal point of view, ABNJ fall squarely within the principles of no-harm and due diligence discussed above, as articulated in Principle 2 of the Rio Declaration:

> *States have, in accordance with the Charter of the United Nations and the principles of international law, the sovereign right to exploit their own resources pursuant to their own environmental and developmental policies, and* ***the responsibility to ensure that activities within their jurisdiction or control do not cause damage*** *to the environment of other States or of* ***areas beyond the limits of national jurisdiction.*** *(emphasis mine)*

As analysed above, the ICJ has many times confirmed that this principle is now part of the corpus of international law, and is thus applicable to all states of the world. It has to be admitted that even if this principle is *lex lata*, the obligation that states are required to ensure that no significant harm is caused to ABNJ has gained its main support from international treaties and other instruments rather than actual state practice.

The most developed EIA regime for a *sui generis* type of ABNJ is functioning in the Antarctica.[70] Yet there are also some interesting developments as regards space activities and the deep seabed.[71] Articles 204–206 of the UN Convention on the Law of the Sea (UNCLOS) outline clear but general legal obligations to conduct an EIA over the potential effects of planned activities under their jurisdiction or control that may cause substantial pollution of or significant and harmful changes to the marine environment in general. In a similar vein, the Convention on Biological Diversity (CBD) in its Article 14(1) obligates states to conduct TEIAs over impacts to biological diversity in ABNJ.[72] The most recent developments have taken place in international discussions over how the biodiversity in ABNJ should be managed.[73] Even if no consensus has yet emerged, it is interesting that both main processes that have tackled the issue – United Nations Ad Hoc Open-ended Informal Working Group to study issues relating to the conservation and sustainable use of marine biological diversity in ABNJ and the thematic programme of marine and coastal biodiversity of the CBD – have endorsed TEIA as an important means of protecting biodiversity in ABNJ. The UN Working Group recommended to the UN General Assembly (16 March 2010) the role of TEIA in the following way:

> ***Environmental impact assessments***
>
> *14. The General Assembly should recognize the importance of environmental impact assessments, in particular for the implementation of ecosystem and precautionary approaches.*
>
> *15. It should request the Secretary-General to include, in the annual report on oceans and the law of the sea, information on environmental impact assessments undertaken with respect to planned activities in areas beyond national jurisdiction, including capacity-building needs, on the basis of information requested from States and competent international organizations.*
>
> *16. It should recognize the importance of further developing scientific and technical guidance on the implementation of environmental impact assessments with respect to planned activities in areas beyond national jurisdiction, including consideration of the assessment of cumulative impacts.*[74]

Very similar normative guidance was provided by the ninth meeting of the Conference of the Parties (COP) to the CBD in 2008.[75]

Transboundary EIA and international financial institutions

The development of environmental policy in multilateral development banks predates many of the well-known international environmental agreements.[76] This early interest of development banks in environmental issues may be explained by the strong interrelationship between economic development and environmental issues. While certain environmental concerns may in part be addressed by promoting economic development, development projects may cause significant adverse environmental consequences.[77]

Probably a combination of various factors explains why, since the end of the 1980s, development banks have paid increasing attention to the environmental aspects of their lending activities. These factors include the growing environmental consciousness worldwide, awareness of environmental concerns of development projects, attention to these issues by international commissions and NGOs, and the development of declarations and agreements regarding environmental protection. In particular, EIA for development projects has become a central instrument of the environmental guidelines and policies of development banks. The World Bank has played a central role in this trend with its 2007 revised Operation Policy Statement 4.01 and Best Practice 4.01 on Environmental Assessment.[78] On a regional level, the European Bank for Reconstruction and Development (EBRD) has played a strong role in EIA, explicitly using the Espoo Convention as the main standard in projects likely to involve transboundary impacts.[79]

Governments play a key role in developing and implementing TEIA procedures. Although non-state actors have been involved in the negotiating process of certain TEIA systems, it is the state governments that remain the main actors: they have reached agreement on the various components of TEIA, such as the threshold for determining the need to conduct an EIA and various procedural arrangements, and they must ensure that legal entities under their jurisdiction are subjected to the EIA requirements in accordance with the TEIA system and that the system is being applied in practice. As development banks are primarily owned by governments, governments have had a great deal of influence on the banks' use of EIA as an instrument of environmental policy.

However, anthropogenic transboundary impacts on the environment or other values and interests (e.g. human rights and cultural values) are certainly not the exclusive concern of governments. With the globalization of the economy, prevention and control of adverse social and environmental impacts by multinational companies operating in developing countries has become a topical issue internationally. Particularly in the last 15 years, corporate social responsibility (CSR) in industry has become an important subject of research and practice. Among the many developments that illustrate this trend are the establishment of the World Business Council for Sustainable Development,[80] the adoption of the OECD Guidelines for Multinational Enterprises[81] and the emphasis on the need to develop partnerships in order to involve business in searching for solutions to sustainability issues. While in the past CSR was often considered important for the prevention of damage to a company's reputation, environmental performance is now increasingly and explicitly considered part of a company's core areas of concern.

The special position of the financial sector has been widely acknowledged in developments towards increased responsibility for industry in sustainability issues, among these the prevention of adverse transboundary environmental effects. Financial institutions may have great influence on the environmental performance of industries and even governments through the stock market, direct investments and insurance practices. This influence may be established by banks' choices of which companies they wish to do business with and what terms they set for their clients. With respect to investment activities, this is called 'Socially Responsible Investment' (SRI): expanding stock market practice to evaluate the social records of companies, including their record of environmental compliance and performance, for instance, the one in use in the Johannesburg Stock Exchange.[82]

Commercial banks have also responded to this trend, for instance, by developing sustainable funds. With the adoption of the Equator Principles, 65 commercial banks have committed themselves to make an effort to ensure that the projects these banks finance are developed in a manner that is socially responsible and reflect sound environmental management practices. EIA takes a central place in the principles, thereby confirming that it is a fundamental component of SRI. Currently (October 2010), the Equator Principles are put through a Strategic Review process, the intention being to produce a multi-year strategic vision to make sure that the 2006 Equator Principles[83] retain their viability in environmental and social risk management for project finance within the financial sector.

Conclusion

TEIA has clearly become an important instrument of environmental policy and law in many parts of the world and has been prescribed by a number of international conventions.[84] Certain conventions apply to activities that may cause significant impacts on the environment of a neighbouring country, while only a few regulate EIA for activities in certain ABNJ. In areas where such agreements do not apply, TEIA has increasingly been implemented through the application of national EIA legislation to activities with potential significant transboundary impacts. Parallel to the development and implementation of these systems initiated and adopted by state governments, financial institutions have adopted and continuously improved policies to limit the risk of getting involved in a project that would be inconsistent with the aim of sustainable development, for instance, because the project would cause environmental degradation. The instrument of EIA constitutes a central component of these policies, given that many of the World Bank financed investment projects require an EIA report and the EBRD is committed to applying the Espoo Convention when a project it finances is likely to have transboundary impacts, even if the host country is not a party to the Convention.

It is also likely that the spread of TEIA will gradually continue. Even if the ICJ in the 2010 *Pulp Mills* case was hesitant in using the Espoo Convention as establishing standards for due diligence, this was mainly due to the fact that the Court needs to be careful in not applying conventions that are not in force for the disputing parties. It is important that the ICJ has now confirmed that TEIA – in cases where there is a risk that the proposed industrial activity may have a significant adverse impact in a transboundary context – can now be considered a requirement under general international law and thus be binding on all states.[85] States that are currently applying national EIA law to activities with potential transboundary impact in a certain region may, at a certain moment in time, decide to harmonize their practices by the adoption of a TEIA agreement or guidelines for that particular region, or to become part of an already existing TEIA agreement, such as the Espoo Convention.[86]

Financial institutions will most likely continue to promote this process: by requiring EIA in the process of financing a project, a state government becomes familiar and experienced with applying the instrument of EIA and may thereby be prompted to adopt national EIA legislation, to apply such legislation to transboundary impact situations, to become a party to an existing TEIA convention, or to initiate a process of developing a new TEIA agreement for a region. In respect of state governments that are already parties to a TEIA agreement, financial institutions may stimulate governments to implement such agreements adequately. Moreover, financial institutions may promote the spreading of TEIA directly in their relations with private operators.

Finally, TEIA will probably be developed further for ABNJ. In view of the recent developments regarding ABNJ, TEIA for these areas will require serious attention in the near future. A particularly important development is the emerging consensus on having a strong role for TEIA in managing difficult issues related to marine biological diversity in ABNJ.

TEIA has evolved most clearly within the UNECE region. The Espoo Convention and its SEA Protocol are clearly the leading legally binding treaties on TEIA, the Convention having 45 parties in North America, Europe and Central Asia and its SEA Protocol recently having entered into force (11 July 2010). If states are serious about observing their customary law TEIA obligations, it seems fairly obvious that they should make use of the standards of the Espoo

Convention for TEIA – an observation that applies also to multilateral development and commercial banks when they grant loans for projects likely to have significant adverse transboundary impacts. The first meeting of the Espoo Convention opened up the membership of the Convention to any UN member state; unfortunately, it will probably still take quite some time for this amendment to enter into force.[87] The same applies to multilateral development and commercial banks when they grant loans for projects likely to have significant adverse transboundary impacts.

Even if it can be concluded that the Espoo Convention and other international TEIA standards are a best guarantee for states to follow their customary law obligations, it is also the case that TEIA procedures are currently and increasingly taking place mainly within the UNECE region in general and between Member States of the EU in particular. In many instances, customary law obligations are not followed as standard practice, in contrast with the situation where states have negotiated, implemented and continuously develop a specific international convention for TEIA, such as the Espoo Convention. The EU has implemented the requirements of the Espoo Convention in a very rigorous manner, as will be shown in Chapter 3, which has further increased existing TEIA procedures between the EU Member States.

Notes

1 Council Directive of 27 June 1985 on the assessment of the effects of certain public and private projects on the environment (85/337/EEC) (OJ L175, 5.7.1985, p40).

2 Senate resolution 49 passed by the Senate on 21 July 1978. It is reproduced in 4 *ILM* 1082 (1978). In the following year, the Council of the OECD made a recommendation to the member governments on the 'Assessment of Projects with Significant Impacts on the Environment'. However, this recommendation only referred to 'environmental assessment procedures for actions that might have significant transboundary effects' (paragraph 8). The recommendation adopted on 8 May 1979 (C (79) 116). The recommendation can be found in Hohmann, H. (ed) (1992) *Basic Documents of International Environmental Law: The Important Declarations*, Graham & Trotman Ltd, vol 1, pp400–401, available also at http://sedac.ciesin.org/entri/texts/oecd/OECD-4.15.html

3 The 1974 Nordic Environmental Protection Convention (NEPC). The NEPC was signed on 19 February 1974 by Finland, Sweden, Norway and Denmark and came into force on 5 October 1976. It is still in force. The NEPC contains a Protocol, paragraph 1 of which provides: 'In connection with the signing today of the Nordic Environmental Protection Convention the duly authorized signatory agreed that the following comments on its application shall be appended to the Convention'. For an overview of the Convention, see Broms, B., The Nordic Convention on the Protection of the Environment, in Flinterman, Kwiatkowska, Lammers (eds) (1986) *Transboundary Air Pollution: International Legal Aspects of the Co-operation of States*, Martinus Nijhoff Publishers, pp141–152.

4 Ibid., Article 1 of the NEPC ends its enumeration of pollution effects with 'etc.'.
5 Ibid., Article 5.
6 Ibid., Article 11.
7 Ibid., Article 5.
8 Ibid. Article 4 defines the concept of 'supervisory authority': 'Each State shall appoint a special authority (supervisory authority) to be entrusted with the task of safeguarding general environmental interests insofar as regards nuisances arising out of environmentally harmful activities in another Contracting State'.
9 Ibid., Article 5.
10 Ibid.
11 Ibid., Article 6.
12 Ibid., Article 7.
13 Ibid., Article 10.
14 See Broms 1986, Ibid. note 3, p 148.
15 Article 11 of the NEPC.
16 Ibid., Article 12. The Espoo Convention contains a similar provision which is discussed later in this chapter and also in Chapters 3 and 5.
17 Ibid.
18 Senate resolution 49 passed the Senate on 21 July 1978. It is reproduced in 4 *ILM* 1082 (1978).
19 Article I of the resolution.
20 Ibid., Article II(1).
21 Ibid., Article II(2).
22 Ibid.
23 Section 1 of the Executive Order, Executive Order 12114 – Environmental effects abroad of major Federal actions, available at www.archives.gov/federal-register/codification/executive-order/12114.html
24 Reproduced in Basic Documents of International Environmental Law, vol 1, 1992, pp187–190, Ibid. note 2, also available at www-penelope.drec.unilim.fr/penelope/library/Libs/Int_nal/unep/unep.htm
25 Principle 11 of the Goals and Principles.
26 Ibid., Principle 12.
27 Ibid.
28 For an analysis of this connection, see Connelly, R., The UN Convention on EIA in a Transboundary Context: a Historical Perspective (1999) *Environmental Impact Assessment Review*, vol 19, pp37–46.
29 The Convention on Environmental Impact Assessment in a Transboundary Context, printed in 30 *ILM* 800 (1991), and can be accessed via the home page of the Convention at www.unece.org/env/eia/documents/conventiontextenglish.pdf The Convention entered into force on 10 September 1997.
30 Espoo Convention, Article 2 (2, 3 and 7).
31 Ibid., Article 3 (1, 2 and 3).
32 Ibid. The expression 'The public' was defined in Article 1(x) of the Espoo Convention simply as 'one or more natural or legal persons'. Due to the influence of the Aarhus Convention on Access to Information, Participation in Decision-making and Access to Justice in Environmental Matters (25 June 1998, Aarhus, Denmark, in force

30 October 2001, 38 *ILM* 517 (1999)), the second MOP, held in Sofia in 2001, adopted an amendment to the Espoo Convention, adding the following to Article 1(x): 'and, in accordance with national legislation or practice, their associations, organizations or groups'.

33 Ibid., Article 3 (1, 2 and 3), Article 3 (4-8) and Article 2(6) on non-discrimination.

34 Ibid., Articles 4 and 5.

35 Ibid., Article 6.

36 Ibid., Article 7 and Appendix V.

37 Churchill and Ulfstein argued in 2000 that a special feature of recent multilateral environmental agreements has been that states' parties not only conclude an agreement with binding obligations but also establish institutional frameworks, such as COPs, to oversee the development of the treaty system rather than setting up traditional inter-governmental organizations for this purpose. They also perceive that such AIAs, as they call them, can be considered as inter-governmental organizations (IGO) for the purpose of applicable law even though they are of a less formal nature and less permanent than the traditional IGOs. It seems clear that the system of the Espoo Convention qualifies as such an AIA. See Churchill, R. and Ulfstein, G., (2000) Autonomous Institutional Arrangements in Multilateral Environmental Agreements: A Little-Noticed Phenomenon in International Law, *American Journal of International Law*, vol 94, pp623–659.

38 There is a significant difference between the Protocol and the Espoo Convention where the definition of environmental effect is concerned. The Protocol states explicitly that 'environmental effect' includes effects on the state of biodiversity, an element of the environment that was left out of the Espoo Convention's definition of 'impact' in Article 1(VII). The Protocol also places greater emphasis on impacts to human health than the Convention does. This focus manifests itself in many parts of the Protocol and is partly attributable to the influence of the World Health Organization (WHO) on the drafting of the instrument. On the other hand, in contrast to the Protocol, the Convention's definition of 'impact' covers changes in socio-economic conditions where such effects are caused by changes in physical factors.

39 The main difference is that the European EIA Directive procedures are triggered if objective criteria are fulfilled, notably in relation to a list of large scale environmentally harmful activities. If the procedure is initiated, it normally continues until the licencing decision is made. The North American EIA procedures contain more flexibility, as was noted above. Especially in the federal EIA procedure of the US, the agency in charge can decide to make a preliminary impact study after which it may decide whether a full-blown impact study is required or whether the project can proceed. See Gilpin, A. (1995) *Environmental Impact Assessment (EIA): Cutting Edge for the Twenty-First Century*, Cambridge University Press.

40 See further, Craik, N., Transboundary Environmental Impact Assessment in North America: Obstacles and Opportunities in Bastmeijer, K. and Koivurova, T. (eds) (2008) *Theory and Practise of Transboundary Environmental Impact Assessment*, Martinus Nijhoff, Leiden pp93–117.

41 Article 1 of the draft articles. Available at http://untreaty.un.org/ilc/texts/instruments/english/draft%20articles/9_7_2001.pdf

42 Ibid., Article 2(b).
43 Ibid., Article 2(d).
44 Ibid., Article 2(e).
45 Ibid. Article 8 provides: 'Any decision in respect of the authorization of an activity within the scope of the present draft articles shall be based on an evaluation of the possible transboundary harm caused by that activity'.
46 Ibid., Article 7.
47 Ibid., Article 10(1).
48 Ibid.
49 Ibid., Article 10(2).
50 Ibid., Article 13(1).
51 Ibid., Article 13(2).
52 Ibid., Article 11(1).
53 Ibid., Article 13(3).
54 Ibid., Article 17(2).
55 Ibid., Article 9.
56 Ibid.
57 The ICJ first stated this view in its Advisory Opinion in the Legality of the Threat or Use of Nuclear Weapons, Advisory Opinion, ICJ Reports 1996, p226, paragraph 29. The Court also repeated this passage in the *Case Concerning the Gabcíkovo-Nagymaros Project* (Hungary/Slovakia), Judgment, ICJ Reports 1997, p7, paragraph 53.
58 See Birnie, P., Boyle, A. and Redgwell, C. (2009) *International Law & the Environment*, Oxford University Press, in particular pp143–150.
59 *Case Concerning Pulp Mills on the River Uruguay* (Argentina v. Uruguay), 20 April 2010, Judgment, available at www.icj-cij.org/docket/files/135/15877.pdf. For recent commentary, see De Mulder, J., (2010) Case Note, *Review of European Community and International Environmental Law*, vol 19, no 2, pp263–267.
60 Ibid., paragraph 101.
61 Ibid., paragraph 204.
62 Ibid., paragraphs 203–205. The Court did emphasize another instrument taken up by Argentina, namely the above-mentioned UNEP Guidelines on EIA. Yet, it stated that these principles as non-binding international technical standards should be taken into account when interpreting and applying Article 41(a) of the Statute of the River Uruguay, which referred to such standards (paragraph 205).
63 Ibid., paragraph 205.
64 After a careful examination, the Court came to the conclusion that Uruguay had acted with the requisite degree of due diligence in conducting its transboundary EIA over the harmful effects of permitting the construction of the Botnia pulp mill, even if it also found Uruguay to have violated procedural obligations of the 1975 Statute.
65 *Gabcíkovo-Nagymaros Project (Hungary/Slovakia), Judgment, ICJ Reports* 1997, p7. Paragraph 140 of the judgment reads: 'In order to evaluate the environmental risks, current standards must be taken into consideration. This is not only allowed by the wording of Articles 15 and 19, but even prescribed, to the extent that these articles impose a continuing – and thus necessarily evolving – obligation on the parties to maintain the quality of the water of the Danube and to protect nature [...]. The Court is mindful that, in the field of environmental protection, vigilance and

prevention are required on account of the often irreversible character of damage to the environment and of the limitations inherent in the very mechanism of reparation of this type of damage [...]. Such new norms have to be taken into consideration, and such new standards given proper weight, not only when States contemplate new activities but also when continuing with activities begun in the past. This need to reconcile economic development with protection of the environment is aptly expressed in the concept of sustainable development [...].' *Gabcíkovo-Nagymaros Project (Hungary/Slovakia),* Ibid. note 65, p111.

66 In his separate opinion, Weeramantry provided, under the heading 'The Principle of Continuing Environmental Impact Assessment' (at 88–119) the following: 'I wish in this opinion to clarify further the scope and extent of the environmental impact principle in the sense that environmental impact assessment means not merely an assessment prior to the commencement of the project, but a continuing assessment and evaluation as long as the project is in operation. This follows from the fact that EIA is a dynamic principle and is not confined to a pre-project evaluation of possible environmental consequences. As long as a project of some magnitude is in operation, EIA must continue, for every such project can have unexpected consequences; and considerations of prudence would point to the need for continuous monitoring'. Ibid. p111.

67 The Helsinki Rules can be obtained in vol. 1, Basic Documents of International Environmental Law, p227. These rules were adopted by a non-governmental expert body of international lawyers, the ILA.

68 Article I of the Helsinki Rules declares: 'The general rules of international law as set forth in these chapters'. Thus, the intent was to codify the law relating to transboundary watercourses. Although the term 'international river' is used in the name of these rules, they apply to 'international drainage basin', a term defined in Article II of the Helsinki Rules as follows: 'An international drainage basin is a geographical area extending over two or more States determined by the watershed limits of the system of waters, including surface and underground waters, flowing into a common terminus'.

69 Articles 5 and 6 of the 1997 Watercourses Convention. In the 1966 Helsinki Rules, equitable utilization is defined in the following manner: Article IV provides: 'Each basin State is entitled, within its territory, to a reasonable and equitable share in the beneficial uses of the waters of an international drainage basin'; the beginning of Article V adds: 'What is a reasonable and equitable share within the meaning of Article IV is to be determined in the light of all the relevant factors in each particular case'; and finally, Article VI states: 'A use or category of uses is not entitled to any inherent preference over any other use of category of uses'. Article 6 of the 1997 Watercourses Convention is the most relevant here. Paragraph 1 of the Article provides: 'Utilization of an international watercourse in an equitable and reasonable manner within the meaning of Article 5 requires taking into account all relevant factors and circumstances'. Paragraph 3 adds: 'The weight to be given to each factor is to be determined by its importance in comparison with that of other relevant factors. In determining what is a reasonable and equitable use, all relevant factors are to be considered together and a conclusion reached on the basis of the whole'. See also the ILC produced 2008 Draft Articles on the Law of Transboundary Aquifers, available at http://untreaty.un.org/ilc/texts/instruments/english/draft%20articles/8_5_2008.pdf

70 Bastmeijer, K. and Roura, R., Environmental Impact Assessment in Antarctica in Bastmeijer, K. and Koivurova, T. (eds) (2008) *Theory and Practise of Transboundary Environmental Impact Assessment*, Martinus Nijhoff, Leiden, 175–219.

71 See Part II in *Theory and Practise of Transboundary Environmental Impact Assessment*, Ibid. note 70.

72 Article 14(1) of the Convention on Biological Diversity prescribes that: 'Each Contracting Party, as far as possible and as appropriate, shall: ... (c) Promote, on the basis of reciprocity, notification, exchange of information and consultation on activities under their jurisdiction or control which are likely to significantly affect adversely the biological diversity of other States or areas beyond the limits of national jurisdiction, by encouraging the conclusion of bilateral, regional or multilateral arrangements, as appropriate; (d) In the case of imminent or grave danger or damage, originating under its jurisdiction or control, to biological diversity within the area under jurisdiction of other States or in areas beyond the limits of national jurisdiction, notify immediately the potentially affected States of such danger or damage, as well as initiate action to prevent or minimize such danger or damage ...', available at www.cbd.int/convention/conventions.html

73 For a comprehensive overview, see Warner, R. (2009) *Protecting the Oceans Beyond National Jurisdiction: Strengthening the International Law Framework*, Martinus Nijhoff Publishers, in particular pp207–235.

74 The Working Groups recommendation to the General Assembly can be found at the beginning of the Co-Chairs document, available at http://daccess-dds-ny.un.org/doc/UNDOC/GEN/N10/277/20/PDF/N1027720.pdf

75 The COP provided that: '8. *Invites* Parties, other Governments and relevant organizations, including in the context of the United Nations Ad Hoc Open-ended Informal Working Group to study issues relating to the conservation and sustainable use of marine biological diversity beyond areas of national jurisdiction, to cooperate in further developing scientific and technical guidance for the implementation of environmental impact assessments and strategic environmental assessments for activities and processes under their jurisdiction and control which may have significant adverse impacts on marine biodiversity beyond national jurisdiction, taking into consideration the work of Food and Agriculture Organization of the United Nations, the International Maritime Organization, and other relevant organizations, with a view to ensuring such activities are regulated in such a way that they do not compromise ecosystem integrity, and to report to the COP at its Tenth Meeting on progress made in that regard; 9. *Notes* the need for capacity-building for developing countries, in order to fully implement existing provisions of environmental impact assessment, as well as the challenges and difficulties in carrying out environmental impact assessment in areas beyond national jurisdiction; 10. For the purpose of paragraphs 8 and 9 of the present decision, taking into account the relevant provisions of the United Nations Convention on the Law of the Sea and the Convention on Biological Diversity, *decides* to convene an expert workshop, including experts from different relevant organizations, with balanced regional and sectoral representation, to discuss scientific and technical aspects relevant to environmental impact assessment in areas beyond national jurisdiction with a view to contributing to the development of such scientific and technical guidance,

building on ongoing relevant sectoral, regional and national environmental impact assessment efforts'; Decision adopted by the Conference of the Parties to the Convention on Biological Diversity at its Ninth Meeting (IX/20. Marine and Coastal Biodiversity), UNEP/CBD/COP/DEC/IX/20, 9 October 2008, available at www.cbd.int/doc/decisions/cop-09/cop-09-dec-20-en.doc

76 See, among others, Steer, A. and Mason, J. (1995) The Role of Multilateral Finance and the Environment: A View from the World Bank, *Indiana Journal of Global Legal Studies,* vol 3, pp35–45.

77 See Guyett, S. C. (1991–92) Environment and Lending: Lessons of the World Bank, Hope for the European Bank for Reconstruction and Development, *New York University Journal of International Law and Politics*, vol 24, p892.

78 See the OP, available at http://web.worldbank.org/WBSITE/EXTERNAL/PROJECTS/EXTPOLICIES/EXTOPMANUAL/0,,contentMDK:20064724~menuPK:64701637~pagePK:64709096~piPK:64709108~theSitePK:502184,00.html and BP available at http://web.worldbank.org/WBSITE/EXTERNAL/PROJECTS/EXTPOLICIES/EXTOPMANUAL/0,,contentMDK:20064614~menuPK:64701637~pagePK:64709096~piPK:64709108~theSitePK:502184,00.html

79 See the EBRD's Environmental and Social Policy (May 2008), available at www.ebrd.com/downloads/about/sustainability/2008policy.pdf

80 Available at www.wbcsd.org/templates/TemplateWBCSD5/layout.asp?MenuID=1

81 The text is available at www.oecd.org/dataoecd/56/36/1922428.pdf

82 Available at www.jse.co.za/about-us/media/press-releases/full-story/06-05-04/Launch_of_JSE_Socially_Responsible_Investment_Index.aspx

83 Available at www.equator-principles.com/documents/Equator_Principles.pdf

84 For comprehensive studies, see Craik, N. (2008) *The International Law of Environmental Impact Assessment: Process, Substance and Integration*, Cambridge University Press and Theory and Practise of Transboundary Environmental Impact Assessment, Ibid. note 70.

85 *Case Concerning Pulp Mills on the River Uruguay* (Argentina v. Uruguay), 20 April 2010, Judgment, paragraph 204, available at www.icj-cij.org/docket/files/135/15877.pdf

86 This is, however, not yet possible, given that the first amendment to the Convention, which makes it potentially possible for any UN member to become a party outside of the UNECE region, has not yet entered into force. See the current status of ratification of the first amendment, available at http://treaties.un.org/Pages/ViewDetails.aspx?src=TREATY&mtdsg_no=XXVII-4-a&chapter=27&lang=en

87 Article 17(2) addition: 'Any other State, not referred to in paragraph 2 of this Article, that is a Member of the United Nations may accede to the Convention upon approval by the Meeting of the Parties. The Meeting of the Parties shall not consider or approve any request for accession by such a State until this paragraph has entered into force for all the States and organizations that were Parties to the Convention on 27 February 2001', available at www.unece.org/env/eia/about/amendment.html. For a proposal on how to speed up this process, see Knox, J. (2003–05) Assessing the Candidates for a Global Treaty on Transboundary Environmental Impact Assessment, *New York University Environmental Law Journal*, vol XII, pp153–168, at p168, available at www1.law.nyu.edu/journals/envtllaw/issues/vol12/1/12n1a6.pdf

3

Transboundary Environmental Impact Assessment in European Law

Simon Marsden

Chapter outline

This chapter considers the legal development and implementation of transboundary environmental impact assessment (TEIA) in European Union (EU) or European law, from the introduction of brief provisions in the original EIA Directive to their expansion in the amended EIA Directive to implement the Espoo Convention. More recently, directives for public participation and strategic environmental assessment (SEA) have also contained provisions relating to TEIA, partly to implement aspects of the Aarhus Convention. In addition to the substantive law, the chapter examines the role of the European Commission and European Court of Justice (ECJ) in reporting on, implementing and enforcing these directives where relevant to TEIA.

Introduction

Article 7 of EC Directive 85/337 (the EIA Directive[1]) introduced provisions for significant transboundary environmental effects to be assessed under European law. The first two Implementation Reports on the EIA Directive, however, contained little evidence of the application of this provision. Subsequently, after the Espoo Convention[2] came into force,[3] an amended EIA Directive 97/11/EC[4] introduced a new Article 7, making the Member States responsible for the detailed arrangements for implementing the requirements of the Espoo Convention. More recently, Directive 2003/35/EC has further amended the transboundary procedure in respect of public participation,[5] Directive 2009/31/EC has amended Annexes I and II by adding projects related to the transport, capture and storage of carbon dioxide (CO_2),[6] and Directive 2001/42/EC (the SEA Directive) makes provision for transboundary consultations in respect of plans and programmes where there are effects on another Member State.[7]

As the original EIA Directive was amended to take account of the coming into force of the Espoo Convention, and the amended EIA Directive was again revised by the Public Participation Directive to do the same for the Aarhus Convention,[8] so EU law will need to be further amended to account for the coming into force of the Kiev Protocol in 2010,[9] which contains slightly different provisions than the SEA Directive.[10] However, EU law has also influenced the development of the Espoo Convention and Kiev Protocol. In respect of the Convention, the original EIA Directive was adopted six years before it, providing impetus to its development although not in respect of its content; the SEA Directive was adopted two years before the Protocol, the latter which, as an instrument primarily operating in the context of individual states, was based substantially on it. The role of the Aarhus Convention has also been influential in producing change to EU law and, in addition to the Public Participation Directive, it was responsible for the new Directive 2003/4/EC on access to information,[11] which also contains provisions relevant to EIA.[12]

This chapter begins by analysing the substantive secondary EU law together with its relationship with the Convention and Protocol. The results of the reviews of implementation of the EIA, Public Participation and SEA Directives are evaluated where comment is made in relation to the TEIA provisions. The most recent EIA Directive Review was reported on in 2009 and, as a follow-up exercise, public consultation was carried out in mid-2010 to collect opinions on the overall view on its functioning and effectiveness, the need for amendment, possible policy options for review, and the areas to be improved or amended. The first review of the Public Participation Directive was reported on in 2010, and the first SEA Directive Review in 2009 and a comparable consultation exercise in respect of the SEA Directive can be anticipated now that the Kiev Protocol has come into force. Treaty reporting obligations in respect of the Conventions and Protocol and implementation experience by EU law, is also discussed. Enforcement matters are thereafter considered, specifically the limited ECJ case law on TEIA and participation. Throughout the chapter, the terminology of the Espoo Convention is used loosely to refer to the proponent state as the 'origin' Member State, and the state likely to be affected as the 'affected' Member State.

EIA DIRECTIVE

Recitals (12) and (13) of the amended EIA Directive refer to the desirability of strengthening provisions concerning EIA in a transboundary context, and the signature of the Community to the Espoo Convention. As such it was recognized that initially the EU was not compliant with the Convention, and changes were required. The EIA Directive as amended now implements almost all of the provisions of the Espoo (and, as will be seen, Aarhus) Convention(s) regarding TEIA. As argued in Chapter 5 (Marsden), provision for the Inquiry Procedure in the Espoo Convention has, however, not been transposed by the amended EIA Directive, although Article 7(1) allows an affected Member State to request the application of the transboundary consultation procedure if it is likely to be significantly affected.[13] The EIA Directive contains two more obligations than the Espoo Convention. It contains stronger wording that the results of the TEIA are to be taken into account, and it requires the affected states to inform their own public when they receive the final decision, which the Convention does not.

Substantive obligations

Article 7(5) of the amended EIA Directive, delegates to the concerned Member State the 'detailed arrangements for implementing' all of the aspects of the transboundary consultation procedure, which is triggered under Article 7(1) when the origin Member State is aware that the project is likely to have significant effects on the environment of another Member State, or if the latter requests its application. If triggered, Article 7(1) requires documentation to be forwarded to the affected Member State describing the project, possible transboundary impact, and the nature of the decision to be taken, giving the affected Member State a reasonable time to indicate whether it wishes to participate.

If it does, under Article 7(2) the origin Member State must send all the information gathered and information on the procedure, including the request for development consent. Article 7(3) requires both or all involved Member States (if more than one) to arrange for the information to be made available within reasonable timeframes to relevant authorities and the public concerned in the affected Member State. Those authorities and the public concerned must be given an opportunity before the granting of the development consent to forward their opinion. Article 7(4) stipulates that consultations within an agreed timeframe must be entered into regarding the potential transboundary effects and measures proposed to mitigate or prevent them.

Implementation reviews

The European Commission is tasked with carrying out periodic reviews of specific directives, as required by the instruments themselves.[14] The Commission also prepares annual reports on the application of EU law generally.[15] With reference to the two first EIA Directive implementation reports, it was concluded that 'there was scanty evidence on the use of Article 7' but that 'several states without a formal provision did in fact notify neighbouring states of projects'.[16] The 2003 Commission Report comments on the variety of issues that have arisen in applying Article 7 across the EU, including those of a procedural nature, such as establishing points of contact and obtaining translation; ways of overcoming such issues are also discussed.[17] At the time, few bilateral agreements had been entered into between Member States concerning these issues, but this would seem to have developed since then.[18] Following a detailed examination of the substantive obligations of the EIA Directive relating to TEIA, the most recent Commission review is considered below together with its response to the Espoo Convention Implementation Committee.

The most recent review of the EIA Directive was concluded in 2009.[19] The Report comments on the need to generate synergy between the EIA Directive and Espoo Convention, and the EIA Directive has been identified as a potential instrument for a simplification exercise concerning transboundary projects. Specifically, the Report draws attention to the fact that despite increasing public participation in the decision-making process, there is still no standard practice across the EU, with the meaning of early and effective participation tending to be interpreted as public consultation once the information has been collected; timeframes for consultation also vary considerably, as do arrangements for informing and consulting the public and criteria for standing in a review procedure, which are determined by the Member States.[20]

Difficulties regarding TEIA procedures point to the differences between national EIA procedures, suggesting the need for further harmonization; language barriers also continue to cause concern. Projects that physically cross Member State boundaries are singled out, with risks of duplication, inconsistency and cumulative effects highlighted; the need to improve consultation arrangements including coordinated, joint or single procedures is therefore emphasized, with the Report recommending:

> *Improved formal and informal arrangements are needed for consultation on transboundary impacts with neighbouring countries. This could be achieved by developing guidance or by strengthening the Directive's provisions (e.g. definition of minimum timeframes for transboundary consultation, coordinated or possibly joint EIA procedures for multi-country projects; explore the possibility for having a single EIA procedure).*
>
> *Furthermore, following approval by the Council of the SEA Protocol to the Espoo Convention, the EIA Directive could be reviewed to ensure better implementation of the requirements of the Protocol.*

Since the Report, the European Commission has carried out an internet-based public consultation exercise in respect of the review, looking at options for reform.[21] The target group is deliberately broad and comprises all citizens, stakeholders and organizations involved in EIAs of plans, programmes or projects. The Commission is equally interested in getting the views of national parliaments,

Public Administrations (at national, regional and local level), the European Parliament and the Council, the Social and Economic Committee and the Committee of the Regions, as well as other institutions such as the European Investment Bank and the European Bank for Reconstruction and Development. The Commission will take all contributions into account and will publish a summary paper on its environment public consultation website, which will influence future reform proposals.

As indicated in Chapter 1 of this book, the Commission on behalf of the EU and other Parties to the Convention has also responded to the Implementation Committee, which has recently been compiling completed questionnaires from Parties to the Convention in order to prepare a report for the period to 2009, with the third Review of Implementation expected to be published in 2011.[22] Since the questionnaire is directed towards states, the EU has responded by way of a brief one page letter repeating its competence from previous reviews, and indicating the extent of the EU's compliance with its obligations with specific reference to measures in the directives that implement provisions of the Convention, together with the recent EU Commission reports on effectiveness.[23]

Public Participation Directive

Transboundary procedure was amended further under Directive 2003/35/EC to comply with the Aarhus Convention. This related UNECE treaty has a close relationship with the Espoo Convention, as discussed in Chapter 6 (Marsden), and the Espoo Convention is one of only three conventions cited in the recitals to the Aarhus Convention. Recital (5) of the Directive cites the need for EU law to be compliant with the Convention. Recital (7) emphasizes that Article 6 of the Aarhus Convention is concerned with public participation in relation to EIA of specific activities, which is confirmed by recital (11) which refers to the EIA Directive; recital (8) further states that Article 7 is concerned with public participation in relation to SEA concerning plans and programmes. Recital (10) states the general objective of this and related directives:

> *Provision should be made in respect of certain Directives in the environmental area which require Member States to produce plans and programmes relating to the environment but which do not contain sufficient provisions on public participation, so as to ensure public participation consistent with the provisions of the Aarhus Convention, in particular Article 7 thereof. Other relevant Community legislation already provides for public participation in the preparation of plans and programmes and, for the future, public participation requirements in line with the Aarhus Convention will be incorporated into the relevant legislation from the outset.*

Substantive obligations

Article 3 amends the EIA Directive generally. Definitions of 'the public' and 'the public concerned' have been added, the latter which includes not only the public affected or likely to be affected by the decision but also those having an interest in environmental decision-making, specifically NGOs. The complete exemption from assessment of projects for national defence has been removed, and information is now required on alternative assessment procedures to be made public when a Member State decides to exempt specific projects from the Directive. Detailed provisions establish which information is to be made available, and that it must be released early in the process. Early and effective opportunities for public participation within reasonable timeframes are required, and the public must be informed of the main reasons for the decision and of the consultation process.

Article 3(4) of the Public Participation Directive requires the public to be informed early in the environmental decision-making procedures, and at the latest as soon as the information can be provided, about the fact that the project is subject to an EIA procedure and, where relevant, that Article 7 applies, and hence there are potential significant transboundary environmental effects. This is as a result of the insertion of a new Article 6(2)(b) in the EIA Directive. Article 3(6) amends Article 9(2) of the amended EIA Directive requiring the competent authority or authorities to inform any Member State that has been consulted pursuant to Article 7, forwarding to it the information referred to in Article 9(1).

The detailed amendments to the transboundary procedures are found in Article 3(5) of the Public Participation Directive, which amends Article 7 paragraphs (1), (2) and (5) of the amended EIA Directive to make specific reference to the environmental decision-making procedures in Article 2(2) of the Public Participation Directive. These are as follows:

> *Article 7 shall be amended as follows:*
>
> *(a) paragraphs 1 and 2 shall be replaced by the following:*
>
> > *'1. Where a Member State is aware that a project is likely to have significant effects on the environment in another Member State or where a Member State likely to be significantly affected so requests, the Member State in whose territory the project is intended to be carried out shall send to the affected Member State as soon as possible and no later than when informing its own public, inter alia:*
> >
> > *(a) a description of the project, together with any available information on its possible transboundary impact;*

> *(b) information on the nature of the decision which may be taken,*
>
> *and shall give the other Member State a reasonable time in which to indicate whether it wishes to participate in the environmental decision-making procedures referred to in Article 2(2), and may include the information referred to in paragraph 2 of this Article.*
>
> *2. If a Member State which receives information pursuant to paragraph 1 indicates that it intends to participate in the environmental decision-making procedures referred to in Article 2(2), the Member State in whose territory the project is intended to be carried out shall, if it has not already done so, send to the affected Member State the information required to be given pursuant to Article 6(2) and made available pursuant to Article 6(3)(a) and (b).'*
>
> *(b) paragraph 5 shall be replaced by the following:*
>
> *'5. The detailed arrangements for implementing this Article may be determined by the Member States concerned and shall be such as to enable the public concerned in the territory of the affected Member State to participate effectively in the environmental decision-making procedures referred to in Article 2(2) for the project.'*

Article 3(7) of the Public Participation Directive has also inserted a new Article 10a into the EIA Directive to implement the third pillar of the Aarhus Convention on access to justice in specific regard to EIA, which has application in a transboundary context. It recognizes the significance of this area for environmental decision-making across the EU, and operates in the absence of the proposed Access to Justice Directive.[24] This provides access for the public concerned to the courts or other independent and impartial bodies, to challenge 'the substantive or procedural legality' of decisions, acts or omissions, which must be 'fair, equitable, timely and not prohibitively expensive'; the public concerned must, however, have a 'sufficient interest' or 'maintain the impairment of a right'.

Member States retain the power to decide in accordance with their national legal system at what stage decisions and acts of omissions may be challenged, and what constitutes a sufficient interest or impairment of a right; this limits the potential for challenges, although 'wide access to justice' is specified as an objective in determining what a 'sufficient interest' or 'impairment of a right' means, which restricts Member State discretion. Member States must also provide the public with practical information on access to administrative and judicial review procedures.[25] Article 10a is likely to result in increased litigation with the EIA Directive requiring interpretation under Article 6 of the Aarhus Convention regarding specific activities,[26] and also in accordance with the principles of effective judicial protection and consistent interpretation (indirect effect).

Article 2 of the Public Participation Directive provides for public participation in relation to the preparation of plans and programmes. Article 2(2) obliges Member States to ensure that the public is given early and effective opportunities to participate in the preparation and modification or review of the plans or programmes required to be drawn up under the provisions listed in Annex I. Article 2(5) confirms that this Article shall not apply to plans and programmes set out in Annex I for which a public participation procedure is carried out under the SEA Directive or under the Water Framework Directive.[27] The obligations require the public to be informed about such proposals, the right to participate in decision-making, and the competent authority to whom comments and questions are to be submitted. Comments and opinions can be expressed at a time when all options are open and before any decisions are taken, and such decisions must take due account of these comments and opinions. Reasonable efforts must thereafter be taken to inform the public of the decisions taken and the reasons and considerations on which they are based including the public consultation.

Implementation reviews

On 14 April 2010, the Commission adopted a report on the application and effectiveness of Directive 2003/35/EC, compiled pursuant to its Article 5.[28] As indicated, the aim of this Directive is to give fuller effect to the Aarhus Convention. Although the Directive does not explicitly stipulate that Member States are to submit to the Commission national reports on the experience acquired, the Commission felt that it needed to be able to draw on information gained by them. The level of information and detail in these contributions differs considerably. The Report is limited to an examination of Article 2, due to the fact that aspects dealing with Article 3 were examined in the EIA Directive Report (see above discussion of relevant provisions for TEIA).

It is the first Commission report on the Directive, hence its significance, although there is no specific reference to implications for TEIA. The Report examines issues of relevance to both the EIA and SEA Directives, including issues relating to the necessity or appropriateness of submitting proposals for amendments and the possibility of extending the scope of the Directive to plans and programmes other than the ones provided for in Annex I. The Report discusses the potential extension of Article 2 beyond the categories of plans and programmes listed in Annex I, and comments on the relationship with the SEA Directive as follows (p5):

> *Following the legislative changes made after the adoption of the Directive, the plans and programmes have concerned and will concern*

> *from end-2010 in particular waste-management and air-quality plans. In view of their nature and scope the SEA Directive might appear to be a more suitable instrument in reality than Article 2 for dealing with these matters, including the public participation aspect. Moreover, in its Article 6, the SEA Directive provides for public participation on a scale comparable to that of Article 2 of the Directive. A similar consideration emerges from the experience acquired in some States. At this stage, this is no more than a trend. The future will tell whether this trend is built upon and becomes a general practice.*

In the same way that Parties to the Espoo Convention are required to report to the Implementation Commission on measures taken to implement the Convention and practical implementation, so Parties to the Aarhus Convention must also report to the Secretariat in advance of the MOP.[29] The Aarhus Clearinghouse on Environmental Democracy contains all of these reports, including the first by the EU in 2008.[30] The EU Report not surprisingly refers to the legislative amendments and EU reports discussed above, and adds nothing further to these.

SEA Directive

The SEA Directive, although adopted and in force before the Kiev Protocol, implements most of its obligations. Like the Protocol, it is not limited to the assessment of proposed plans and programmes having potential significant environmental effects within a state. It also applies in respect of proposed plans and programmes that may have effects in other Member States, whether the plan or programme relates to the territory of the Member State itself, or, less commonly, the territory of more than one Member State. The general obligations of the SEA Directive and Kiev Protocol and the relationship with the transboundary procedure are described and analysed in detail in Chapter 10 of the book (Marsden and De Mulder).

Recital (7) of the SEA Directive refers to both the Espoo Convention encouraging application of its principles to plans and programmes and to the Kiev Protocol, at the time not yet adopted. The recital draws attention to the distinction between Member States and non-EU Member States, recognizing existing discretionary provisions in the Espoo Convention and inevitable obligations that the Kiev Protocol would bring to the EU upon adoption and entering into force for the EU itself. Reference to a 'reciprocal and equivalent basis within an appropriate legal framework' in the recital for these consultations to take place therefore recognizes the absence of the harmonizing role of EU directives in respect of these relationships unless the non-EU Member State complies with such directives.[31] The recital states:

The systems operating within the Community for environmental assessment of plans and programmes should ensure that there are adequate transboundary consultations where the implementation of a plan or programme being prepared in one Member State is likely to have significant effects on the environment of another Member State. The information on plans and programmes having significant effects on the environment of other States should be forwarded on a reciprocal and equivalent basis within an appropriate legal framework between Member States and these other States.

Recital (16) also makes reference to the need for transboundary consultations in respect of plans and programmes prepared by one Member State on another Member State, only adding to Recital (7) in referring specifically to the public as well as relevant authorities. What is missing from this recital is any reference to effects on non EU-Member States and existing provisions in the Espoo Convention and anticipated obligations from the Kiev Protocol to involve relevant authorities and the public in non-EU Member States in the procedure of the origin EU Member State. Recital (17) comments that the results of any transboundary consultations are to be taken into account during the preparation of the plan or programme before adoption or submission to the legislative procedure.

Substantive obligations

The relevant substantive obligations of the SEA Directive are found in Articles 7 and 8, which relate closely to Articles 10 and 11 of the Kiev Protocol. Article 7(1) sets out the procedure whereby the origin Member State must forward a copy of the draft plan or programme and related environmental report to the affected Member State where it considers the implementation of the proposal is likely to have significant environmental effects upon the other state, or where that state requests the information. The information must be submitted before the adoption of the proposal or its submission to the legislative procedure.

Article 7(2) states that where the affected Member State is sent a copy of both documents, it must indicate to the origin Member State whether it wishes to enter into consultations and, if so, this should be in accordance with agreed, detailed arrangements providing the public in the affected Member State with the opportunity to be informed and make their views known within reasonable times frames, which are required to be decided upon in advance in accord with Article 7(3). In 2003, the European Commission published a guidance document on the SEA Directive. In relation to what is 'reasonable' the report comments:

Compared with non-transboundary situations, these will need to be sufficient for contact to be established between the States concerned, the

> *identification and consultation of the public and environmental authorities in the affected State, and consideration of the resulting comments by the appropriate authorities in the State of origin. Practical matters such as the need to prepare translations may also lengthen the process.*[32]

Article 8 contains the obligation that the environmental report and opinions from consultation in the origin and affected Member States must be taken into account during the preparation of the plan or programme and before its adoption or submission to the legislative procedure. Whether this is a weaker provision than that in the Protocol that refers to 'due account' is unlikely, given ambiguity of the meaning of the additional word. The Directive also contains a requirement in Article 9 for information on the decision to be provided to the public and any Member State consulted as a result of potential transboundary effects, a requirement that is not included in the Protocol. The guidance document from the European Commission comments that the obligations in Article 8:

> *... reflect the obligation in Article 7 of the Aarhus Convention which, in conjunction with Article 6(8) of that Convention, requires that in decisions on plans and programmes due account is taken of the outcome of the public participation.*[33]

Implementation reviews

The Commission review of the SEA Directive was also completed in 2009, delayed three years as experience within the Member States was limited, and new Member States acceded in 2004 and 2007.[34] The Report notes many cases of transboundary consultations were carried out with a high degree of satisfaction other than in regard to language and translation issues. NGOs, however, report that in practice the consultation does not always take place at an early stage, but rather at a time when the proposal is already well advanced.[35] The Report emphasizes that the Kiev Protocol is not limited to transboundary impacts from plans and programmes, but that it is also concerned with impacts from plans and programmes within a state.[36] It further stresses that the Protocol addresses policies and legislation, which is not the case for the SEA Directive.

In regard to the relationship between the SEA and EIA Directives, comment is made on the unclear boundaries as to what constitutes a project, plan or programme, and therefore which of the two directives applies. This may have implications for TEIA. As experience in the application of the SEA Directive is limited, however, the Commission does not consider it is appropriate at the present time to merge the two processes. Instead, better coordination and coherence by correcting the inconsistencies between the provisions and clarifying definitions of problematic project categories in the EIA Directive (by amending the latter and/or through the development of guidance) is recommended.[37]

The Kiev Protocol contains a reporting mechanism in Article 14(7) which requires each Party to report on measures that it has taken to implement the Protocol, at intervals to be decided on by the MOP. Article 14(4) also states that the MOP must keep under regular review the implementation of the Protocol. There are six obligations in this paragraph, including review of policies for and methodological approaches to SEA improving procedures, sharing information regarding experience, and setting up subsidiary bodies.

ENFORCEMENT BY THE ECJ

The ECJ interprets EU law including directives, for example, when references for preliminary rulings are made by Member State courts concerning specific provisions and their application, or where actions are commenced by the Commission against a Member State for failure to transpose directives correctly or at all into national law. There have been numerous cases under the original and amended EIA Directives, and there is a growing case law in relation to the Public Participation and SEA Directives. In regard to the latter, a recent case has determined that nitrogen management programmes relating to designated vulnerable zones are programmes within the meaning of the SEA Directive.[38] Transboundary matters specifically have only been considered twice by the ECJ, in relation to proposals originating in Belgium and Austria. Member State implementation of the Public Participation Directive has also been considered in a Commission case against Ireland and a preliminary ruling in relation to Sweden, which may have implications for TEIA.

Screening

That the first TEIA case related to Belgium was perhaps appropriate as Belgium recognizes three territory-related regions under constitutional law (Flanders, Wallonia and Brussels), and allows implementation of TEIA between them as well as in the relationship with other states, which was the concern of the ECJ case. The Belgium situation is evaluated in Chapter 12 of this book (De Mulder), along with an examination of the role of transboundary SEA. In this case, the Commission took infringement proceedings against Belgium, succeeding in various arguments including that the Flemish Region and the Region of the Capital City of Brussels had failed to transpose Article 7 and comply with requirements relating to works with a transboundary impact.[39] In rejecting the position put by the Kingdom of Belgium, the Court held (at paragraph 53):

> *The argument based on the geographical situation of the Region of the Capital City of Brussels is based on the assumption that only projects located in frontier regions are capable of affecting the environment in*

> *another Member State. Yet, as the Commission has rightly observed, that assumption is wrong since it ignores the possibility of air- or water-borne pollution.*

The second case, regarding a proposal for overhead powerlines from Austria to Italy also concerned a failure to conduct a transboundary EIA, and was a request for a preliminary ruling.[40] The substantive question was whether there was an obligation to conduct a TEIA further to the threshold for overhead electric power cables in Annex I of the amended EIA Directive, specifically whether the length of more than 15kms applied to the territory of the origin Member State alone, or to the territory of both the origin and affected Member States. The Court was asked to provide the national court with criteria for interpreting Directive 85/337 so as to permit it to assess whether, under Community law, the project at issue was subject to the procedural obligations laid down in the Directive, despite national law not containing such obligations. In its judgement, not surprisingly the Court made reference to the case law on splitting projects defeating the purpose of the Directive, particularly the issue of cumulative effects. It held (paragraphs 54–55):

> *It follows that projects listed in Annex I to Directive 85/337 which extend to the territory of a number of Member States cannot be exempted from the application of the Directive solely on the ground that it does not contain any express provision in regard to them.*
>
> *Such an exemption would seriously interfere with the objective of Directive 85/337. Its effectiveness would be seriously compromised if the competent authorities of a Member State could, when deciding whether a project must be the subject of an environmental impact assessment, leave out of consideration that part of the project which is located in another Member State.*

Concerns relating to 'screening' for effect significance were only noted generally in the most recent review of the EIA Directive by the European Commission, with a tendency among many Member States to set thresholds to avoid application of the Directive. However, ECJ jurisprudence has traditionally emphasized the need for a wide scope and broad purpose in interpreting the EIA Directive,[41] and the two TEIA decisions above accord with this approach.

Participation and review

In a case against Ireland, the ECJ determined the question of how much flexibility is available to a Member State in deciding the 'detailed arrangements' governing public participation in the EIA Directive, including the question of fees payable, which would also apply in respect of the Article 7 transboundary requirements. The Court held that the action of a Member State must not undermine the effectiveness of a directive in practice,[42] especially the role of the public, which was acknowledged in an earlier case.[43] However, in the more recent decision, the Court held that a Member State is free to impose a participation fee provided it does not constitute an obstacle to the exercise of participation rights.[44]

In a case against Sweden, the ECJ considered the right of access to a review procedure[45] to challenge decisions authorizing projects likely to have significant effects on the environment and the relationship with public participation in environmental decision-making procedures.[46] The ECJ was asked three questions: the first in relation to the interpretation of Annex II of Directive 85/337, which permits Member States to consider whether projects need to be assessed on the basis of their characteristics; the second in relation to Article 10a of the EIA Directive as amended by the Public Participation Directive, as discussed above; and the third in relation to national requirements regarding the public concerned and the review procedure. The ECJ held first that Annex II did apply to the activity in question, irrespective of its purpose. Second, the ECJ held that the fact that the decision within the meaning of Article 10a was given by a court exercising administrative powers did not prevent the NGO exercising its right to a review procedure in order to challenge the decision, and participation in the decision-making procedure had no effect on the conditions for access to the review procedure. Third, decisions as to minimum numbers of members required to access the review procedure ran counter to the objectives of the Directive.

CONCLUSION

The main conclusion of this chapter is that European law has to a significant extent implemented treaty obligations in respect of TEIA, and the absence of ECJ rulings in this area to an extent confirms this. The Espoo Convention has been implemented through the amended EIA Directive, and the participation and access to justice obligations of the Aarhus Convention have been implemented through further amendments to the EIA Directive provided for by the Public Participation Directive. However, Commission reports supplementing reports to the Convention Implementation Commission confirm that several issues continue to challenge implementation in practice, in particular differences in national procedures, such as participation and review.

Other specific omissions identified in this book include the failure to transpose the Inquiry Procedure under the Espoo Convention, and the absence of implementing provisions for the Aarhus Convention to provide for direct access to the ECJ for individuals and NGOs, which remain subject to the restricted standing provisions under Article 263(4) of the TFEU,[47] or the ability to convince a national court to refer questions to the ECJ, the latter of which is also dependent upon access provisions in a domestic context.[48] While the focus of this book is upon TEIA in the EU, it must be remembered that the application of the Espoo Convention and Kiev Protocol outside the EU may trigger compliance and dispute settlement mechanisms of an entirely different nature, and that this will also be the case for EU Member States in relation to non-EU Member States, as with the Danube Delta case cited in several chapters of this book.[49]

Notes

1 Council Directive of 27 June 1985 on the assessment of the effects of certain public and private projects on the environment (85/337/EEC) (OJ L175, 5.7.1985, p.40).

2 Convention on Environmental Impact Assessment in a Transboundary Context (Espoo) 30 *ILM* 802 (1991).

3 The EU signed the Convention on 26 February 1991 and ratified on 24 June 1997; it came into force on 10 September 1997. For EU ratification, see Council Decision of 27 June 1997 on the conclusion, on behalf of the Community, of the Convention on environmental impact assessment in a transboundary context (Espoo Convention) (proposal OJ C 104, 24.4.1992, p5; decision not published).

4 Council Directive 97/11/EC of 3 March 1997 (97/11/EC) (OJ L73, 14.3.1997, p5).

5 Directive 2003/35/EC of the European Parliament and of the Council of 26 May 2003 providing for public participation in respect of the drawing up of certain plans and programmes relating to the environment (2003/35/EC) (OJ L156, 25.6.2003, p17).

6 Directive 2009/31/EC of the European Parliament and of the Council of 23 April 2009 on the geological storage of carbon dioxide (09/31/EC) (OJ L140, 5.6.09, p114).

7 Directive 2001/42/EC of the European Parliament and of the Council of 27 June 2001 on the assessment of the effects of certain plans and programmes on the environment (Directive 2001/42/EC) (OJ L197, 21.7.01, p30).

8 Convention on Access to Information, Public Participation in Decision-Making and Access to Justice in Environmental Matters (Aarhus) (1998) 38 ILM, 517. The EU signed the Convention on 25 June 1998 and ratified on 17 February 2005. See Council Decision 2005/370/EC of 17 February 2005, OJ L 124, 17.5.2005.

9 Protocol on Strategic Environmental Assessment to the Convention on Environmental Impact Assessment in a Transboundary Context (Kiev). The Protocol was signed by the EC on 21 May 2003 and ratified on 12 November 2008, subsequently approved by Decision 2008/871/EC, OJ L 308, 19.11.08, p33.

10 The 2009 EIA and SEA Directive implementation reviews report and comment on the potential for such changes at paras 3.5.1 (EIA Report) and paragraph 7 (SEA Report).

11 Directive 2003/4/EC of the European Parliament and of the Council of 28 January 2003 on public access to environmental information and repealing Council Directive 90/313/EEC, (OJ L41, 14.2.03, p26).

12 See Article 7(2) which requires information to be made available and disseminated and specifies (b) policies, plans and programmes relating to the environment; (c) progress reports on their implementation; (e) data or summaries of data relating to the monitoring of activities affecting, or likely to affect, the environment;

(f) authorizations with a significant impact on the environment; (g) environmental impact studies. Under Article 9(1) Member States were due to report on the implementation of this no later than February 2009. The Commission Report summarizing responses, with possible references to the relationship to the other directives discussed, is awaited.

13 The implication is that where the Member States disagree, there is no independent scientific procedure established to resolve the matter, which would need determining by the ECJ in the absence of agreement. See Koivurova, T. (2002) *Environmental Impact Assessment in the Arctic: A Study of International Legal Norms*, Ashgate, Aldershot, p201.

14 Article 11(3) of the EIA Directive requires a review of application and effectiveness every five years; Article 12(3) of the SEA Directive and Article 5 of the Public Participation Directive contain similar provisions.

15 See Commission of the European Communities, Commission Staff Working Document accompanying the 25th Annual Report on Monitoring the Application of Community law (2007) Situation in the Different Sectors, Brussels, 18.11.2008 SEC(2008) 2854, paragraph 6.2.3. One of the further challenges identified in this report is to ensure that citizens and NGOs can seek reviews at the Member State level, which means full implementation of Article 10a of the EIA Directive (paragraph 6.2.3.4).

16 Woodliffe, J., Environmental Damage and Environmental Impact Assessment, in Bowman, M. and Boyle, A. (eds) (2002) *Environmental Damage in International and Comparative Law: Problems of Definition and Valuation*, pp133–147.

17 Commission of the European Communities, Report from the Commission to the European Parliament and the Council on the Application and Effectiveness of the EIA Directive (Directive 85/337/EEC as amended by Directive 97/11/EC), How Successful are the Member States in Implementing the EIA Directive?; section 4.5 deals with transboundary impacts, pp78–82.

18 See Albrecht, E. (2008) Transboundary Consultations in Strategic Environmental Assessment, *Impact Assessment and Project Appraisal*, vol 26, p289 for example, who discusses the implementation of the Espoo Convention on TEIA between Germany and Poland, making explicit reference to the requirements of Article 7 of the EIA Directive.

19 Commission of the European Communities, Report from the Commission to the Council, the European Parliament, the European Economic and Social Committee and the Committee of the Regions on the Application and Effectiveness of the EIA Directive (Directive 85/337/EEC, as amended by Directives 97/11/EC and 2003/35/EC), Brussels, 23.7.2009 COM(2009) 378 final. See p7 in particular.

20 See also Annex 2 (at p13) entitled: 'Preliminary observations on the implementation of the new provisions introduced by Directive 2003/35/EC', which states in respect to Article 10(a) 'it is not possible to draw any conclusions at this stage on the implementation of these provisions. However, the Commission has already identified transposition problems in some MS and initiated infringement procedures, in accordance with Article 226 EC Treaty.'

21 See European Commission Environment, Public consultation on the Review of the Environmental Impact Assessment (EIA) Directive (Directive 85/337/EEC on the

assessment of the effects of certain public and private projects on the environment, as amended), available at http://ec.europa.eu/environment/consultations/eia.htm

22 See Article 14 bis, adopted by Decision III/7, which provides a legal obligation on Parties to report on the implementation of the Convention, and Decision IV/1 on the Second Review of Implementation. Article 14 bis is part of the Second Amendment to the Convention which requires ratification by at least three quarters of the Parties, followed by the expiry of the stipulated time period. Currently, only 16 of the 45 Parties have done so, and it is therefore not in force.

23 See Espoo Convention Review of Implementation 3rd Review, Report of the European Union for 2006–2009 on the implementation of the Convention on Environmental Impact Assessment in a Transboundary Context (1991 Espoo Convention), available at www.unece.org/env/eia/documents/Review_2006_2009/Questionnaire2006_09_EU_reply_en.pdf

24 Proposal for a Directive of the European Parliament and of the Council on access to justice in environmental matters, Brussels, 24.10.2003 COM(2003) 624 final, 2003/0246 (COD).

25 See Ryall, A. (2007) *Effective Judicial Protection and the Environmental Impact Assessment Directive in Ireland*, Hart, Oxford.

26 Ryall, A. (2007) EIA and Public Participation: Determining the Limits of Member State Discretion, *Journal of Environmental Law*, vol 19, pp247, 256.

27 2000/60/EC of the European Parliament and of the Council of 23 October 2000 establishing a framework for Community action in the field of water policy.

28 European Commission, Report from the Commission to the Council, the European Parliament, the European Economic and Social Committee and the Committee of the Regions on the application and effectiveness of Directive 2003/35/EC of the European Parliament and of the Council of 26 May 2003 providing for public participation in respect of the drawing up of certain plans and programmes relating to the environment and amending with regard to public participation and access to justice Council Directives 85/337/EEC and 96/61/EC, Brussels, 14.4.2010 COM(2010)143 final.

29 See Article 10(2) of the Convention, and Decision I/8, which effectively established National Implementation Reports (NIRs), produced by the majority of Parties with a full Convention Report in 2005 and 2008, and the next anticipated in 2011.

30 For a more user friendly PDF version, see Commission of the European Communities, Aarhus Convention Implementation Report European Community, Brussels, 7 May 2008, SEC (2008) 556. Available at http://ec.europa.eu/environment/aarhus/pdf/sec_2008_556_en.pdf

31 States such as Iceland and Norway do this under agreement between the European Free Trade Association (EFTA) and the EU, as they are Parties to the European Economic Area (EEA) Agreement, which is explained in Chapter 8 of this book in relation to the Arctic (Koivurova).

32 Representatives of the Member States and the Environment Directorate-General of the European Commission, Implementation of Directive 2001/42/EC on the assessment of the effects of certain plans and programmes on the environment, 23 September 2003, paragraph 7.27.

33 Paragraph 7.30.

34 Commission of the European Communities, Report from the Commission to the Council, the European Parliament, the European Economic and Social Committee and the Committee of the Regions on the application and effectiveness of the Directive on Strategic Environmental Assessment (Directive 2001/42/EC), Brussels, 14.9.2009 COM(2009) 469 final. See paragraphs 3.7 (transboundary consultations), and 4.3 (SEA Protocol) in particular.

35 Paragraph 3.7, p5.

36 Paragraph 4.3, p7.

37 Paragraph 4.1.

38 Joined Cases C-105/09 and C-110/09, *Terre Wallonne ASBL, Inter-Environnement Wallonie ASBL v. Région Wallonne,* 17 June 2010.

39 Case C-133/94, *Commission v. Belgium* [1996] ECR I-2323.

40 Case C205/08, Reference for a preliminary ruling under Article 234 EC from the Umweltsenat (Austria), made by decision of 2 April 2008, received at the Court on 15 May 2008, in the proceedings *Umweltanwalt von Kärnten v. Kärntner Landesregierung,* 10 December 2009.

41 Case C-2/07, Abraham and Others, ECR I-1197; Case C-72/95, *Kraaijeveld and others,* ECR I-5403, paragraph 31; Case C-227/01, *Commission v. Spain,* ECR I-8253, paragraph 46.

42 Case C-216/05, *Commission v. Ireland* [2006] ECR I-10787.

43 Case C-332/04, *Commission v. Spain* [2004] ECR I-40.

44 Case C-216/05, *Commission v. Ireland* [2006] ECR I-10787.

45 Note pending Case C240/09 *Lesoochranárske zoskupenie VLK v. Ministerstvo životného prostredia Slovenskej republiky,* and Opinion of Advocate General Sharpston delivered on 15 July 2010, in regard to whether Article 9(3) of the Aarhus Convention can be regarded as having direct effect within the Member State's legal order.

46 Case C-263/08, *Djurgården-Lilla Värtans Miljöskyddsförening v. Stockholms kommun genom dess marknämnd,* 15 October 2009.

47 Case C-355/08 P, *WWF-UK v. Council of the European Union and Commission of the European Communities,* 5 May 2009.

48 Milieu Ltd, *Summary Report on the Inventory of EU Member States' Measures on Access to Justice in Environmental Matters,* Report for the European Commission, DG Environment, Brussels, p18.

49 See Koyano, M., The Significance of the Convention on Environmental Impact Assessment in a Transboundary Context (Espoo Convention) in International Environmental Law: Examining the Implications of the Danube Delta Case (2008) *Impact Assessment and Project Appraisal,* vol 26, no 4, p299.

Part II

The Relationship between International and European Law

4

The Case of Vuotos: Interplay Between International, Community and National Environmental Law

Timo Koivurova

INTRODUCTION

The *Vuotos Case* has been one of the longest and best-known environmental struggles in Finland.[1] It has highlighted the changed legal environment in the country, inasmuch as international and European Community (EC) law has very much influenced how the case was finally resolved. Although the Finnish Water Act was invoked by the courts as the main basis for their decision, *Vuotos* is a good example of how much the legal environment of Finland has changed, for the case hinged on not only the Finnish Water Act but also international and Community law.

The *Vuotos Case* concerns the proposed construction of a reservoir and dam on the upper course of the Kemijoki River in Finland in 1992. Swedish officials and civil society were concerned about the possible effects of the construction of the reservoir on the Gulf of Bothnia, particularly with regard to Swedish maritime areas, which prompted Sweden to participate in the Finnish licensing procedure. Moreover, the Commission of the EU was very worried about the Vuotos project,[2] so much so that it initiated two supervisory procedures against Finland. The Commission took the position that the Kemihaara mires, which formed part of the planned Vuotos reservoir area, should have been designated by Finland as a special protection area (SPA) in the Natura 2000 network.

The first goal of this chapter is to examine how the different legal procedures concerned with the Vuotos project proceeded over time and how they interacted. This will provide the reader with a general context in which the individual procedures can be more fully understood. The second goal is to examine, in depth, the two main legal procedures influencing the outcome of the case – Sweden's participation in the Finnish licensing procedure and, more importantly, the licensing procedure itself. The analysis here will focus on the arguments used by the courts, especially when interpreting international, EC and national law, and will explore why the courts reasoned as they did. Finally, a look at the latest developments in the case is in order, after which general conclusions are drawn. Before proceeding to the main goals of the chapter, however, it will be useful to briefly outline the background of the case.

Background to the *Vuotos Case*

The debate on the construction of the Vuotos reservoir has been a protracted one. In 1960, Parliament passed the Kemijoki River Regulation Act,[3] on the basis of which permits were granted for the construction of the Lokka and Porttipahta reservoirs. The initial plans for the Vuotos reservoir, or, as it was known then, 'the Kemihaara project', were completed in the 1960s, with the reservoir to be located in the municipalities of Savukoski, Salla and, most extensively, Pelkosenniemi. The project was suspended for environmental reasons in 1974 and in 1982, the Finnish Government made an informal decision not to proceed with the construction of the reservoir. The project was resumed as the 'Vuotos project' in the mid-1980s, and the necessary preparatory work was completed by the time that Kemijoki Ltd applied to the Water Court of Northern Finland on 25 September 1992 for a permit to construct the Vuotos reservoir and dam.

According to Kemijoki Ltd, the purpose of the reservoir was to function as a water reserve for all the existing and future power plants on the main course of the Kemijoki River. The Vuotos reservoir would:

> *... increase power generation of all power plants located below it. Furthermore, it would make economical sense to build additional machinery at the existing plants and two new power plants. The addition to the energy production would be about 560 GWh/a.*[4]

The size of the reservoir 'would be about 0.25 per cent of the area of the province of Lapland'.[5]

The applicable legal procedures in the *Vuotos Case*

It will be useful to examine the legal procedures pertaining to the Vuotos project from two perspectives in order to grasp the complexity of these procedures. The principal procedure comprised the inspection proceedings and the licensing procedure under the Finnish Water Act. The latter came to involve Sweden when it turned out that the construction of the Vuotos project could result in harmful effects on Swedish maritime areas.

The other relevant procedure was the Natura 2000 process stipulated in the Wild Birds Directive[6] and Habitats Directive[7] of the EU. The European Commission was convinced that the Kemihaara mires, which would have been partly immersed by the planned reservoir, should be designated by Finland as an SPA in the Natura network.

The procedure stipulated in the Finnish Water Act

As no national environmental impact assessment (EIA) legislation was in force when Kemijoki Ltd applied for the permit for the Vuotos project, the Finnish Water Act was applied in the permit procedure.[8] Kemijoki Ltd first filed its application with the Water Court of Northern Finland (Water Court) on 25 September 1992. According to the Water Act, all large-scale projects have to be studied to determine the environmental and social impacts of the project, and to draw up recommendations for the Water Court. To this end, the Water Court appoints a panel consisting of an engineer and two trustees.[9] In the case of Vuotos, this occurred on 16 March 1993.

Inspection proceedings include several studies of the impacts of the activity and the panel draws conclusions as to whether a permit should be granted for the activity and under what conditions. After the inspection proceedings, private individuals and governmental authorities have the opportunity to make comments on, and objections to, the results of the inspection through final public hearings. In the present case, these took place in July 1996.[10] After these meetings, the public had the opportunity to submit its comments and objections to the Water Court, up until 11 September 1996.

The engineers proposed to the Water Court that Article 2(5) of the Finnish Water Act, which defines the criteria by which a permit may be denied where important public interests are involved, did not present any obstacles to the construction of the project. The engineers also stated that the project would have no significant transboundary impacts on the maritime area of Sweden. They endorsed the Vuotos project because they viewed that its benefits greatly exceeded its costs, a criterion in keeping with Article 2(6) of the Water Act.[11]

The studies on likely environmental impacts of the Vuotos project identified possible impacts on the Gulf of Bothnia, a development that led to a second procedure in addition to the inspection proceedings. In a letter dated 1 July 1993, the Finnish Water and Environment Council informed the Water and Environment District of Lapland that the Vuotos project might cause environmental impacts on Swedish maritime areas. It also noted that, pursuant to the Espoo Convention,[12] the Finnish Ministry of the Environment (FME) was the authority responsible for providing notification to Sweden.[13] When Kemijoki Ltd applied for the permit from the Water Court, there was only one international treaty that applied to the case. This was not the Espoo Convention but the Nordic Environmental Protection Convention (NEPC),[14] which had been in force since 1976. The

Espoo Convention had been signed in 1991, but did not enter into force until long after the Vuotos application, in September 1997.

Since the Finnish Water Act and the NEPC provide that any procedure to be carried out by the origin and potentially affected states may only be commenced after the inspection proceedings have been completed, the FME proposed that Finland apply the Espoo Convention, even though it was not binding on the country at the time. The FME initiated the procedure by first discussing informally with the Water Court how the potential notification to Sweden should be arranged. The Water Court, the competent authority to notify by virtue of the NEPC, had some doubts about this procedure, but saw no obstacle in light of the Espoo Convention and the resolution of the signatories to the FME notifying Sweden.[15]

According to the EIA Act, which entered into force on 1 September 1994,[16] the FME in these circumstances must consult with the Finnish Ministry of Foreign Affairs.[17] This consultation took place before the FME notified Sweden of the potential transboundary impact of the Vuotos project. On the basis of this informal consultation, it was agreed that the Espoo Convention should be invoked provided that the obligations it imposed did not exceed the level of those set forth in a valid international treaty, i.e. the NEPC.

Notification of the potential transboundary impacts of the project was given to Sweden on 22 June 1995, long before the conclusions of the inspection proceedings. In the notification, the FME informed the Swedish Environmental Protection Agency (SEPA) that the construction of the Vuotos project might result in an overload of phosphorus in the maritime area of Sweden in the Gulf of Bothnia. While the FME referred primarily to Article 3 of the Espoo Convention and the resolution of the signatories as the legal bases for the notification, it also referred to the NEPC. SEPA was asked to respond before 1 September 1995 and did so, on 30 August 1995, requiring that environmental studies be done covering the possible impact on Swedish maritime areas. The FME responded to the Swedish concerns by collecting expert opinions on the matters identified.[18] Even though SEPA was well informed about the potential impacts, as well as the inspection proceedings and the procedure in the Water Court, it failed to send its objections regarding the Vuotos project before the deadline set for these comments and objections by the Water Court on 11 September 1996; they were received two days later.

Although the FME had collected expert opinions on the likely impacts of the project on the Gulf of Bothnia, the Water Court based its decision of 29 February 2000 very much on the findings and recommendations of the inspection proceedings. As the proceedings' engineers had recommended, the Water Court concluded that Article 2(5) of the Finnish Water Act, which defines the criteria by which a permit may be denied where important public interests are involved, did not present any obstacle to the construction of the reservoir and dam. The Vuotos project was granted a licence by the Water Court because the anticipated benefits of the project greatly exceeded the costs by an estimated 1800 million Finnish marks.[19] The Court also determined that there would be no significant transboundary impact on the maritime area of Sweden and, accordingly, that as the examining authority of the NEPC, it was not obligated to notify the Swedish supervisory authority, SEPA, under Article 5 of the NEPC.[20]

Quite unexpectedly, on 14 June 2001, the Administrative Court of Vaasa, to which several parties had appealed the Water Court's decision, overturned the licensing decision made by the Water Court and rejected Kemijoki Ltd's application. The Court decided that the criteria for one of the three absolute prohibitions enumerated in Article 2(5) of the Finnish Water Act had been fulfilled (prohibiting the causing of widespread and harmful consequences to the environment) and cited the requirements of EC law and changed environmental values as grounds for its decision.[21] This was the first time in the history of the Finnish Water Act that Article 2(5) had prevented the construction of a proposed activity. The Administrative Court also found that the construction of the Vuotos project might entail damaging effects on the Swedish side of the Gulf of Bothnia[22] and, accordingly, that SEPA, as a supervisory authority under Article 4 of the NEPC, could appeal the licensing decision of the Water Court.[23]

Kemijoki Ltd appealed to the Supreme Administrative Court of Finland, which handed down its decision on 18 December 2002. The Court upheld the decision made by the Administrative Court of Vaasa, namely that no permit was to be granted, but gave different reasons for its decision. The Court argued that although Article 2(5) had never prevented the construction of a proposed activity, neither had there been a case in which the impacts on the environment would have been as devastating; the construction of the reservoir would have altered the entire ecosystem. Unlike the Administrative Court, the Supreme Administrative Court did not cite changed environmental values as a relevant factor in interpreting the Water Act; rather, it argued that the Water Act must be interpreted within the bounds of legal discretion.[24]

The Natura process in Finland

While the Vuotos litigation was on-going, parallel proceedings were taking place relating to the EU's Natura process. Article 3 of the Habitats Directive established an ecological network to protect biodiversity in Europe – Natura 2000. The network consists of the natural habitat types listed in Annex I of the directive and the habitats of the species listed in Annex II, the two sets forming sites of community importance (SCI). In addition, the network includes the SPAs classified under the Wild Birds Directive. Each EU Member State is obligated to notify the Commission of SPA sites in its territory under the criteria of the Wild Birds Directive and propose SCI sites to the Commission in accordance with the Habitats Directive.

The Habitats Directive obligated Member States to send their SCI proposals to the Commission by the end of June 1995. (The deadline for Finland to submit its list of SPA areas coincided with the date on which it became a member of the EU – 1 January 1995).

Finland faced problems in notifying the Commission of the SPAs and SCIs on its territory. On 11 October 1996, it notified the Commission of 15 SPAs and informed the Commission that it would provide a definitive list of SPAs by the beginning of 1997,[25] which it subsequently failed to do. Even though Finland was late in submitting its Natura proposal, it managed to lay down the basic legislative framework for implementing the network through the Finnish Nature Protection Act, which came into force on 1 January 1997.[26]

Due to Finland's failure to provide a definitive list of SPAs, the Commission commenced a supervisory procedure against the country by a letter of formal notice, dated 10 July 1998. In this notice, the Commission referred to Finland's failure to classify certain areas as SPAs, which in light of ornithological criteria should have been so classified; one of these areas was the Kemihaara mires.[27]

On 20 August 1998, the Government of Finland decided which areas should be included in Natura. The Government did not propose that the Kemihaara mires should be included in the network even though the FME recommended it on 21 March 1997. The FME was not solely responsible for preparing the Natura proposal, however. Rather, it was the work of an inter-ministerial Natura 2000 committee, which did not find consensus on including the mires.[28]

The European Commission was not satisfied with Finland's Natura decision because, in its opinion, Finland had still not fully and definitively classified its SPAs; it gave Finland a reasoned opinion on 15 December 1998, after the Government's decision. Even though Finland had classified 439 sites as SPAs, it reported that it could not as yet send the final list of sites to the Commission, because there were several appeals against the Government's Natura decision pending before the Supreme Administrative Court. The Commission reminded Finland that several sites were missing from the Finnish proposal that under the ornithological criteria should have been designated as SPAs, one of these being the Kemihaara mires.[29] On 1 July 1999, the Commission decided to bring an action against Finland in the European Court of Justice (ECJ),[30] with the Commission alleging that Finland had not fully and definitively notified the Commission of its SPA sites. The Commission mentioned the Kemihaara mires as an example of an area that should have been included as an SPA.

From the Commission's point of view, the decision by the Water Court on 29 February 2000 to permit the construction of the Vuotos project was an unfortunate turn of events. Even though the Commission had clearly expressed in its supervisory procedure that the Kemihaara mires should have been designated as an SPA, the Water Court of Northern Finland had now allowed the construction of a reservoir, which would flood the mires. In addition, Kemijoki Ltd had applied for extensive rights to go ahead with necessary work related to the construction of the reservoir. The Water Court permitted the company to clear the forests under its ownership but provided that 'the permit to initiate work does not apply to the clearing of the shores of lakes'.[31] It thus seemed that Vuotos would proceed.

The events following the judgment by the Water Court on 29 February 2000 signalled a new direction in the *Vuotos Case*. First of all, the right of Kemijoki Ltd to start clearing the forests in the Vuotos area was revoked by the Administrative Court of Vaasa on 19 April 2000, pursuant to several appeals against the Water Court's decision in an initial ruling pending its final decision on the matter.[32] Second, as mentioned above, the Government's decision of 20 August 1998 establishing the Finnish Natura areas was challenged on several grounds, one of which was the need to protect the Kemihaara mires. Individual persons, the Finnish Nature Protection Association and Birdlife Finland had lodged a complaint with the Supreme Administrative Court. On 14 June 2001, the Supreme Administrative Court decided that there was not enough information for it to evaluate the legality of the Government's decision concerning the Kemihaara mires and it remanded the Kemihaara mires case to the Government for additional preparation.[33]

Third, the Commission commenced another supervisory procedure against Finland, this time focusing directly on the Kemihaara mires. In its letter of formal notice of 12 September 2000, the Commission argued that, on the basis of information available to it, the construction of the Vuotos project would breach Article 4(4) of the Wild Birds Directive and Article 6(2), (3) and (4) of the Habitats Directive because it would significantly harm the ecosystems and birds there. The Commission argued that, under the settled case law of the ECJ, Member States must observe the requirements of Article 4(4) of the Wild Birds Directive, even though an area has not been designated an SPA, if the area, such as the Kemihaara mires, was an ornithologically suitable site for a SPA. In fact, the Commission was of the opinion that the Kemihaara mires would be, in its ornithological assessment, one of the most suitable sites for a SPA in Finland. The Commission also argued that the Kemihaara mires contain habitat types listed in Annex I of the Habitats Directive as well as habitats of the species listed in Annex II, some of which are priority natural habitat types.[34]

In its opinion delivered on 10 April 2001, the Commission focused on Finland's breach of the Wild Birds Directive by permitting the construction of the Vuotos reservoir, pointing out that the area also contains priority natural habitat types. It found that Finland had not observed the requirements in Article (4) of the Wild Birds Directive, because it had approved an activity that would significantly harm the habitats and species in the Kemihaara mires area. The Commission also urged Finland to implement the measures required in the reasoned opinion within two months of its notification and to inform the Commission when these measures had been taken.[35] Some two months after the receipt of this reasoned opinion, the Vaasa Administrative Court, on 14 June 2001, overturned the licensing decision made by the Water Court and rejected Kemijoki Ltd's application for the construction of the Vuotos reservoir and dam. Since the Supreme Administrative Court upheld the decision made by the Vaasa Administrative Court on 18 December 2002, there was no need for the Commission to continue its supervisory procedure.

ANALYSIS OF THE SWEDISH PARTICIPATION TO THE FINNISH LICENSING PROCEDURE

As mentioned above, the relevant entities in the inter-State procedure between Finland and Sweden – the FME, SEPA and the three Finnish courts – approached the possible impacts to the Gulf of Bothnia from the perspective of two international treaties: the Espoo Convention and the NEPC. One of the peculiarities of the *Vuotos Case* was that the FME took the view that SEPA should have been notified even before this was required by the NEPC. The FME invoked a resolution that was made in connection with the signing of the convention in 1991. In this resolution, all the parties 'resolve to strive for the entry into force of the Convention as soon as possible and to seek to implement the Convention to the maximum extent possible pending its entry into force'.[36] But could this be considered a legal obligation for Finland?

The resolution can be considered binding but only in requiring an attempt to observe it, for instance, by establishing a working group to consider its application.[37] Most scholars would probably define this declaration as a soft-law instrument, non-binding in international law but requiring compliance in some other manner. In any case, even though Finland can be seen as having been under an obligation to try to observe its provisions, there was no legal obligation on its part to notify Sweden. Why did it then do so?

The FME invoked the NEPC as one basis for notification, although the NEPC did not require any notification before the case had proceeded to the examining authority, which in this case was the Water Court. It thus seems that both the FME and SEPA were operating on the assumption that it was the resolution related to the Espoo Convention that was the basis for the early notification; the two agencies did not contemplate any legal bases other than this resolution and the NEPC, and the latter did not require the notification at this stage of proceedings. It can be argued that the FME involved SEPA simply because it needed more resistance to the Vuotos project. Clearly, the resistance to the Vuotos project would seem considerably stronger when another state, its authorities and civil society opposed it. This argument can be buttressed by referring to the fact that the FME had consistently opposed the construction of the Vuotos project. It was the FME that proposed in 1997 to the Government of Finland that the Kemihaara mires be included in Natura. Yet, the Natura 2000 Committee and the Government rejected this proposal. The FME also consistently argued against the construction of the reservoir during the licensing procedure. In fact, in a comment on the Natura preparation procedure by Kemijoki Ltd, published on its website and dated 15 May 2003, the company accuses the FME of being in breach of principles governing the preparation of administrative decision-making, for example the principle of objectivity, partly because of the above-mentioned factors.[38] Yet, even though it is clear that the FME opposed the Vuotos project, it seems too far-fetched an explanation that it used notification to SEPA as part of its overall strategy to have the project rejected.

First, the FME did not apply the Espoo Convention before its entry into force only in the case of Vuotos, but also in two other cases.[39] Indeed, it seems that the FME took the declaration related to the convention very seriously. Second, since the system set out in the NEPC initiates the transboundary procedure at a very late stage of the national procedure, it was probably considered necessary to involve the Swedish governmental authorities before all environmental impact studies had been made. All in all, it seems likely that, even though the actions of the FME could be interpreted as advancing its general goal of seeing the Vuotos project rejected, the more likely explanation is that Finland, and the competent authorities, were merely trying to follow the requirements of the resolution.

Although the Espoo Convention and its related resolution were invoked by the FME as the initial bases for notification to Sweden, the rights of the Swedish participants in the licensing procedure itself were dealt with in the context of a valid international treaty, the NEPC. At the outset, the Water Court made a decision not to notify the supervisory authority under Article 5 of the NEPC because the effects to the Gulf of Bothnia would have been, in its assessment, minor.[40]

The Administrative Court was not as convinced of the insignificance of the potential effects on Swedish maritime areas as the Water Court had been. For the Administrative Court, the model calculations used by the inspection proceedings and the Water Court, which showed that the potential effects extended no further than the waters near the Finnish town of Kemi, were not credible. The Administrative Court asserted that the effects on the Gulf of Bothnia would extend to Swedish maritime areas as well, but noted that the ecological impacts on those areas would, in all likelihood, be obscured by natural variation and would therefore be hard to measure.[41] The Administrative Court decided that SEPA had a right to appeal because it was a supervisory authority under the NEPC; hence, it possessed the rights stipulated in Article 4 of the NEPC. Yet, Kemijoki Ltd argued that SEPA had not lodged its objection or comment on the Vuotos by the deadline, 11 September 1996, which was a precondition for challenging the Water Court's judgment. Indeed, the Administrative Court observed that the written communication by SEPA arrived at the Water Court on 13 September 1996, two days after the deadline. Moreover, this had not been a direct written communication from SEPA, but an appendix to the statement by the FME. The Administrative Court nevertheless took the view that, despite these errors, it would be unreasonable to have SEPA lose its right to participate in the procedure. It gave SEPA this right because, in its view, the Water Court and the FME had agreed on a special procedure for how the opinions and views of the Swedish officials would be transmitted to the Water Court.[42]

There were also other Swedish parties arguing that they had a right to participate in the Vuotos licensing procedure, i.e. the Municipality of Haparanda, the County Administrative Board of Norrbotten, Nature Protection Association of Norrbotten and a private fisher. The Administrative Court first determined that the Municipality of Haparanda and the County Administrative Board of Norrbotten did not have a right to appeal, because they were not supervisory authorities under Article 4 of the NEPC. However, the Court decided that the views and complaints of these officials could be taken into account as part of SEPA's submission. The private fisher was given the status of an injured party even though his complaint had come in late.[43] This was due to the fact that Article 3 of the NEPC gives private parties the same rights as they have in the state of origin, and the Administrative Court found it unreasonable to exclude the fisher's rights simply when he had not been properly informed under the Water Act.[44]

In its appeal to the Administrative Court, SEPA stated that the permit could only be granted if heavy metal and phosphorus emissions were controlled through the permit conditions in such a way that no danger would be caused to Swedish territory and the Swedish environment. SEPA also noted that if the permit were granted, a separate permit condition had to be included that Kemijoki Ltd regularly submit monitoring information to the Country Administrative Board of Norrbotten. Since the Supreme Administrative Court did not change the negative permit decision by the Administrative Court, the threat to the Gulf of Bothnia posed by the Vuotos project never materialized.

Irrespective of the reasons why the Swedish parties were involved in the *Vuotos Case*, it is clear that the case took on a new dimension when private and public parties from another State participated in the Finnish national licensing procedure. All the Swedish parties argued that the Vuotos project should not be constructed because of the significant environmental damage the project would cause to Swedish maritime areas in the Gulf of Bothnia. Their participation in the case clearly had an impact on the way the *Vuotos Case* was decided. As discussed above, the Administrative Court decided that the criteria under one of the three absolute prohibitions enumerated in Article 2(5) of the Finnish Water Act was fulfilled (prohibiting widespread and harmful consequences to the natural environment and water ecosystems), and it was this criterion that ultimately prevented construction of the project. It was also this element that was the basis of the Swedish submissions to the Finnish administrative courts.

Evidently, one part of this overall assessment of whether the natural environment and ecosystems were in danger of being seriously damaged was the assessment, by the Administrative Court, that Swedish maritime areas might in fact be affected by the construction of the Vuotos project, although the changes to the Swedish maritime areas would be hard to measure. Thus, since the Supreme Administrative Court agreed with the way the Administrative Court had evaluated the potential effects of the Vuotos project on the Gulf of Bothnia, it appears that these had an influence on the Courts' rejection of Kemijoki Ltd's application.

Analysis of the Finnish Licensing Procedure

The most important procedure in the *Vuotos Case*, at least from the perspective of the construction of the Vuotos project, was the Finnish licensing procedure. As shown above, the initial judgment by the Water Court of Northern Finland was appealed both to the Administrative Court of Vaasa and to the Supreme Administrative Court of Finland. Since all three courts ended up grounding their decisions in national and EC law in a different manner, it is useful to examine each of the respective court decisions in order to look at what basic approach each court was using in interpreting the law and whether its reasoning was consistent with the approach chosen. The analysis in this section then proceeds to clarify the problems in the approach chosen by the two lower courts in particular, and to highlight the strengths of the decision by the Supreme Administrative Court. The Finnish Water Act was enacted in 1961.[45] The main focus in the licensing procedure is Article 2(5), which sets out three separate criteria, the fulfilment of any one of which prohibits a permit being granted for the proposed activity. According to the original version of Article 2(5), a permit shall not be given if:

1. the construction work endangers public health;
2. causes significant and wide-ranging negative changes in the natural conditions of the environment; or
3. seriously impacts the residents' livelihoods or living conditions. The 1987 amendment to Article 2(5) of the Water Act, which applied to the case, provides that:

> *If the construction work endangers public health,* causes significant and wide-ranging negative changes in the natural conditions of the environment or the aquatic environment and its functioning, *or if it seriously impacts the livelihood or settlement conditions, no permit shall be granted.*[46] *(emphasis added)*

As noted above, prior to the *Vuotos Case*, Article 2(5) of the Water Act had never prevented the construction of a proposed activity.

Other relevant norms can be found in Article 2(6). According to the 1987 version of this article under the Water Act, the licensing authority may also grant a permit if a project is required by significant public interests. In addition, and more importantly, the provision also stipulates that if the benefits of a proposed activity clearly outweigh its negative costs, a permit may be granted.

THE WATER COURT OF NORTHERN FINLAND

As outlined above, the Water Court gave Kemijoki Ltd permission to proceed with the Vuotos project on the basis of Article 2(6). It determined, as had the engineers, that the benefits of the project clearly outweighed the drawbacks. The Water Court first examined whether any of the absolute prohibitions enumerated in Article 2(5) applied; when none was found, it studied whether the project could be seen as one required by the public interest, which it also found was not the case. Finally, the Court looked at whether the benefits of the project greatly exceeded its costs and concluded that they did; in fact, as noted above, it estimated that the benefits in monetary terms to be 1800 million Finnish marks greater than the costs.[47]

The reasoning of the Water Court as to whether the Vuotos project was prohibited by Article 2(5) arose from a basic examination of the amendments made by the Finnish Parliament to Article 2(5) and how these had changed the article's content and applicability. After a careful study of these legislative amendments, the Court concluded that, even though Article 2(5) had been amended in 1987 and in 1994, its basic content had remained much the same; that is, the legislature consistently upheld a high threshold for applying the article. Indeed, as mentioned above, Article 2(5) had never prevented the construction of any proposed activity. Moreover, because the article had not changed much in the course of time, the Court argued that the events close to the enactment of the Water Act, in the 1950s and 1960s, were relevant in interpreting the content and applicability of the provision. The principal question in the Water Court's view was whether the only reservoirs in Finland, Lokka and Porttipahta, could be regarded as precedents in the case.[48]

The Water Court found that decisions relating to the Lokka and Porttipahta projects could be regarded as relevant case law, even though permits were granted for these reservoirs on the basis of the Kemijoki River Regulation Act,[49] not the Water Act. The Water Court was of the opinion that this special law on the Kemijoki River, in its Article 3, in fact contained a provision similar to Article 2(5) of the Water Act. Article 3 of the Kemijoki River Regulation Act states:

> *The regulation of the river is to be seen as representing a public interest and its benefits clearly override its negative impacts,* if the regulation does not negatively affect the climate, public health or cause any other kind of danger.[50] *(emphasis added)*

According to the Water Court, this view was supported by the fact that the Government Bill terminating this special law in 1968 had stated that, although the two articles were not identical in content, they basically meant the same thing.[51]

When the Water Court handed down its judgment on 29 February 2000, the Supreme Administrative Court had not yet remanded the matter of the Kemihaara mires area to the Government in the Natura process. This was probably one reason why the Water Court considered the influence of the Wild Birds and the Habitats Directives on this case irrelevant. According to the Water Court: The nature protection values in the area of Vuotos have been examined very carefully during the inspection proceedings, in all likelihood better than for most areas in Finland. When making its Natura 2000 network decision, the Government has had very extensive material at its disposal, including the letter of formal notice by the Commission of the Council of Europe [sic] referring to the Kemihaara mires. In addition, even though the Ministry of the Environment has proposed that the Vuotos area should be included in the Natura 2000 network, these areas have been excluded from the Government's decision.[52]

The conclusion of the Water Court was that the directives had no relevance in this case because Finland had, after careful scientific studies, excluded the Kemihaara mires from its Natura proposal.

Analysis The Water Court's approach was to examine the provisions of the Water Act applicable to the case, study the *travaux preparatoires* of these provisions and look for relevant case law. It is this strong focus on the Water Act that underlies its decision, which, on its own premises, is quite convincing. Indeed, it is likely that although the Lokka and Porttipahta reservoirs were approved on the basis of a special law, they would have been licensed on the basis of the Water Act as it stood in the 1960s. If the Water Act had not changed much from the 1960s, as the Water Court argued, it is quite certain that reservoirs like Vuotos would be permitted as well.

The Water Court also gave reasons why the Wild Birds and the Habitats Directives had no relevance in this case. The inspection proceedings had carefully examined the environmental impacts of the Vuotos project and still proposed to the Water Court that it should be constructed. Also, the Government of Finland decided that the Kemihaara mires were not a suitable site for Natura. Even though the EU Commission had commenced a supervisory procedure against Finland for failing to implement the Wild Birds Directive, mentioning the Kemihaara mires as one site that should have been designated a SPA, the focus of the procedure was on having Finland implement the Wild Birds Directive, not on making the mires a SPA. The Water Court also believed that the Commission would not initiate a supervisory procedure focusing directly on the Kemihaara mires. It stated explicitly – and erroneously – that this was the view of the Commission;[53] this turned out to be a misunderstanding on the part of the Court but the Court was clearly convinced that the only supervisory procedure Finland faced was the Commission's effort to have Finland implement the Wild Birds

Directive in an appropriate manner. Thus, the Water Court had its reasons for not applying the Wild Birds and the Habitats Directives, particularly as the Government of Finland and the inspection proceedings had opined that the construction of the reservoir in the area of the Kemihaara mires was legally permissible and socially desirable.

THE ADMINISTRATIVE COURT OF VAASA

The Administrative Court of Vaasa overturned the licensing decision of the Water Court and rejected Kemijoki Ltd's application on the basis of Article 2(5). As noted above, it took the view that one of the absolute prohibitions in Article 2(5) – that the proposed activity shall not cause significant and wide-ranging negative changes in the natural conditions of the environment or in the aquatic environment and its functioning – was triggered and, accordingly, prevented the construction of the Vuotos project. According to the Administrative Court, the Vuotos project was not prevented by the other two prohibitions in Article 2(5), i.e. it neither endangered public health in the area nor seriously affected residents' livelihoods or living conditions.[54]

The Administrative Court examined Article 2(5) from a totally different perspective than the Water Court had. According to the Administrative Court, Article 2(5) is a flexible legal norm, the interpretation of which is dependent on the social values prevailing in the society at the material time. It stated:

> *Society and its values have significantly changed from when the Water Act was enacted in 1961. Also, the environmental sciences have developed considerably during this period. The societal and legal relevance of environmental and nature protection values has increased nationally as well as internationally in the recent decades, especially during the last 10 years.*[55]

It is precisely this focus on changed values that underlie the decision by the Administrative Court. But how could the Court prove that societal values had changed? The Court referred primarily to the legislative activity, both in Finland as well as in the EU. The Administrative Court reviewed the development of Finnish legislation aimed at protecting the environment. Among other things, it pointed out that Article 20 of the Finnish Constitution guarantees protection of the environment.[56] With regard to the EU, the Court noted the change in both primary and secondary law towards environmental protection.[57]

Interestingly, even though the Administrative Court's decision differed greatly from that of the Water Court on a general level, it came to the same conclusion in one respect: the content and applicability of Article 2(5) had not changed essentially during the lifetime of the Water Act. Yet, because societal values have changed so clearly since the time the Water Act was enacted, the Court did not consider it justifiable to use the *travaux preparatoires* of that Act. The Court also opined that there had not been an activity under the licensing procedure of the Water Act comparable to Vuotos, in terms of either extent or potential impacts. Yet, it also made clear that no category of activities is excluded from the licensing procedure in the Water Act. According to the Court, it was the effects of the proposed activity that were crucial in determining whether an activity was permissible or not.

One distinct factor in the approach of the Administrative Court was its emphasis on how much our knowledge of the environment has increased. It used this factor to argue that the Lokka and Porttipahta decisions should not be taken into account today. Whereas the Water Court relied on the results of the inspection proceedings, the Administrative Court:

> *... base[d] its conception regarding the nature potentially affected by the activity and the impacts potentially caused by the activity on the studies made in the inspection proceedings, on the additional studies done during the procedure in the Administrative Court and on its own on-site observations.*[58]

Significantly, the approach of emphasizing increased knowledge on the environment influenced the way the Administrative Court evaluated the information on the potential effects of the Vuotos project. In many instances where the Water Court deemed the results of the inspection proceedings trustworthy, the Administrative Court applied the precautionary approach and stated that there was not enough information on certain effects of the Vuotos project to make a conclusive judgment on the issue. Where there was enough information, the Court took the view that the models used in studying the likely impacts of the Vuotos project contained many uncertainties.[59] It also understood the effects of the project in a broader manner than the Water Court. The Water Court had primarily studied the area that would be flooded by the Vuotos reservoir, whereas the Administrative Court understood that the reservoir was part of an ecosystem interconnected with the Luiro and Kilpiaava mires, with the correspondingly broader scope of possible negative impacts.[60]

European Community law contributed to the reasoning of the Administrative Court in one principal way. The Court argued that:

> *Article [2(5)] must also be interpreted through the loyalty principle enshrined in the Founding Treaty of the European Community, meaning that the protection objectives of the above-mentioned directives [the Wild Birds and the Habitats Directives] are to be taken into account.*[61]

In contrast to the Water Court, which had excluded the requirements of the directives because the Kemihaara mires were not included in Natura by the Government, the Administrative Court stated that the conservation value of the area did not depend on whether it had been included in Natura.[62] According to the Administrative Court, the requirement of taking into account the protection objectives of the directives derived from the principle of loyalty as laid down in Article 10 of the Treaty Establishing the European Community.

Analysis The reasoning of the Administrative Court is quite logical: what was permissible in the 1960s may not be permissible today. The Administrative Court did not focus only on the developments in the Finnish Water Act, as the Water Court had done, but tried to observe the changes in the legal environment of Finland as a whole. It seems safe to argue that this legal environment, which after 1995 also included Community law, had adapted itself to be more protective of environmental values. Indeed, the whole environmental movement, both internationally and nationally, did not start its path into the corridors of power until after the 1972 Stockholm Conference on Human Environment. It was with this movement that knowledge on the environment, how it functions and knowledge of the environmental threats started to increase immensely.

But could this change in the societal value system be used in interpreting Article 2(5) of the Water Act? The Administrative Court answered this in the affirmative because Article 2(5), as a flexible legal norm, leaves much room for interpretation; it is an article whose interpretation depends on the values prevailing in society. It was this approach that was heavily criticized as interfering with the traditional view of the task of judiciary in the Finnish understanding of the separation of powers and, in the end, was corrected by the Supreme Administrative Court. Yet, it should be borne in mind that the Court did not simply 'discover' the change in societal values but firmly based its view on the development of the Finnish and EU legal systems.

The Administrative Court also gave interpretational value to the protection objectives of the Wild Birds and the Habitats Directives, as required by the principle of loyalty. This doctrine of indirect effect or interpretative effect is well established in the case law of the ECJ. Its essence is that national law should be interpreted in such a manner that it observes the requirements of Community law, in this case the directives. The Administrative Court thus ascribed relevance to the Wild Birds and the Habitats Directives when interpreting the content of Article 2(5), which was an approach that was in direct contrast to the reasoning by the Water Court.

It can be seen how different the position of the Administrative Court was to that of the Water Court with regard to the requirements of the directives. After the Water Court's decision on 29 February 2000, there were three significant events urging the Administrative Court to give relevance to the directives. First, on 14 June 2000 the Supreme Administrative Court remanded the Kemihaara mires case to the Government, arguing that there was still not enough information about the area. Second, on 15 June 2000, the ECJ registered the Commission's case against Finland for failing to classify SPAs fully and definitively; in the case, the Commission took up the Kemihaara mires as one example of an area that should have been designated a SPA. Third, and most importantly, the Commission started a new supervisory procedure against Finland on 12 September 2000 because, in its view, Finland was in danger of breaching the Wild Birds and the Habitats Directives by permitting the construction of the Vuotos project. It would seem fair to argue that these three events had at least some influence on how the Administrative Court reasoned in the case.

The Supreme Administrative Court

The Supreme Court did not change the judgment of the Administrative Court: it overturned the permit decision made by the Water Court and rejected Kemijoki Ltd's application on the basis of the same criterion of Article 2(5) as the Administrative Court had used. However, the Supreme Court found it essential to justify its decision in a different manner.

Like the Administrative Court, the Supreme Court found that Article 2(5) is a flexible legal norm. However, it did not accept the reasoning of the Administrative Court that interpretation in such cases should be based primarily on changes in societal values. In the view of the Supreme Court, it was the semantic openness of Article 2(5) that left discretion to the legal decision maker. For the Supreme Court this discretion meant legal discretion – the interpretation of an individual norm through other sources of law.

The Supreme Court first determined whether Article 2(5) had changed in the course of time. Both the Water Court and the Administrative Court were of the opinion that no essential changes had been made to the article over the years. The Supreme Court, in contrast, strongly emphasized the amendment made to Article 2(5) in 1987 – 'in the aquatic environment and its functioning' – which, according to the Court, had inserted ecological values into the article. This argument was supported by the fact that the *travaux preparatoires* to the 1987 amendment showed that the legislator had included ecological values in Article 2(5) at that time.

The Court also asserted that Article 2(5) must be interpreted in light of Article 20 of the Finnish Constitution, which guarantees the right to a decent environment. The Supreme Court submitted that the constitutional clause protecting the environment was not only a prescription to the legislators but also a principle guiding the application of the law and its interpretation, whereby it should also guide the interpretation of Article 2(5) in this case. Other considerations that the Court deemed relevant to the decision were the constitutionally protected right to property and a general value underlying the legal system – the predictability of legal decision-making.[63]

The Water Court and the Administrative Court had had differing views about the relevance of EC law in this case. In one sense the Supreme Court went further than the lower courts and stated that at least certain norms found in the directives are so clear and unconditional that they may have direct binding effect in some cases. Yet, in this case the Supreme Court argued that it was reasonable to apply national law, the Finnish Water Act. The Supreme Court did refer generally to the importance of the interpretative effect of directives, but this was made as a response to the argument of the applicant, Kemijoki Ltd, concerning the influence of directives in EC law, rather than an application by the Court of the interpretative effect doctrine to the case through giving Article 2(5) the content required by the directives.

Analysis The judgment by the Administrative Court of Vaasa caused uproar in Finland because it was felt that the Court had exceeded its authority as a court of law. According to the separation of powers doctrine, as it is understood and accepted in Finland, courts of law are to apply rather than make the law; the determination of how changed values influence society is a task for Parliament, not the courts.

As indicated above, the reality is more complex than this. The Administrative Court did not make its decision on the basis of changed values but on the basis of Article 2(5) of the Water Act, with the changed values and the directives assisting it in interpreting this article. But so strongly did the Administrative Court base its reasoning on changed values that the Supreme Court must have thought it best to correct the Administrative Court's approach.

The reasoning of the Supreme Court was based on the strict separation of the use of legal and political discretion. This was required because Kemijoki Ltd had invoked Article 5 of the Act in its appeal, a provision that stipulates that when a case involves political discretion, the Supreme Court is to refer the matter to the Government of Finland, or to at least consult the Government. The Supreme Court clearly avoided taking a position on the changes in societal values, an argument that had been so prominent in the reasoning of the Administrative Court.

The Supreme Court used the approach of strict legal discretion, i.e. interpretation of an individual flexible norm with the help of other norms that are formally part of Community law and the Finnish legal system, respectively. The Supreme Court also reasoned quite convincingly that Article 2(5) had been amended in 1987 in an essential manner to this particular case, because ecological values had been inserted into the article. This reasoning had been overlooked by the lower courts.

Also important was the reasoning by the Supreme Court on the relevance of the Wild Birds and Habitats Directives. The Administrative Court had used the doctrine of indirect effect and interpreted Article 2(5) in terms of the protection objectives of the directives. The Supreme Court seemingly went further, stating that:

> *... even though Article 4, paragraphs 1 and 2, are sufficiently clear and unconditional in content for them to have at least in some cases a direct effect binding on the officials of the Member States, the Supreme Administrative Court views that ... it is not necessary to determine in the context of this permit application whether the obligations of the Wild Birds Directive mentioned would form an obstacle to granting a permit.*[64]

The Supreme Court was implicitly saying that it was best to apply national law in this case. The fact that the European Commission very clearly wanted Finland to include the Kemihaara mires in Natura did not obviate the difficult national situation that the Kemihaara mires case was still being prepared in the FME and had not yet been accepted by the Finnish Government for inclusion in Natura.

Even though the Commission had a good case against Finland, and it had invoked the direct effect doctrine to prevent the construction of the Vuotos project, it was still wise for the Supreme Court to apply national law. In this way, the Court could distance itself from the Natura process while, at the same time, maintain the importance of the relevant directives, and their direct and indirect effects. The Court was justified in preventing the construction of the Vuotos project by national law alone, simply because the impacts of the Vuotos project were massive enough to trigger Article 2(5). By doing so, the Court avoided any conflict with the requirements of the directives, which clearly seem to require that the Kemihaara mires be included in Natura.

The consequences of the Supreme Court Judgment

The Supreme Court was able to steer its way quite successfully in a very difficult case on which various legal procedures had an impact. First of all, it managed to avoid any conflicts with the Wild Birds and Habitats Directives by preventing the construction of Vuotos on the basis of Article 2(5) of the Water Act. Moreover, because it applied a Finnish legal norm, there was no need to have recourse to the preliminary ruling system of the EU. This was important given that the *Vuotos Case* had already seen too many legal procedures.

On a general level, the Supreme Court's decision relied on sources of law, not changed societal values, and it could thus institutionally support the Finnish version of the separation of powers doctrine. Even though the line between applying and enacting the law can be shown to be rather thin, the acceptance of this doctrine in Finland tends to lead to conservative, rather than creative, application of the law. The Court was also able to send constructive signals to the Finnish judiciary as to how EU law is to be interpreted in the country, following closely the established case law of the ECJ. It was significant that the Supreme Court argued that the constitutional provision on the environment, Article 20 of the Finnish Constitution, is not only an abstract legal principle but one that guides the application of law. Since Article 2(5) had not prevented the construction of a proposed activity before the *Vuotos Case*, the Supreme Court clarified its function as a provision that can deny a permit to a proposed activity. Clearly, however, the construction of the reservoir is an activity of such magnitude and impact that, even though the Supreme Court has now confirmed that the article may prevent an activity being implemented, the threshold for applying the article remains high.

The latest developments

Soon after the judgment by the final court of appeal, the Supreme Administrative Court, the advocates of the Vuotos project started to plan other options for implementing the activity. The Governor of Lapland, as well as a number of Members of Parliament (MPs) from the province, suggested in public that the next government should start to prepare new policies to implement the project. The Governor of the Province of Lapland proposed in a local newspaper that it was time for a special legislative act on Vuotos, which would, as *lex specialis*, override the Finnish Water Act.[65]

Parliamentary elections were held in March 2003 and a new government was formed in April. The main parties that formed the government, especially the Prime Minister's Centre Party, strongly advocated for the construction of the Vuotos project, among other things, as a measure of regional policy. Indeed, after the elections, the Vice-Chairman of the Centre Party, an MP from Lapland, Hannu Takkula, proposed in a local newspaper that the Vuotos project should be implemented. His suggestion differed somewhat from that of the Governor in that he advocated amendments to the Water Act to enable the implementation of the project.[66]

While the parliamentary elections were receiving all the attention in Finland, the ECJ (on 6 March 2003), decided whether Finland had breached its obligations under the Wild Birds Directive.[67] The Commission had accused Finland of not having fully and definitively notified the Commission of its SPAs, mentioning the Kemihaara mires as one example that should have been designated. On 25 April 2002, the Advocate-General issued his opinion on the case, arguing that Finland had not fully and definitively notified its SPAs, a position that was upheld by the ECJ.[68]

Since the ECJ did not mention the Kemihaara mires in its judgment, the local newspaper, *Lapin Kansa*, welcomed the Court's judgment as proof that Community law did not require the mires to be included in Natura.[69] Yet, this action by the Commission was brought before the ECJ because Finland had not notified enough SPAs in general and in due time, with the mires only cited as an example that certain obvious SPAs had not been included. In this action, the ECJ could not have decided that a certain area should be designated as a SPA. It was, however, in a position to specify which sources of ornithological information

Member States should use when deciding on the SPAs in their territory, but chose not to enter into this discussion.[70]

After the judgment by the Supreme Administrative Court on 18 December 2002, preparations to include the Kemihaara mires in the Natura started anew. As explained above, on 14 June 2000 the Supreme Administrative Court had remanded the matter of the Kemihaara mires to the Government because there was not enough information about the area. In mid-April 2003, the FME proposed that the Kemihaara mires should be designated both as a SPA and SCI in Natura. The deadline for comments on, and objections to, the new Natura proposal, which included quite a number of new areas in addition to the Kemihaara mires, was 15 May 2003.

The *Vuotos Case* returned to the headlines when the first meeting of the inter-ministerial committee convened on 17 October 2003. Several Centre Party ministers took a stance against including the Kemihaara mires in Natura, voicing their opinion that the Kemihaara mires can be excluded from Natura also from the perspective of Community law and that at least some kind of mini-Vuotos can still be constructed.[71] At the same time, Kemijoki Ltd reported that it had abandoned the Vuotos project, even a mini-Vuotos, but was now planning to construct hydropower plants, possibly even in the area of the Kemihaara mires. The company insists that the Kemihaara mires should be excluded from Natura.[72] The inter-ministerial committee, responsible for making the Natura proposal to the Finnish Government, disagreed on the status of many areas, including the Kemihaara mires. The decision by the Government in this matter should, in all likelihood, be completed in 2004.

It seems very likely that the new Government will propose that the Kemihaara mires be included in Natura. Two principal reasons may be cited. First, the Commission is pressing hard for inclusion of the mires: it has invoked two supervisory procedures against Finland, arguing that the Kemihaara mires should be included in both. Second, the Commission's view that the Kemihaara mires should be designated as at least an SPA is well supported by ornithological knowledge, the mires clearly being one of the most suitable sites in Finland for SPA status. Presently, the Kemihaara mires are proposed for inclusion in Natura as both a SPA and as SCI.

Conclusion

The *Vuotos Case* shows quite well the effects of the complex regulatory environment that exists today, especially for EU Member States. Almost all aspects of environmental protection are covered by international law, EC law, as well as each Member State's own legal system. International law outlines the basic universal principles and regulates transboundary and global environmental problems; Community law focuses on the minimum standards for harmonization of Member States' environmental protection systems; and national laws stipulate the details of required behaviour and provide the institutional machinery to monitor and enforce the applicable norms. Even though in some cases the division of labour among the three systems works well, it is still often the case that the three overlap. Where they overlap, the uncertainty of applicable legal standards is exacerbated by the very different nature of these three levels of legal systems – frequently only the specialist in European Community law is aware of all the particulars of the case law of the ECJ. This is a formidable challenge to legal decision-makers in all Member States, as was shown by the *Vuotos Case.*

It is the influence of EC law in particular that has brought with it problems for those making and applying national law. As seen in the case of environmental protection, the Commission supervises Member States to ensure that EC law is correctly implemented and applied and, if necessary, brings proceedings against the State before the ECJ. This was arguably one of the factors influencing the way the Finnish administrative courts made their decisions in the *Vuotos Case,* especially the Administrative Court and the Supreme Court. In the end, the Supreme Court decided the case on the basis of Finnish law, but it also made clear that the decision had to be consistent with the requirements of EC law. As seen above, the Commission was convinced that the Kemihaara mires should be designated as a SPA and SCI, and even initiated two supervisory procedures to convey this view to Finnish decision-makers. The Commission had a good case against Finland, given that the ornithological facts indicate that the Kemihaara mires area should be included in Natura as at least a SPA area.

It can convincingly be argued that the Finnish legal system was able to apply both international law and Community law in a legally correct manner in its national licensing procedure. Finland notified Sweden of the potential negative effects on the latter's maritime area even before the legally valid convention, the NEPC, required it to do so. Finland also allowed Swedish private and public parties to the *Vuotos Case* to participate in its national licensing procedure on the basis of the NEPC. Moreover, even if the Supreme Court made its decision on the basis of Article 2(5) of the Finnish Water Act, the influence of the directives to the decision was, arguably, strong. The Supreme Court also clarified the way directives should be applied in the Finnish legal system, following closely the established case law of the ECJ in its reasoning. In this perspective, the *Vuotos Case* can be seen as a good example of how the three legal systems can be accommodated in national legal decision-making.

NOTES

This chapter was originally published as: Koivurova, T. (2004) 'The Case of Vuotos: Interplay between International, Community and National Environmental Law', *Review of European Community and International Environmental Law,* vol 13, no 1, pp47–60.

1 The main legal materials cited in this chapter can be found from the judgment of the Administrative Court of Vaasa, *Vuotoksen tekojärven ja voimalaitoksen rakentaminen sekä vesistön säännöstely*, author's copy from the Court, in Finnish only. The copy can be ordered from the Administrative Court of Vaasa (Rantakatu/Kalarannanpuisto, PB 204, 65101 Vaasa, Finland; neither the decision of the Water Court of Northern Finland, nor the judgment of the Administrative Court of Vaasa is reported). The Administrative Court's judgment contains the judgment of the Water Court (*Vuotoksen tekojärven ja voimalaitoksen rakentaminen sekä vesistön säännöstely Pelkosenniemen, Sallan ja Savukosken kunnissa sekä töiden aloittaminen ennen lupapäätöksen lainvoimaiseksi tulemista*), the findings of the inspection proceedings (the inspection statement) and Kemijoki Ltd's application. The document is divided into two parts, the first containing the Administrative Court of Vaasa judgment (Administrative Court Decision), the second the application by Kemijoki Ltd, the inspection statement and the judgment of the Water Court of Northern Finland (Water Court Decision). Since the numbering of each part has been done separately, the two parts of the document will be referred to as 'part I' and 'part II', respectively. The judgment of the Supreme Administrative Court (KHO 2002:86, taltionumero 3339) (Supreme Administrative Court Decision) can be found on the website – only in Finnish – available at www.kho.fi/paatokset/17859.htm. As this judgment, as it is on the website, contains no numbering, its component chapters only will be referred to.

2 'Vuotos project' is used here to denote all the aspects of the Kemijoki Ltd application, i.e. the construction of the Vuotos reservoir and dam as well as the company's rights to regulate the water conditions in the nearby area.

3 Kemijoki River Regulation Act (62/1960). The Finnish Acts and Decrees from the year 1734 up to the present can be found – only in Finnish and in Swedish – from the Finlex website, available at www.finlex.fi/lains/index.html. The Government Bills can also be found from the same website, albeit covering only Government Bills proposed after 1991 – and only in Finnish – more specifically at www.finlex.fi/esitykset/index.html.

4 Information obtained from the Kemijoki Ltd website, available at www.kemijoki.fi/vuotos_eng/vuotose.htm. The home page of the Kemijoki Ltd website contained a separate section on Vuotos but it was removed after the rejection of Kemijoki Ltd's Vuotos application by the Supreme Administrative Court.

5 Ibid.

6 Council Directive 79/409/EEC of 2 April 1979 on the Conservation of Wild Birds, (1979) OJ L103/1.

7 Council Directive 92/43/EEC of 21 May 1992 on the Conservation of Natural Habitats and of Wild Fauna and Flora, (1992) OJ L206/7.

8 Kemijoki Ltd proposed that a voluntary EIA of Vuotos should be carried out. However, this suggestion was rejected by the Finnish Water and Environment Council after consultation with the FME (1 July 1993).

9 An additional proceedings engineer was appointed for the Vuotos procedure because of the project's size. See Water Court Decision, note. 1 above, at 22–23 (part II) reviewing the inspection proceedings.

10 According to the Water Act, the public may present comments through inspection meetings. The regular inspection meetings were held between 7 and 17 February 1994.

See Water Court Decision, n. 1 above, at 23 (part II). The final inspection meetings were organized in the communities concerned between 2 and 11 July 1996.

11 See Water Court Decision, ibid., at 84–85 (part II).

12 The Convention on Environmental Impact Assessment in a Transboundary Context (Espoo, 25 February 1991), printed in 30 ILM (1991), 800. The convention entered into force on 10 September 1997.

13 Letter from the Finnish Water and Environment Council to the Water and Environment District of Lapland. Letter obtained from Kemijoki Ltd (dated 1 July 1993).

14 The Nordic Environment Protection Convention (Stockholm, 19 February 1974), printed in 13 ILM (1974), 352. The convention entered into force on 5 October 1976.

15 See Administrative Court Decision, note 1 above, at 50–51 (part I).

16 Act on Environmental Impact Assessment Procedure (468/1994) (EIA Act).

17 Section 14(3) of the EIA Act reads: 'The Ministry of the Environment shall request a Ministry for Foreign Affairs opinion on the matter'; ibid.

18 See Administrative Court Decision, note 1 above, at 50–51 (part I) for a review of the procedure between the two countries.

19 This was approximately US$371 million.

20 See Water Court Decision, note 1 above, at 251 (part II).

21 See Administrative Court Decision, note 1 above, at 55–58 (part I).

22 Ibid., at 68.

23 Ibid., at 49–53.

24 See Supreme Administrative Court Decision, note 1 above, at chapter 6.2.2.1.

25 Author's personal copy of the Commission's reasoned opinion, dated 15 December 1998, received from the Ministry for Foreign Affairs.

26 Nature Protection Act (1096/1996).

27 Author's personal copy of the letter of formal notice, dated 10 July 1998. The copy was provided by the Ministry for Foreign Affairs.

28 The process is reviewed in the Supreme Administrative Court judgment of 14 June 2000 (1675, Natura-2000 – *ehdotusta koskevat valitukset*), at 5. The copy was provided by the Supreme Administrative Court.

29 The reasoned opinion of 15 December 1998. The copy was provided by the Finnish Ministry of Foreign Affairs.

30 The Commission's action against Finland was registered at the ECJ on 15 June 2000.

31 See Water Court Decision, note 1 above, at 420 (part II) (author's translation).

32 See Administrative Court Decision, note 1 above, at 9 (part I).

33 See Supreme Administrative Court judgment, note 28 above, at 39–40.

34 Author's personal copy of the letter of formal notice of 12 September 2000, obtained from the Finnish Ministry for Foreign Affairs.

35 Author's personal copy of the reasoned opinion 10 April 2001 obtained from the Finnish Ministry for Foreign Affairs.

36 Paragraph 1 of the resolution of the signatories of the Espoo Convention (ECE/ENVWA/19, 26 February 1991).

37 For a closer analysis, see T. Koivurova (2002) *Environmental Impact Assessment in the Arctic: A Study of International Legal Norms*, Ashgate Publishing, at 296–297.

38 See Kemijoki Ltd's website, available at www.kemijoki.fi/kemijoki/kemijoki.nsf/TD/F1D6C3950CD3705242256D28002E6061/$File/huomautus%20Ymn%20uuteen%20ehdotukseen.doc?Open Element
39 The cases were the extension of the steelmaker Outokumpu's ferrochrome and stainless steel works near the border of Sweden and the planned gas power plant near Imatra, close to the Russian border.
40 See Water Court Decision, note 1 above, at 251 (part II).
41 See Administrative Court Decision, note 1 above, at 68 (part I).
42 Ibid., at 49–53.
43 The Nature Protection Association also failed to lodge its complaint on time and thus the complaint was rejected.
44 See Administrative Court Decision, note 1 above, at 49–53 (part I).
45 Finnish Water Act (264/1961).
46 Amendment to the Finnish Water Act (467/1987) (author's translation). Article 2(5) was also amended in 1994 (553/1994) by removing the qualification 'wide-ranging', which slightly lowered the threshold for applying the provision. The amendment was not applicable to the *Vuotos Case* because it did not apply retroactively.
47 See Water Court Decision, note 1 above, at 253–273 (part II). As noted above, this was approximately US$371 million.
48 Ibid., at 253–256.
49 Kemijoki River Regulation Act, note 3 above.
50 Ibid., author's translation.
51 Government Bill on terminating the Kemijoki River Regulation (Act 45/1968).
52 See Water Court Decision, note 1 above, at 264 (part II) (author's translation). The Water Court uses the incorrect term 'Euroopan neuvoston komissio', which in Finnish means the Council of Europe, even though it was clearly referring to the European Commission.
53 The Water Court noted that (see Water Court Decision, note 1 above, at 263 (part II)): '[The] Commission has stated that it does not commence a separate supervisory procedure, which would focus specificly on the Vuotos area ... and has referred to its above-mentioned decision to bring an action against Finland' (author's translation).
54 See Administrative Court Decision, note 1 above, at 71–72 (part I).
55 Ibid., at 56 (author's translation).
56 Ibid., at 56–57.
57 The Court also notes (at 57) that the recent case law related to the water management in Finland has placed much emphasis on environmental values; ibid.
58 Ibid., author's translation.
59 See Administrative Court Decision, note 1 above, at 59–60 and 68–69 (part I).
60 Ibid., at 61 and 64.
61 Ibid., at 58 (author's translation).
62 Ibid., at 61
63 The Court stated that: 'The Article manifests a stance on behalf of environmental values, at the level of the Constitution and for its part guides interpretation and application of the law. In assessing the significance of the Article in an individual application of the law, the other fundamental rights shall also be taken into account, such as the constitutional protection of property rights. The scope of discretion is

also influenced by the predictability of application of the law'; see Supreme Administrative Court Decision, note 1 above, at chapter 6.2.2.4.(a) (author's translation).

64 See ibid., at chapter 6.2.2.2 (author's translation).

65 See the newspaper article by the Governor of Lapland, Hannele Pokka, 'Vuotospäätökseen ei pitäisi tyytyä', *Kaleva* (3 January 2003) at 2.

66 See the newspaper article 'Keskusta pitää Vuotosta mukana', *Lapin Kansa* (4 April 2003), at 3.

67 ECJ 6 March 2003, Case C-240/00, *Commission v. Finland.* The judgment and the Advocate-General's opinion can be accessed through the EUROPA website, available at http://europa.eu.int/jurisp/cgi-bin/form.pl?lang=en.

68 Opinion of Advocate-General Philippe Léger, 25 April 2002, ibid.

69 See the newspaper article 'EY-oikeus ei vaadi Kemihaaraa Naturaan', *Lapin Kansa* (7 March 2003), at 3.

70 The Advocate-General reasoned in this direction, arguing that the publications *Birdlife Report* and *Important Bird Areas* must be seen as having some scientific value in evaluating which sites should be designated as SPAs. See Advocate-General's Opinion, note 67 above, paragraph 40.

71 See the newspaper articles 'Vuotoksen tekoaltaasta kytee taas uusi kiista', *Helsingin Sanomat* (16 October 2003), at A3; 'Hallitus kiistelee Kemihaaran Natura-soista', *Lapin Kansa* (17 October 2003), at 3.

72 See the newspaper article 'Kemijoki Oy ei halua Kemihaaran soita Naturaan', *Lapin Kansa* (18 October 2003), at 1.

5

MOX Plant and the Espoo Convention: Can Member State Disputes Concerning Mixed Environmental Agreements Be Resolved outside EC Law?

Simon Marsden

INTRODUCTION

In 2006, the European Court of Justice (ECJ) effectively brought an end to a trio of proceedings begun in 2001 by Ireland against the UK[1] pursuant to two treaties they were parties to.[2] It concluded that since both were EC Member States, disputes concerning treaties to which the EC was also a party should be determined by the ECJ, not arbitral tribunals.[3] The proceedings highlighted the complexities of self-contained regimes in international law,[4] emphasized the proliferation and overlapping nature of courts and tribunals, and clarified the relationship between international and EC law.[5] While these issues have received the most attention in the literature,[6] there has been no detailed analysis of whether Ireland's choice of these treaties was the most appropriate litigation strategy. Suggestions have been made that EC directives on access to environmental information[7] and environmental impact assessment (EIA)[8] should have been utilized,[9] yet there has been no discussion concerning the broader application of the underlying international law to the latter directive. Indeed, while one author has proposed a protocol on marine EIA to the UN Convention on the Law of the Sea (UNCLOS) as a means of preventing disputes of this kind again,[10] it is somewhat surprising that the Espoo Convention has not been discussed,[11] since it deals with many of the matters that featured in the proceedings, notably procedures for transboundary environmental impact assessment (TEIA) including access to information, consultation and participation of 'affected parties' in relation to proposals being considered by the 'party of origin'.[12]

The object of this chapter is therefore to consider the *MOX Plant* dispute in the light of the Espoo Convention, making specific reference to relevant provisions including the Inquiry Procedure, which is used to determine effect significance when disputed by the parties. It will firstly summarize the dispute and the litigation that ensued, highlighting the implications for those unfamiliar with it. It will, secondly, examine the provisions in the EIA Directive that were relevant to the dispute, and since it only partly implements the Espoo Convention, will, thirdly, outline the Inquiry Procedure, which was relevant to the dispute and was recently used in the *Danube Delta* case.[13] Although there are examples of the Espoo Convention being applied directly between EC Member States,[14] other than where non-EC Member States are also involved[15] (as in the example of the application to the Cernavoda nuclear power development, described below), it is doubted whether this remains a viable option. This chapter concludes that since the ECJ decision in 2006, disputes between Member States are now clearly matters for the European Commission and ECJ rather than arbitral tribunals and other institutions. While this may further the autonomy of the European legal order, without comparable procedures for determining disputes over effect significance, it may not be the best outcome for the environment.

Background to the dispute

Sellafield is the largest, oldest civilian nuclear complex in the UK, situated on the English coast 112 miles away from Ireland across the Irish Sea. Since the 1950s, it has been the UK centre for reprocessing spent fuel rods from nuclear power stations. At the time of the dispute, the site contained a number of facilities, including a mixed oxide fuel (MOX) Demonstration Facility.[16] The owner/operator, British Nuclear Fuels Limited (BNFL), a UK company owned by the government, wanted to build a plant to make MOX from reprocessed plutonium dioxide and uranium dioxide. MOX was to be used as a fuel in nuclear reactors; since there were none of these in the UK it was therefore likely to be used for export, mainly by sea.

In 1992, BNFL applied for consent from the UK regulators to construct the MOX plant. An EIA was only required under English regulations if the project was likely to have significant environmental effects and although it had already been determined that the likely discharges were insignificant, an Environmental Statement was nonetheless prepared at Ireland's suggestion.[17] This was deficient, as it was limited in scope, containing, for example, no section explaining the potential impacts on the Irish Sea or on the transport of nuclear materials. It also lacked procedures for inter-State consultation and information exchange.[18] It concluded that there were no significant environmental effects, allowing the construction to proceed. The site was permitted in 1994 and constructed in 1995.

In 1996, BNFL applied for consent to operate the MOX plant, which required compliance with national radioactive waste provisions and provisions in the Euratom Treaty.[19] BNFL had to show first that it would not release unacceptable discharges, and, second, that it would bring economic and social benefits. There was no consultation with Ireland concerning the discharges, and it was concluded by the UK regulator that they would not exceed those already permitted. Consultation on the latter was needed under Article 37 of the Euratom Treaty; an opinion of the European Commission was required and, as part of this process, notification of plans for disposal of nuclear waste, such as at Sellafield, took place with Ireland between 1997 and 2001. In October 2001, it was concluded that there would be economic and social benefits and the Commission agreed that discharges would be insignificant. It was therefore permitted in December 2001.

Litigation and outcomes

In June 2001, Ireland began proceedings under the OSPAR Convention for the Protection of the Marine Environment of the North-East Atlantic claiming the UK had failed to make information available during consultations on operation. In October 2001, Ireland began proceedings under UNCLOS claiming there had been an inadequate EIA, a failure to cooperate with Ireland over the protection of the Irish Sea and a failure to take measures to protect and preserve the Irish Sea. Both treaties were 'mixed environmental agreements' to which Ireland, the UK and the EC were parties.[20] In October 2003, the EC Commission began proceedings against Ireland for breaches of the Treaty of the European Community (TEC), for submitting a dispute to another tribunal and for not consulting the Commission.

OSPAR Arbitral Tribunal

The OSPAR Convention contains access to information provisions modelled on EC Directive 90/313.[21] Article 9(1) states that information set out in Article 9(2) should be made available on reasonable request. Article 9(2) includes available information on the state of a maritime area and activities or measures adversely affecting or likely to adversely affect it; Article 9(3) allows refusals on grounds of commercial confidentiality; and Article 32 stipulates that disputes may be referred to arbitration. As a result of the more limited nature of these provisions and the narrowness of the approach taken to them (including a refusal to consider EC Directive 90/313 on which they were based),[22] the Arbitral Tribunal held that the information did not relate directly to activities or measures adversely affecting or likely to affect a maritime area, but rather to economic justification, so there was no breach by the UK.[23]

UNCLOS: ITLOS and Annex VII Tribunal

Two tribunals were involved under the UNCLOS proceedings: the International Tribunal on the Law of the Sea (ITLOS) and the Annex VII Tribunal. The first body is a permanent institution which may be called upon to decide provisional measures where the urgency of the situation so requires; the second is an institution constituted where parties have not chosen the same institution to determine disputes under Part XV of UNCLOS. In November 2001, Ireland asked ITLOS to prescribe provisional measures under Article 290(5) to suspend the permit and prevent the operation of the plant; to prevent movement of radioactive substances through its waters; and for the UK to take no action that would aggravate the dispute. In December 2001, ITLOS made its decision following a consideration of its jurisdiction and whether the matter was sufficiently urgent so as to protect the parties' positions or serious harm to the environment. It decided it had jurisdiction but rejected the request for provisional measures outlined because of a lack of urgency. ITLOS prescribed a measure of its own, however, that the parties should cooperate in exchanging information on risks and the effects of the plant's operation and in devising ways of dealing with them.[24]

The Annex VII Tribunal was later established for a full hearing of the issues to take place because Ireland had not made any choice between the tribunals available to it (International Court of Justice (ICJ), ITLOS or Annex VIII) to determine disputes when it ratified the treaty.[25] Two months after the ITLOS decision, the Annex VII Tribunal was constituted to consider a number of matters: further to Article 206 that there had been no proper EIA of the MOX (and the related Thermal Oxide Processing Plant (THORP) plants; that additional waste was to be stored at Sellafield; and the matter of the international movement of waste. Further to Articles 123 and 197, it was argued that there had been a failure to cooperate adequately regarding the protection of the marine environment of the Irish Sea. Finally, in accordance with Articles 192–194, 207, 211 and 213, it was argued that there had been a failure to take the necessary measures to protect and preserve the marine environment of the Irish Sea.

In June 2003, the Tribunal handed down its statement on jurisdiction. The issue of EC law was regarded by the Tribunal as a 'serious difficulty'. Because UNCLOS was a mixed agreement, it was considered quite possible that the ECJ may well conclude that the provisions of UNCLOS were a matter for it to determine. If the Tribunal were to have decided the matters in dispute, there could therefore be two decisions on the same issue. The proceedings were hence suspended for the ECJ to consider; like ITLOS however, the Annex VII Tribunal still held it had *prima facie* jurisdiction to consider provisional measures.[26]

ECJ

The Commission commenced legal proceedings against Ireland under the Treaty of the European Community (TEC) (which were mirrored in the Euratom Treaty).

These concerned alleged breaches of Article 10 of the TEC, which required 'loyal cooperation', and Article 292 on the exclusive jurisdiction of the ECJ over Treaty matters. The Commission took the view that even if only part of a dispute falls within EC law, then the ECJ was competent to determine it. In 2006, the ECJ agreed, supporting the Opinion of the Advocate-General on the matter;[27] and following the earlier *Commission v. Ireland* decision,[28] where it had held that jurisdiction extends to all mixed agreements and does not depend on whether it has competence for each provision within an international agreement. As a result, Ireland should have consulted the Commission regardless of whether it believed it had competence itself (Article 10). UNCLOS was part of the EC legal order because the EC was a party to it and thus Ireland should have referred the matter to the ECJ (Article 292).[29] This was supported by UNCLOS being adopted under the environment provision in the TEC, and by the implementation of it by secondary law, such as the EIA Directive.

Analysis of jurisdiction

In analysing the jurisdiction issue, different positions were reached by each of the arbitral tribunals. The OSPAR Arbitral Tribunal had no difficulty in pronouncing on the issues before it, and ITLOS had rejected any attempt to side-line or subsume the significance of UNCLOS. It held that each treaty was an individual identifiable legal manifestation worthy of respect, with the dispute-settlement procedures of each treaty intended to deal with the disputes under each. For the Annex VII Tribunal, the question was whether the ECJ's exclusive jurisdiction extends to the interpretation and application of UNCLOS 'as such and in its entirety', although the ongoing suspension of its proceedings indicated it was willing to allow for such a finding and give way to its jurisdiction in the matter. Although the question of the ECJ's competence needed to be determined, the ECJ held there was no need to look at each provision to see whether it fell within the competence of the ECJ, because in *Commission v. Ireland* (2001) the ECJ said it was not necessary.[30]

Therefore, despite the *pacta sunt servanda* central concept that treaties must be complied with, the ECJ litigation indicated that, in some instances, self-contained regimes may claim jurisdiction to determine disputes to the exclusion of other regimes and more specific treaties. This is despite the fact that the problem may remain that if a tribunal of more than one international treaty with a specialist interest in the matter (*lex specialis*) claims competence, there will always be the issue of who has the better claim to decide the case. Despite the ECJ's ruling, there therefore remains great potential for forum shopping, simultaneous litigation and successive proceedings, with inconsistent, conflicting decisions possible.[31] Although this is complicated for mixed agreements, it is practically more straightforward following the ECJ's 2006 decision, for while a tribunal may have a different point of view to the ECJ as to whether it has jurisdiction, the Member States appearing before the tribunal must withdraw or face action under EC law.

What might have happened?

Given the unfavourable outcome to Ireland, it may have been in its interests to have first considered the application of EC primary law for ECJ adjudication and Commission consultation, further to Articles 10 and 292 of the TEC (and Euratom equivalents), and then, second, considered the application of EC secondary law implementing the Espoo Convention: EC Directive 85/337 (EIA) as amended by Directive 97/11 (EIA)[32] and also EC Directive 90/313, concerned with access to environmental information.[33] Although Ireland had not ratified the Espoo Convention at the time and it was only a signatory, since the UK and EC had ratified and it was in force it is possible that it could have been considered to have direct effect in the EC legal order whether or not its provisions had been fully transposed by the EIA Directive.[34] This could have enabled the Commission on behalf of the EC to request the establishment of an Inquiry Commission under the Espoo Convention to consider the significance of the environmental effects that formed the basis of Ireland's complaints.[35] This 'Communitarization', or 'Europeanization', of international law results in international treaties being placed between primary and secondary EC law with directives needing to be interpreted with reference to the relevant international treaty.[36]

Although arguments for direct effect (or, alternatively, that the EC may have been in breach of its international treaty obligations by not having given full effect to the Espoo Convention) were not made, the Espoo (and Aarhus)[37] Convention(s) was in fact cited by Ireland in its Memorial to both the OSPAR and UNCLOS tribunals as relevant contextual law to which the specific law at issue should have regard.[38] Article 31(3)(c) of the Vienna Convention on the Law of Treaties states that in the interpretation of treaties 'there should be taken into account ... any relevant rules of international law applicable in the relations between the parties'.[39] However, when Ireland commenced proceedings against the UK in June 2001 under the OSPAR Convention and in October 2001 under UNCLOS, the fact that it was not a party to the Espoo Convention (nor indeed to the Aarhus Convention) would have weakened its position.[40] Whether direct effect or EC non-compliance was a matter for Ireland to have argued substantively or for the Commission to have initiated, is purely speculative. Yet the establishment of an Inquiry Commission may well have pre-empted the dispute before Ireland commenced litigation.[41] The closest reference in the literature to the application of the Espoo Convention is by Tanaka, who comments:

> *Although the differences between the opposite sides of the Irish Sea might ultimately be resolved in consultation with EC and other*

relevant regional instruments with more specific criteria, such a solution might not be possible in a similar dispute in a different region without instruments establishing common standards for environmental impact assessments.[42]

EC Directives

Since the focus of this chapter is on the application of TEIA via the Espoo Convention and implementing the EIA Directive (as amended), the Environmental Information Directive is not considered here in detail. However, since similar arguments apply to its role in the dispute and because of its comparatively close relationship with the Aarhus Convention, it is also discussed briefly below. Another related provision, Directive 2003/35/EC,[43] has strengthened provisions in the EIA Directive in relation to public participation. While it was not in force at the time of Ireland's dispute, which in any event was directed to the information and consultation needs of the Irish government rather than the public, it is, however, briefly discussed in relation to the EIA Directive because the principle of non-discrimination is an important component of TEIA.[44]

Churchill and Scott draw attention to the failure of Ireland to challenge the UK under the EC Environmental Information and EIA Directives, both of which, even in their original versions, contained more detailed obligations for these areas than OSPAR and UNCLOS. If Ireland had based its challenge upon them, it would have more likely increased its chances of success in the dispute, as well as avoiding a challenge by the Commission. The authors draw attention to the broad approach taken by the ECJ in connection with the Environmental Information Directive (especially since it has been replaced by a more detailed directive), which is also very much the case with regard to the EIA Directive. They indicate that review in an English court would also have been possible for Ireland if the request for access to information had been refused by the UK, and, if denied, a complaint to the Commission or direct action by it to the ECJ would have been available.[45]

Environmental Information

Directive 90/313/EEC of 7 June 1990 on the freedom of access to information on the environment initiated, in the words of recital (2) to the directive that replaced it (Directive 2003/4/EC of 28 January 2003): 'a process of change in the manner in which public authorities approach the issue of openness and transparency, establishing measures for the exercise of the right of public access to environmental information which should be developed and continued'.[46] Following a review of the original directive by the Commission and the signature by the EC of the Aarhus Convention, it was decided to replace the original directive with a new legislative text. Other developments, such as the increased use of information technology, had an impact upon the content and procedural provisions of the new directive.

Revisions to the original directive would not have applied at the time Ireland commenced its litigation against the UK,[47] but now include an amended definition of environmental information and public authorities and tighter restrictions on any exceptions[48] that can prevent the release of the information.[49] Article 1 sets out the two objectives of the new Information Directive, which are:

> *(a) to guarantee the right of access to environmental information held by or for public authorities and to set out the basic terms and conditions of, and practical arrangements for, its exercise; and (b) to ensure that, as a matter of course, environmental information is progressively made available and disseminated to the public in order to achieve the widest possible systematic availability and dissemination to the public of environmental information.*

Although traditional exceptions to disclosure remain in Article 4 (which would also have been the case in 2001) – including 'material in the course of completion or unfinished documents or data', 'internal communications', or if the disclosure of the information would affect 'the confidentiality of proceedings of public authorities, where such confidentiality is provided for by law' – the grounds for refusal 'shall be interpreted in a restrictive way, taking into account for the particular case the public interest served by disclosure'. Any refusal is required to be in writing within specified time limits if so requested,[50] and reasons with details of the review procedure are to be provided. Article 6 contains provisions for access to justice for applicants whose requests have been ignored, wrongfully refused, inadequately answered or otherwise not dealt with in accordance with the Directive. The provisions refer to administrative review 'by an independent and impartial body established by law', with such procedure to be 'expeditious and either free of charge or inexpensive'. They also provide for 'access to a review procedure before a court of law, in which the acts or omissions of the public authority concerned can be reviewed and whose decisions may become final' and 'shall be binding on the public authority'.

While many of the above provisions were not in effect at the commencement of Ireland's dispute, future cases of this type would be subject to the more stringent requirements now in place. All of the above suggest a changed culture of access to environmental information, with a focus on information availability as a norm rather than an exception. The underlying weight of the obligations in the Aarhus Convention has permeated EC law to such an extent as to prevent a Member State exercising too great a discretion, or at least to ensure that such discretion is limited and reviewable. It is also a clear demonstration of how international law can apparently have a greater binding effect when also implemented by EC law, because of the legal effect of EC law in the legal system of the Member State. The limitations on enforcing obligations in international law are significant; in contrast the Commission and ECJ are readily able to enforce obligations in EC law.

EIA

EC Directive 85/337 was adopted in 1985 and entered into force on 3 July 1988. It required a number of public and private projects to be assessed as to their significant direct and indirect effects on the environment prior to approval. A second group of projects had to undergo assessment if impacts were expected in view of their nature, size or location. In all cases, it was for the regulator to decide on whether a project had significant effects or not. The Directive laid down principles for assessment, consultation of the general public and participation of environmental administrations. Such requirements could be integrated into existing consent procedures. The original EIA Directive was revised in 1997 by Directive 97/11/EC, to provide greater clarity and supplement and improve the rules on assessment procedure.[51] The amended directive was in force at the time of Ireland's action against the UK. The drivers were the 1993 Commission Implementation Report[52] and, notably, the need for EC law to reflect the obligations of the Espoo Convention.[53] Yet the revised version has also attracted criticism because of its loose drafting and since consent may be granted even when serious negative effects are expected. Directive 2003/35/EC (the Public Participation Directive)[54] made further changes to the EIA Directive, in order to ensure the implementation of the Aarhus Convention, although, as mentioned above, this was not in force, nor was the issue of (Irish) public participation in the UK process raised, at the time of the dispute between the parties in 2001.

The original Directive's procedures were mandatory with respect to nine broad classes of project listed in its Annex I and discretionary with respect to eleven project categories listed in Annex II. The amended Directive significantly added to the number of projects requiring EIA and in the screening provisions clarifies the way in which a Member State can determine whether Annex II projects require assessment; Article 4 requires a formal procedure for this determination, which is to be publicly available; and Annex III stipulates a range of criteria that must be taken into account in setting thresholds or in a case-by-case examination.[55] Article 5 concerns terms of reference for assessment (known as scoping), setting out a procedure to be followed by the regulator in relation to developers. No formal requirement is made for a mandatory scoping procedure that could include a checklist of matters to be covered. A developer must provide an outline of the main alternatives and reasons.

Article 7 contains the relatively detailed provisions for transboundary impacts, triggered when a Member State is aware that the project is likely to have significant effects on the environment of another Member State, or if the latter requests its application (Article 7(1)), and therefore needed to be complied with by the UK in relation to the significant environmental effects that may have been likely from the MOX Plant. Documentation must be forwarded to the affected Member State describing the project, possible transboundary impact, and on the nature of the decision to be taken (Article 7(1)(a) and (b)), giving the affected Member State a reasonable time to indicate whether it wishes to participate in the procedure. If it does, Article 2 requires the proponent Member States to send all the information gathered under Article 5, and information on the procedure, including the request for development consent. Both or all Member States (as more than one may be affected) must arrange for the information to be made available within reasonable timeframes to relevant authorities and the public concerned in the affected Member States. Those authorities and public concerned must be given an opportunity before the granting of the development consent to forward their opinion (Article 7(3)). Consultations within an agreed timeframe must be entered into regarding the potential transboundary effects and measures proposed to mitigate or prevent them (Article 7(4)).

In referring to the two earlier EIA Directive implementation reports, Woodliffe states that 'there was scanty evidence on the use of Article 7' and that 'several states without a formal provision did in fact notify neighbouring states of projects'.[56] In 2003, the Commission reported again on the EIA Directive further to the requirement for a five-yearly review of its application and effectiveness (Article 11(3)).[57] The report comments on the variety of issues that have arisen in applying Article 7 across the EU up to 2003, including those of a procedural nature, such as establishing points of contact and obtaining translation; and ways of overcoming such issues are discussed. At the time, few bilateral agreements had been entered into between Member States concerning these issues,[58] but this would seem to have developed since then, and will no doubt be cited in the report that is currently due. Albrecht, for example, discusses the implementation of the Espoo Convention on TEIA between Germany and Poland, making explicit reference to the requirements of Article 7 of the EIA Directive that implement the treaty.[59]

The amended Directive delegates to the Member States the 'detailed arrangements for implementing' all of the aspects of the transboundary consultation procedure (Article 7(5)), which is provided for in Article 7. Transboundary procedure was also amended further under Directive 2003/35/EC (the Public Participation Directive). Article 3 made a number of additional significant amendments, many to comply with the Aarhus Convention. Article 1(2) has inserted a definition of 'the public' and 'the public concerned', the latter which includes not only the public affected or likely to be affected by the decision but also those having an interest in environmental decision-making, such as non-governmental organizations (NGOs) who are explicitly referred to. Article 1(4) removed the complete exemption from assessment of projects for national defence, and Article 2(3) requires information produced on alternative assessment procedures be made public when a Member State decides to exempt specific projects from the Directive. Article 6(2) and (3) contains detailed provisions on which information is to be made available and a requirement that it be released early in the process. Article 6(4–6) requires early and effective opportunities for public participation, with reasonable timeframes. Article 9(1) requires the public to be informed of the main reasons for the decision and of the consultation process.

Article 10a provides for access for the public concerned to the courts or other independent and impartial bodies, to challenge 'the substantive or procedural legality' of decisions, acts or omissions (Article 10a[2]), which must be 'fair, equitable, timely and not prohibitively expensive' (Article 10a[6]); the public concerned must, however, have a 'sufficient interest' or 'maintain the impairment of a right' (Article 10a[1]). Member States retain the power to decide in accordance with their national legal system at what stage decisions, acts or omissions may be challenged (Article 10a[3]), and what constitutes a sufficient interest or impairment of a right (Article 10a[4]); this limits the potential for challenges, although 'wide access to justice' is specified as an objective in determining what a 'sufficient interest' or 'impairment of a right' means (Article 10a[4]), which restricts Member States' discretion. Member States must also provide the public with practical information on access to administrative and judicial review procedures (Article 10a[7]). Article 10a has been analysed in detail by Ryall, including its application in an Irish domestic context.[60] It is likely to result in increased litigation with the EIA Directive requiring interpretation under Article 6 of the Aarhus Convention: regarding specific activities,[61] and also in accordance with the principles of effective judicial protection and consistent application (indirect effect).[62] Ryall also comments on the minimum standards set out in Article 10a[5], notably that the procedure must not be 'prohibitively expensive'.[63] The original and amended directives have been considered many times by the ECJ,[64] but although in a few cases the question of significance of environmental effects has been examined,[65] transboundary matters have only been considered once, perhaps not surprisingly in Belgium, which recognizes three territory-related regions under constitutional reforms (Flanders, Wallonia and Brussels) and allows implementation of TEIA between them, as well as in the relationship with other States.[66] In *Commission v. Belgium*, the Commission took infringement proceedings against Belgium, succeeding in various arguments including that the Flemish Region and the Region of the Capital City of Brussels had failed to transpose Article 7 and comply with requirements relating to works with a transboundary impact.[67] In rejecting the position put by the Kingdom of Belgium, the Court held:

> *The argument based on the geographical situation of the Region of the Capital City of Brussels is based on the assumption that only projects located in frontier regions are capable of affecting the environment in another Member State. Yet, as the Commission has rightly observed, that assumption is wrong since it ignores the possibility of air-or water-borne pollution.*[68]

The ECJ has also decided on the question of how much flexibility is available to a Member State in deciding the 'detailed arrangements' governing public participation in the EIA Directive, including the question of fees payable, which would also apply in respect of the Article 7 transboundary requirements. In *Commission v. Ireland,* the Court decided that the action of Member States must not undermine the effectiveness of a directive in practice,[69] especially the role of the public, which was acknowledged in the earlier case of *Commission v. Spain.*[70] However, the Court held that a Member State is free to impose a participation fee provided it does not constitute an obstacle to the exercise of participation rights in the Directive.

THE ESPOO CONVENTION

The Espoo Convention, 1991, entered into force in 1997, and Ireland's argument that the terms used in UNCLOS were to be elaborated in light of other international law included it. Since the treaty had not been ratified by Ireland until 2002, however, Ireland was unable to utilize its procedures expressly outside of their incorporation through Article 7 of the EIA Directive. The EC, however, ratified the Espoo Convention in 1997, and as a result it was arguably binding on both the UK and Ireland since then, regardless of the effectiveness of its transposition into the EIA Directive.[71] This section examines the Espoo Convention in some detail, with specific reference to the role of the Inquiry Procedure for determining effect significance. Although the current role of the Implementation Committee and the potential role of the Arbitral Tribunal are also mentioned, they are not dealt with in great detail. Since there is far less doubt that when EC Member States are involved, compliance with TEIA and disputes concerning it will fall to the Commission and ECJ respectively; this is in part also due to the fact that the latter has in fact never been relied upon.

The reasons for the focus on the Inquiry Commission in the context of this article are twofold: first, the significance of the likely environmental effects from the MOX Plant was disputed by Ireland, and although only used once, the Inquiry Commission has recently proven to be an effective means of deciding these matters in an independent, scientific way;[72] second, although the procedure cannot be said to be judicial in nature, it is still, however, likely to be caught by the prohibition in Article 292 of the TEC, preventing disputes being determined outside EC law. The effect of this is considered in the light of the ECJ ruling, and solutions to overcome the problem are suggested; in the absence of a comparable EC procedure, one is that the Commission could perhaps decide to initiate the Inquiry Procedure so that Article 10 of the TEC will also not be breached, although, as stated above, this is in any event unlikely.

General procedure and effectiveness

The Espoo Convention is concerned with harm prevention and cooperation. The Preamble sets out the objective to 'ensure environmentally sound and sustainable development' and to 'enhance international cooperation'. Article 1 defines 'party of origin', 'affected party', 'proposed activity', 'environmental impact assessment', 'impact', 'transboundary impact', 'competent authority' and 'public'. Article 2 contains general provisions for the management of effects, legal measures required, anticipation and notification, consultation and participation, exemptions and requirements for more stringent measures. The detailed specific procedures are found in Articles 3 (notification), 4 (documentation), 5 (consultation), 6 (final decision), 7 (post-project analysis) and 8 (bilateral and multilateral cooperation). Article 2(10) states that 'the provisions of this Convention shall not prejudice any obligations of the parties under international law with regard to activities having or likely to have a transboundary impact'.[73] Article 6(1) contains the substantive obligation to give 'due account' to the outcome of the EIA and Article 6(2) is the requirement for the reasons and other considerations on which the decision is based.

The Espoo Convention contains a number of innovative substantive elements. Sands draws attention to the detailed information to be submitted to the competent authority of the party of origin, for onward transmission to the affected party which, before revision of the EIA Directive, was more comprehensive than EC law required.[74] Provisions for post-project analysis and follow-up (monitoring) are also good examples, as are references to human health and safety.[75] Although Ebbesson[76] and Knox[77] emphasize that many previous treaties have made provision for TEIA and that many of the core concepts of the Convention are drawn from these, Ebbesson indicates that because it operates to change the behaviour of 'non-State actors' (such as individuals and business organizations), as well as States, formal compliance with the Convention is dependent upon the establishment of implementing provisions in domestic law.

It is also dependent upon the principle of non-discrimination, which Gray emphasizes: '[t]he novel provisions of Espoo relate to public participation. The public in the affected area has a right to be informed of and to participate in the EIA procedure, even though the procedure takes place in another country'.[78] Knox identifies this as the 'principal innovation of the Convention',[79] with Ebbesson commenting that if the affected public is not given the opportunity to participate, 'the state in question does not comply with international law'.[80] Schrage also draws attention to this feature, commenting that the Convention 'reflects new trends in international environmental law, which allow all those who are likely to be affected by an environmental impact to be involved in the decision-making process'.[81] Ebbesson identifies two other innovative elements: it provides more specific procedures for notification and consultation regarding transboundary effects, and it advances the use and meaning of EIA in international law.[82]

Implementation of the Espoo Convention has been reviewed twice by the United Nations Economic Commission for Europe (UNECE), following detailed reports submitted to the Secretariat by the parties to the Convention in response to a standard questionnaire.[83] The most recent was prepared for the period mid-2003 to the end of 2005, in which the relationship with EC law is only superficially discussed.[84] The 2008 review reveals a number of weaknesses in the implementation of the Convention, including deficiencies in information exchange, public participation and reporting obligations, and a lack of monitoring. Many of these issues are also common to the enforcement of other treaties.[85] There also remains a continuing need for bilateral and multilateral agreements to identify contacts, and address differences in procedural steps, timing of participation, the interpretation of certain terms and the content of the EIS. With respect to the clarity of the Convention for example, Switzerland considered there to be a large margin for interpretation, the meaning of the word 'likely' was unclear to Hungary, and the terms 'major', 'large' and 'close to an international frontier' were also unclear to Kyrgyzstan. The Czech Republic further sought guidance on post-project analysis.

Screening and the inquiry procedure

As indicated in the 2003 EC Commission Report, the implementation of the Espoo Convention through Article 7 of the EIA Directive included application to several nuclear power developments, which provide a point of comparison with the *MOX Plant* case.[86] In the same year that the ECJ handed down its decision on jurisdiction, the Espoo procedure was activated at Cernavoda in Romania, very close to the Danube Black Sea Canal.[87] Two years before, in 2004, the Espoo Convention Inquiry Procedure was used for the first time in relation to the Bystroe Canal Project, coincidentally also on the Danube Black Sea Canal.[88] Just a month after the ECJ decision, the opinion of the Inquiry Commission was given to the parties in dispute, with a unanimous conclusion that there was likely to be a significant environmental impact from the proposed works.

Cernavoda

The Cernavoda nuclear power plant is located in South-East Romania, and the Romanian Government proposed to expand this facility. Screening began in 2006 further to Appendix I of the Espoo Convention, which identified the proposal as requiring an EIA.[89] Notification of the five affected countries that are parties to the Convention followed: Austria, Bulgaria, Hungary, Moldova and the Ukraine. It is significant to note that only the last two states, Moldova and the Ukraine, are not EC Member States. Austria joined in 1995 and Hungary in 2004. Bulgaria and Romania both joined in 2007. This is important because while Article 7(1) of the EC Directive would have obliged a transboundary procedure including other Member States, it would not have required consultation with non-EC Member States, in this case Moldova and the Ukraine. For these two countries, membership of the Espoo Convention brings a real advantage, although as will be indicated in the following section, the Ukraine was found to be in breach of its obligations under the Convention subsequently in the Bystroe

Canal Project. It is also notable that should disagreements occur between the states involved in the Cernavoda case, it is only where Moldova and the Ukraine are involved that the disputes would be determined in accordance with the Espoo procedures (below). For disputes involving just Member States, these would have to be determined by the ECJ following its 2006 ruling.

After the completion of the screening process, information was provided to the affected parties on the plant, including the scope, scale, description and rationale for the proposed activity. Information was also provided on the spatial and temporal boundaries, including its location, rationale for location, and the start and duration of construction and operation. Finally, information was provided on the expected environmental impacts and proposed mitigation measures, including the scope of the assessment, and on the potential sources of releases to water, of emissions to the air and soil pollution. The consultation stage followed, with EIA documentation submitted to the Romanian environmental authorities, and posted on the Ministry of the Environment and Sustainable Development and project-owner websites; the Romanian public was also consulted. Information was submitted via the Ministry of Foreign Affairs to the affected parties, in turn, to provide it to the national authorities and their own public. A reasonable time to answer was set at eight weeks. As a result, Austria requested consultation based on both Article 5 of the Espoo Convention and Article 7 of the EIA Directive. This resulted in a site visit and discussions at the Ministry. The decision to be taken will take into account the comments of the affected parties and the reasons for the decision will be given.[90]

The Bystroe Canal Project

As indicated, the Inquiry Procedure is an important part of the screening process and provides a scientific, non-judicial means of determining significance in relation to transboundary impact. Article 3(7) of the Espoo Convention provides that 'when a party considers that it would be affected by a significant adverse transboundary impact of a proposed activity listed in Appendix I', if it has not been notified and '[i]f those parties cannot agree whether there is likely to be a significant adverse transboundary impact, any such party may submit that question to an inquiry commission in accordance with the provisions of Appendix IV ...'. Appendix IV requires notification from the party or parties in order to activate the procedure, which consists of a 'scientific or technical expert' appointed by each of the parties in dispute, with a third expert (the president) to be designated by the common agreement of the parties, or having failed to do so within two months, by the Executive Secretary of the UNECE, who also has the power to appoint the other experts if there is a failure to do so.[91] The remainder of the Appendix deals with practical procedural matters, although there is clearly no intention for it to operate as a tribunal, with no provisions for legal representation or evidential matters, for example.[92]

It was successfully used for the first time in relation to the Bystroe Canal Project[93] proposed for the Danube Delta, which was designated as a Wetland of International Importance under the Ramsar Convention in 1991,[94] was inscribed on the World Heritage List in 1991[95] and was listed as a Biosphere Reserve under UNESCO's Man and Biosphere Programme in 1992.[96] Full details of activation and operation of the procedure are outlined by Koyano.[97] Essentially, Romania requested an inquiry under Appendix IV believing there were likely significant adverse effects, and after the scientific evaluation the Inquiry Commission agreed, and thus the provisions of the Espoo Convention were held to apply. The Ukraine was therefore obliged to send a notification about the canal to Romania to commence the procedure, including communication between the parties and public participation. During the Fourth MOP in Bucharest, May 2008, the Ukrainian delegation committed to fulfil its obligations under the Espoo Convention, and this is apparently now being done,[98] with the Implementation Committee (see below) also monitoring the situation closely.[99]

Since the conclusion of the first Inquiry Procedure, the Working Group on EIA to the Espoo Convention has been active in related areas. Of particular note is the emphasis given to large-scale transboundary projects, proposing solutions to the challenges of applying the Convention to complex activities, which include nuclear power development and canal project works of the significance of those on the Danube. In a note by the Secretariat to the twelfth Meeting of the Working Group to the Parties to the Convention,[100] further examples such as infrastructure projects that are part of an energy or transport network crossing several parties are given. The strategic dimension of such projects is emphasized,[101] and the need to determine and assess this in advance if at all possible is recognized.[102]

Compliance and the implementation committee

Non-compliance procedures (NCP) are now a common feature of environmental treaties, such as the Implementation Committee of the Espoo Convention. The objective of this body is to review compliance by the parties with their obligations under the Convention, in order to assist them fully to meet their commitments. The Committee was established by the MOP in February 2001,[103] and comprises eight members, nominated by parties that are, in turn, elected by the MOP.

Article 11(2) of the Convention is the trigger for the Implementation Committee, including in the terms of reference of the MOP to require parties continuously to review implementation in the ways stated. Article 14*bis* was adopted at MOP-3 and is the second amendment to the Convention. It deals with review of compliance by regular reporting arrangements and will come into force when ratified by the parties in accordance with Article 14(4). At MOP-4, a report on the activities of the Implementation Committee was presented highlighting a number of topical issues, including those of a specific nature.[104] As indicated, compliance matters concerning two or more Member States would be dealt with by the EC Commission, unless non-EC Member States were also involved.

Dispute settlement and the arbitration procedure

Dispute-settlement procedures are found in many international treaty provisions, some of which are tailored to the specific treaty and some of which refer to outside means such as the Permanent Court of Arbitration or ICJ judgments. Dispute settlement is often distinguished from dispute avoidance, as many provisions are of the latter type and do not result in either a court or tribunal; NCP such as those of the Espoo Implementation Committee can be considered of this type.[105] If the parties remain in dispute, for non-EC Member States or for disputes between EC Member States and non-EC Member States, the Espoo dispute-settlement provisions can be utilized. Article 15(2) provides that disputes arising are to be settled by negotiation or other acceptable methods, which could include submission to the ECJ for EC Member States. Under Article 15(2), submissions to the ICJ or application of the arbitration procedure in Appendix VII can become compulsory if disputes are not resolved under Article 15(1); if both the ICJ and the Appendix VII Arbitral Tribunal are acceptable to the parties, then a dispute can only be submitted to the ICJ if the parties otherwise agree according to Article 15(3); again this could include reference to the ECJ for disputes between two or more EC Member States. The ICJ and Arbitral Tribunal options are only available if the parties have made a declaration in writing accepting them.[106] In the event that both parties are EC Member States, the dispute would therefore have to go to the ECJ and be decided on the basis of the implementing directives, although it would be open to the ECJ to consider whether these directives comply with the Espoo and Aarhus Conventions and, if not, interpret and apply their provisions directly.[107]

CONCLUSION

Given the discussion in the foregoing sections, these concluding comments will be brief. What is now clear is that the ECJ's position on disputes between Member States that are parties to mixed environmental agreements has implications beyond the OSPAR and UNCLOS treaties, and for the Espoo (and Aarhus) Convention the same stance would undoubtedly also apply. When Member States are in dispute with one another, it is therefore for the ECJ to decide whether it has competence to determine provisions of relevant mixed environmental agreements, and in the unlikely event that it does not, conclude that they may be decided upon by another tribunal, such as the Appendix VII Arbitral Tribunal, or the ICJ further to Article 15 of the Espoo Convention. Although the Arbitral Tribunal may disagree with the

ECJ's claims of jurisdiction and claim competence for itself, the decision of the UNCLOS Annex VII tribunal suggests that caution may result with other tribunals following its path and suspending proceedings if similar circumstances occur again. In any event, the disputing Member State simply does not have this freedom to decide whether to challenge another Member State outside of EC law and it will unquestionably be brought to account if Articles 292 and 10 are breached again. In all cases where a Member State has a dispute with another Member State, therefore, consultation with the Commission must first be undertaken prior to the matter being referred by the Commission to the ECJ for a ruling.[108]

The content of the Espoo Convention was broad enough to cover most of the matters in Ireland's dispute with the UK, and it was extremely surprising that a claim was not based upon it, either pursued through the Commission and ECJ under the EIA Directive (as amended), or, in the absence of its own ratification, directly through the Espoo Convention and its Inquiry Procedure based on its direct effect in EC law,[109] through the lens of which the EIA Directive must be interpreted. In the dispute between Romania and the Ukraine, the Inquiry Procedure has proven itself to be an extremely useful addition to the TEIA procedures in the Convention when significant questions are disputed between the parties. There is no similar provision in the EIA Directive, and in the absence of disputing Member States being encouraged by the EC Commission to resolve their grievances through it (which is suggested to be unlikely), judicial procedures under the TEC would be needed, which, because of their very different nature and need to maintain the autonomy of the EC legal order, may not produce the best outcome for the environment. The fact that the ECJ ruling failed to address the substantive matters of Ireland's complaint with the UK and effectively brought to an end international treaty mechanisms that could have done so is an illustration of how little was ultimately achieved in this respect.[110] An independent scientific evaluation of effect significance, albeit that may later be challenged if procedures are not followed, is therefore strongly advocated.

Despite the 2006 ruling, it does not appear to have influenced bilateral arrangements between EC Member States as yet. The 2008 Review of Implementation of the Espoo Convention unfortunately does not comment on this,[111] based as it is upon information provided until April 2006, just before the ECJ's ruling. It is, however, recommended that guidance on the implications of the ruling should be provided to EC Member States (which constitute the majority of the parties to the Convention), to enable them to review their current arrangements and ensure they are compliant with the ruling; this could, of course, be of a general nature as applicable to all mixed environmental agreements. Recent published work referring to TEIA arrangements between Germany and Poland further to a 1994 Agreement, for example, suggests dispute-settlement provisions between these Member States may not be compliant,[112] and no doubt arrangements between other Member States will also be of a similar nature, especially if concluded prior to the ECJ 2006 ruling.

NOTES

This chapter was originally published as: Marsden, S. (2009) 'MOX Plant and the Espoo Convention: Can Member State Disputes Concerning Mixed Environmental Agreements Be Resolved outside EC Law?', *Review of European Community and International Environmental Law*, vol 18, no 3, pp312–327.

1 ECJ 30 May 2006, Case C-459/03, *Commission v. Ireland*, [2006] ECR I-4635, available at http://curia.europa.eu/jcms/jcms/Jo1_6308/ecran-d-accueil. The first proceedings were begun before the Arbitral Tribunal established under the Convention for the Protection of the Marine Environment of the North-East Atlantic (Paris, 22 September 1992); the second were before the International Tribunal for the Law of the Sea (ITLOS); and the third before the Annex VII Arbitral Tribunal – both the second and third were under the United Nations Convention on the Law of the Sea (Montego Bay, 10 December 1982) (UNCLOS). The ECJ case was begun by the Commission in response to the third proceedings, but it has relevance for all.

2 Since the 2006 decision, the OSPAR, UNCLOS and ECJ proceedings have been extensively discussed by N. Lavranos in the following articles and case note: 'Protecting its Exclusive Jurisdiction: The *MOX Plant* Judgment of the ECJ', 5 *The Law and Practice of International Courts and Tribunals* (2006), 479; 'The *MOX Plant* and *Ijzeren Rijn* Disputes: Which Court is the Supreme Arbiter?, 19 *Leiden Journal of International Law* (2006), 223; 'MOX Plant Dispute – Court of Justice of the European Communities', 2 *European Constitutional Law Review* (2006), 456; and 'The Epilogue in the MOX Plant Dispute: An End Without Findings', *European Energy and Environmental Law Review* (2009), 180.

3 P. J. Cardwell and D. French (2007) 'Who Decides? The ECJ's Judgment on Jurisdiction in the *MOX Plant* Dispute', 19:1 *Journal of Environmental Law*, 121; C. P. R. Romano, 'Commission of the European Communities v. Ireland, Case C-459/03, Judgment', 101:1 *The American Journal of International Law* (2007), 171; C. Semmelmann (2006) 'Forum Shopping between UNCLOS Arbitration and EC Adjudication – And the Winner ... should be ... the ECJ!' *European Law Reporter*, 234; G. Jianjun (2008) 'Comments on *Commission of the European Communities v. Ireland*', 7:2 *Chinese Journal of International Law*, 417.

4 See, e.g., B. Simma and D. Pulkowski (2006) 'Of Planets and the Universe: Self-Contained Regimes in International Law', 17:3 *European Journal of International Law*, 483; R. Higgins (2003) 'The ICJ, ECJ and the Integrity of International Law', 52 *International and Comparative Law Quarterly*, 1; N. Lavranos, 'Concurrence of Jurisdiction between the ECJ and other International Courts and Tribunals', August/September *European Environmental Law Review* (2005), pp213–225, at p224; J. Pauwelyn (2003) *Conflict of Norms in Public International Law: How WTO Law Relates to other Rules of International Law* (Cambridge University Press); Y. Shany (2003) *The Competing Jurisdictions of International Courts and Tribunals* (Oxford University Press).

5 T. Stephens (2009) *International Courts and Environmental Protection*, Cambridge University Press,, at 295–301; N. Craik (2008) *The International Law of Environmental Impact Assessment*, Cambridge University Press, at 115–116 and 127.

6 For readings from the non-English literature, the ECJ website has an 'Annotation of judgments' section to the case law, which lists several articles on the *MOX Plant* case in French, German, Spanish and Russian as well as in English (latest edition 17 February 2009). These are found in Part 2 under the case number (in French). See CJCE-Système MINIDOC, *Notes aux arrêts de la Cour de justice* (Édition du 17/02/2009), 924–925, available at http://curia.europa.eu/jcms/jcms/Jo2_7083/annotation-of-judgments

7 Directive 90/313 on the freedom of access to information on the environment, [1990] OJ L158/56 (repealed by Directive 2003/4 on public access to environmental information, [2003] OJ L41/26).

8 Directive 85/337 on the assessment of the effects of certain public and private projects on the environment, [1985] OJ L175/40 (as amended by Directives 97/11, [1997] OJ L73/5 and 2003/35, [2003] OJ L156/17).

9 R. Churchill and J. Scott (2004) 'The MOX Plant Litigation: The First Half-Life', 53 *International and Comparative Law Quarterly*, at 667, 675 and 676.

10 M. Tanaka (2004) 'Lessons from the Protracted MOX Plant Dispute: A Proposed Protocol on Marine Environmental Impact Assessment to the United Nations Convention on the Law of the Sea', 25 *Michigan Journal of Environmental Law*, 337.

11 Convention on Environmental Impact Assessment in a Transboundary Context (Espoo, 25 February 1991).

12 For a recent overview of TEIA, including a chapter on the Espoo Convention by W. Schrage, see K. Bastmeijer and T. Koivurova (eds) (2008) *Theory and Practice of Transboundary Environmental Impact Assessment*, Martinus Nijhoff, at 29–51. See also 'Special Issue on Environmental Assessment and Transboundary Impact Assessment', W. Schrage and N. Bonvoisin (eds) (2008) 26:4 *Impact Assessment and Project Appraisal*, 234.

13 M. Koyano (2008) 'The Significance of the Convention on Environmental Impact Assessment in a Transboundary Context (Espoo Convention) in International Environmental Law: Examining the Implications of the Danube Delta Case', 26:4 *Impact Assessment and Project Appraisal*, 299.

14 See E. Albrecht (2008) 'Implementing the Espoo Convention in Transboundary EIA between Germany and Poland', 28 *Environmental Impact Assessment Review*, 359, note in particular the section on dispute settlement at 364.

15 For a list of EC Member States, see Europa, *European Countries* (undated), available at http://europa.eu/abc/european_countries/index_en.htm. For a list of the parties to the Espoo Convention, see United Nations Treaty Collection, *Status of Treaties, Chapter XXVII Environment – 4. Convention on Environmental Impact Assessment in a Transboundary Context* (United Nations, undated), available at http://treaties.un.org/Pages/ViewDetails.aspx?src=TREATY&mtdsg_no=XXVII-4&chapter=27&lang=en. There are therefore 16 non-EC Member State parties to the Espoo Convention; Iceland, Russia and the US are signatories but have not ratified.

16 For comprehensive background to the dispute, see M. Tanaka, note 10, at 358–370.

17 See the Town and Country Planning (Assessment of Environmental Effects) Regulations 1988, which implemented the 1985 EIA Directive, and which were triggered by the planning application to Copeland Borough Council in October 1992, and a related application for authorization from Her Majesty's Nuclear Installations Inspectorate.

18 Although the Regulations may have been complied with, they failed to implement the EIA Directive as even the original Article 7 in the 1985 EIA Directive required the latter. This states: 'Where a Member State is aware that a project is likely to have significant effects on the environment in another Member State or where a Member State likely to be significantly affected so requests, the Member State in whose territory the project is intended to be carried out shall forward the information gathered pursuant to Article 5 to the other Member State at the same time as it makes it available to its own nationals. Such information shall serve as a basis for any consultations necessary in the framework of the bilateral relations between two Member States on a reciprocal and equivalent basis'. The amended 1997 EIA Directive, which contains the detailed Article 7 provision intended to comply with the Espoo Convention, was to have been implemented into Member State domestic law by 14 March 1999, and should therefore have been implemented by the UK regardless of the requirements of the 1988 Regulations.

19 Treaty Establishing the European Atomic Energy Community (Rome, 25 March 1957), as amended.

20 J. Heliskoski (2001) *Mixed Agreements as a Technique for Organizing the International Relations of the EC and its Member States*, Kluwer.

21 See Directive 90/313, note 7.

22 McDorman puts Ireland's failure down to putting too great an emphasis on context and evolving practice with access to environmental information. See T. L. McDorman, 'Access to Information under Article 9 of the OSPAR Convention (*Ireland v. United Kingdom*)', 98:2 *The American Journal of International Law* (2004), 330, at 338.

23 In addition to the references already cited at notes 2, 9, 10 and 22, see also Y. Shany (2004) 'The First *MOX Plant* Award: The Need to Harmonize Competing Environmental Regimes and Dispute Settlement Procedures', 17 *Leiden Journal of International Law*, 815; and C. Chamoux, 'Access to Environmental Information in International Law – The Significance of the MOX Plant Case (*Ireland v. United Kingdom*)', unpublished LL.M. thesis, Faculty of Law, University of Western Cape (17 November 2005).

24 See B. Kwiatkowska (2003) 'The *Ireland v. United Kingdom* (Mox Plant) Case: Applying the Doctrine of Treaty Parallelism', 18:1 *The International Journal of Marine and Coastal Law*, 1; C. Brown (2002) 'Provisional Measures before the ITLOS: the MOX Plant Case', 17:2 *International Journal of Marine and Coastal Law*, 267; M. J. C. Forster (2003) 'The Mox Plant Case – Provisional Measures in the International Tribunal for the Law of the Sea', 16:3 *Leiden Journal of International Law*, 611.

25 Although pursuant to Article 287(1) a party may by declaration choose one of the dispute-settlement options set out (ITLOS, ICJ, Annex VII or Annex VIII Tribunal) it was open to EC Member States to make a reservation in favour of the ECJ as part of this process.

26 V. Röben (2004) 'The Order of the UNCLOS Annex VII Arbitral Tribunal to Suspend Proceedings in the Case of the MOX Plant at Sellafield: How Much Jurisdictional Subsidiarity?', 73 *Nordic Journal of International Law*, 223.

27 Opinion of Advocate-General Poiares Maduro delivered on 18 January 2006: Case C-459/03, *Commission of the European Communities v. Ireland*, [2006] ECR I-4635, available at http://curia.europa.eu/jcms/jcms/Jo1_6308/ecran-d-accueil

28 ECJ 19 March 2002, Case C-13/00, *Commission v. Ireland,* [2001] ECR I-2943.

29 See the references in note 3.

30 Some authors have contrasted the approach in the *Iron Rhine Railway* case. See N. Lavranos, 'The *MOX Plant* and *Ijzeren Rijn* Disputes, note 2; and P. J. Cardwell and D. French, note 3 above. See also J. Harrison (2006) 'Signi?cant International Environmental Law Cases: International Arbitrations; *Iron Rhine Railway Arbitration*', 18:3 *Journal of Environmental Law*, 505.

31 See T. Stephens, note 5, who identifies three types, giving examples: two or more dispute-settlement systems with similar general jurisdiction ICJ and Permanent Court of Arbitration, yet to occur); two systems with partially overlapping jurisdiction, *Southern Bluefin Tuna* (UNCLOS and Commission for the Conservation of Southern Bluefin Tuna) and *MOX Plant* (OSPAR, UNCLOS and TEC/Euratom); common facts giving rise to litigation under two specialized systems (World Trade Organization and a Multilateral Environmental Agreement, with trade restrictions imposed to achieve environmental objectives). It is argued in the last instance that there is no true overlap because parallel issues are dealt with.

32 See note 8.

33 It is surprising that Ireland did not pursue the UK directly for breaching EIA obligations in EC law, since it had originally raised this with the UK by letter in 1999, alongside other EIA obligations that it believed the UK was in breach of (such as under the Espoo Convention). Ireland's Memorial, Part II: The Law to the UNCLOS Annex VII Tribunal, refers indirectly to the EIA Directive, see note 38, paragraph 6.19, at 104; Ireland's Memorial to the OSPAR Arbitral Tribunal refers indirectly to the Information Directive and ECJ interpretation of it, see note 38, paragraph 102, at 31. See also B. Kwiatkowska, note 24 at 9–10, and note her view that the legitimacy of the ITLOS Order could have been increased had it defined the dispute as an 'EU-law centered dispute', which would have also brought the precautionary principle into play operationally (at 29 and 55).

34 The direct effect of international treaties on Member States in the absence of secondary EC law is a complex matter involving considerations of monism and dualism, constitutional competence, and compliance with criteria for direct effect, the latter being similar to those applicable to directives (it must be clear, precise and unconditional, in some circumstances allowing individuals to rely on the measure). See K. Lenaerts and P. van Nuffel (2005) *Constitutional Law of the European Union*, Thomson/Sweet and Maxwell, at 702; H. G. Schermers and D. F. Waelbroeck (2001) *Judicial Protection in the European Union*, Kluwer, at 183–185; E. Szyszczak and A. Cygan (2008) *Understanding EU Law*, Thomson/Sweet and Maxwell, at 325–330; G. Conway (2002) 'Breaches of EC law and the International Responsibility of Member States', 13:1 *European Journal of International Law*, 679; G. Betlem and A. Nollkaemper (2003) 'Giving Effect to Public International Law and European Community Law before Domestic Courts: A Comparative Analysis of the Practice of Consistent Interpretation', 14:3 *European Journal of International Law*, 569; and J. Wouters, A. Nollkaemper and E. de Wet (eds) (2008) *The Europeanisation of International Law: The Status of International Law in the EU and its Member States*, TMC Asser Press/Cambridge University Press. See also note 83, where the EC considers the position of the Aarhus Convention to have direct effect.

35 This is however extremely unlikely given the comments made by the European Commissioner for the Environment concerning the Nord Stream Pipeline Project, which is discussed at note 107 below. Indeed, the response of the Commission to a question raised in the Parliament in May 2003 indicates this was unlikely even when it was aware of the dispute at that time. See V. Röben, note 26, at 228, and note 13. What is also extremely unlikely is that Article 292 will be modified to allow for more flexibility for other dispute-settlement fora as suggested by Lavranos. He argues that Article 292 could be changed 'in the sense that the ECJ would be required to take decisions of arbitral tribunals into account – at least when one of the parties involved is an EC member state'. See N. Lavranos, 'The *MOX Plant* and *Ijzeren Rijn* Disputes, note 2, at 246. The autonomy of the EC legal order would be compromised by such an approach if there were a requirement for this, although there is no reason why the TEC cannot be interpreted as Article 31(3)(c) of the Vienna Convention requires as it is an international treaty between the EC Member States (see note 39).

36 See N. Lavranos, 'The *MOX Plant* and *Ijzeren Rijn* Disputes, note 2, at 232–233; V. Röben, note 26, at 242; and J. Wouters et al, note 34.

37 Convention on Access to Information, Public Participation in Decision-Making and Access to Justice in Environmental Matters (Aarhus, 25 June 1998). The Aarhus Convention had therefore just come into force prior to the commencement of Ireland's OSPAR litigation, and although the revised Information Directive was not in force at the time, the same arguments about direct effect applied to it as to the Espoo Convention, which is closely related.

38 Memorial before OSPAR Arbitral Tribunal, paragraphs 100–101 (at 30–31) cites the Aarhus Convention; Memorial before UNCLOS Annex VII Tribunal, paragraph 6.19 (at 104) cites the Espoo Convention; EIA as customary international law is also argued in paragraph 6.29 of the latter memorial. See 1992 Convention on the Protection of the Marine Environment of the North-East Atlantic in the Dispute Concerning Access to Information under Article 9 of the OSPAR Convention and the MOX Plant, *Ireland v. United Kingdom*, Memorial of Ireland (7 March 2002); and see 1982 Convention on the Law of the Sea before an Arbitral Tribunal Established Under Annex VII, in the Dispute Concerning the MOX Plant, International Movements of Radioactive Materials, and the Protection of the Marine Environment of the Irish Sea, *Ireland v. United Kingdom*, Memorial of Ireland (26 July 2002), Part II – The Law, available at www.pca-cpa.org. For the ITLOS decision, see International Tribunal for the Law of the Sea, *Year 2001, 3 December 2001, List of Cases, No 10, The MOX Plant Case, Ireland v. United Kingdom, Request for Provisional Measures, Order*, available at www.itlos.org/

39 C. McLachlan (2005) 'The Principle of Systemic Integration and Article 31(3)(c) of the Vienna Convention', 54:2 *International and Comparative Law Quarterly*, 279.

40 It ratified the Espoo Convention on 25 July 2002; the UK ratified on 10 October 1997. Ireland has still not ratified the Aarhus Convention although it signed on 25 June 1998; the UK ratified on 23 February 2005.

41 The Implementation Committee of the Espoo Convention was established by the MOP in February 2001 (MOP-2) so was in place at the time of Ireland's dispute with the UK.

42 See M. Tanaka, note 10, at 394, note 321.

43 Directive 2003/35/EC providing for public participation in respect of certain plans and programmes relating to the environment and amending with regard to public participation and access to justice Council Directives 85/337/EEC and 96/61/EC, [2003] OJ L156/17.

44 See N. Craik, note 5, at 55–59.

45 See note 9, at 674–675.

46 For commentaries on the original directive, see: R. Hallo (ed.) (1996) *Access to Environmental Information in Europe: The Implementation and Implications of Directive 90/313/EEC*, Kluwer; J. Janssen (1998) 'Recent Developments on Access to Environmental Information: Transparency in Decision Making', October *European Environmental Law Review*, 268; C. Kimber, 'Understanding Access to Environmental Information: The European Experience', in T. Jewell and J. Steele (eds) (1998) *Law in Environmental Decision Making*, Clarendon.

47 Despite the original directive being applicable, Shany indicates that the ECJ had, however, adopted a broad concept of the scope of the information covered by it, and held that information pertaining to consent proceedings of development projects with environmental implications should be disclosed. See note 23, at 822, where ECJ Case C-321/98, *Mecklenburg v. Kreis Pinneberg der Landrat*, [1998] ECR I-3809, is cited. This case was, in fact, referred to by Ireland in its Memorial to the OSPAR Arbitral Tribunal. See note 38, paragraph 102, at 31.

48 See ECJ 26 June 2003, Case C-233/00, *Commission v. France*, [2003] ECR I-6625.

49 For comment on the revised directive, much of which relates it to the Aarhus Convention and its other two pillar rights, see J. Jendroska (2005) 'Aarhus Convention and Community Law: The Interplay', *Journal for European Environmental and Planning Law*, 12; and A. Ryall (2004) 'Implementation of the Aarhus Convention through Community Environmental Law', 6 *Environmental Law Review*, 274.

50 See ECJ 21 April 2005, Case C-186/04, *Pierre Housieaux v. Délégués du conseil de la Region de Bruxelles-Capitale*, [2005] ECR I-3299.

51 See Directive 85/337, note 8.

52 Commission of the European Communities, *Report on the Implementation of Directive 85/337/EEC on the Assessment of the Effect of Certain Public and Private Projects on the Environment*, COM (93) 28 final.

53 The EC made the following Declaration upon signature to the Espoo Convention, which was confirmed upon ratification, indicating that Directive 85/337/EEC was not at the time compliant with it and that it did not cover treaty obligations between Member States and non-Member States. See United Nations Treaty Collection, note 15, which includes the following EU declaration:

> *'It is understood, that the Community Member States, in their mutual relations, will apply the Convention in accordance with the Community's internal rules, including those of the EURATOM Treaty, and without prejudice to appropriate amendments being made to those rules. The European Community considers that, if the information of the public of the Party of*

origin takes place when the environmental impact assessment documentation is available, the information of the affected Party by the Party of origin must be implemented simultaneously at the latest. The Community considers that the Convention implies that each Party must assure, on its territory that the public is provided with the environmental impact assessment documentation, that it is informed and that its observations are collected. In the field covered by the Espoo Convention, Council Directive 85/337/EEC of 27 June 1985, annexed to this Declaration, applies. It enables the Community to comply with most of the obligations under the Espoo Convention. Member States are responsible for the performance of those obligations resulting from the Espoo Convention not currently covered by Community law and more specifically by Directive 85/337/EEC. The Community underlines that Directive 85/337/EEC does not cover the application of the Espoo Convention between the Community on the one hand and non-Member States Party to the Espoo Convention on the other hand. The Community will inform the depositary of any future amendment to Directive 85/337/EEC. From this, it follows that the Community, within the limits indicated above, is competent to enter into binding commitments on its own behalf with non-members countries which are Contracting Parties to the Espoo Convention'.

54 Directive 2003/35/EC of the European Parliament and of the Council of 26 May 2003 providing for public participation in respect of the drawing up of certain plans and programmes relating to the environment and amending with regard to public participation and access to Justice Council Directives 85/337/EEC and 96/61/EC, [2003] OJ L156/17.

55 Guidance has recently been released on these categories, with reference to the ECJ case law. See European Commission, *Directive 85/337/EEC on the Assessment of the Effects of Certain Public and Private Projects on the Environment (EIA Directive) – Interpretation of Definitions of Certain Project Categories of Annex I and II of the EIA Directive* (European Communities, 2008), available at http://ec.europa.eu/environment/eia/pdf/interpretation_eia.pdf

56 J. Woodliffe, 'Environmental Damage and Environmental Impact Assessment', in M. Bowman and A. Boyle (eds) (2002) *Environmental Damage in International and Comparative Law: Problems of Definition and Valuation*, Oxford University Press, 133, at 144.

57 European Commission, *Report from the Commission to the European Parliament and the Council on the Application and Effectiveness of the EIA Directive (Directive 85/337/EEC as amended by Directive 97/11/EC) – How Successful are the Member States in Implementing the EIA Directive?* (European Commission, 2003). Section 4.5 deals with transboundary impacts, at 78–82, available at http://ec.europa.eu/environment/eia/pdf/report_en.pdf

58 Table 15 in the report outlines the treaties and agreements that were in place at the time between a number of the Member States and Table 16 indicates the number and types of transboundary projects that had been subject to the procedure. Several of these have concerned nuclear plants, uranium enrichment and nuclear re-processing, including apparently in relation to the UK. See ibid. at 80–82. It is

also interesting to note the following comment (at 82, paragraphs 4.5.7) in relation to disputes, which most likely related to Ireland's specific concerns at the time:

> *'Questionnaire responses do not often mention problems of disagreement over the final decision, although Ireland sees a need for a more formal arrangement to be developed in due course, particularly in the context of nuclear energy-related developments ...'.*

59 See E. Albrecht, note 14, at 361.

60 A. Ryall, 'Access to Justice and the EIA Directive: The Implications of the Aarhus Convention', in J. Holder and D. McGillivray (eds) (2007) *Taking Stock of Environmental Assessment: Law, Policy and Practice*, Routledge-Cavendish, 191, at 197–198.

61 A. Ryall (2007) 'EIA and Public Participation: Determining the Limits of Member State Discretion', 19:2 *Journal of Environmental Law*, 247, at 255–256.

62 See A. Ryall, note 60, at 215; S. Drake (2005) 'Twenty Years After Von Colson: The Impact of "Indirect Effect" on the Protection of the Individual's Community Rights', 30 *European Law Review*, 329.

63 See A. Ryall, note 60, at 199.

64 See European Court of Justice, *Table Alphabétique des Matières (Alphabetical Table of Subject Matter)*, Volumes 1985–2009, at 456–459, which provides a comprehensive list of cases (only in French).

65 ECJ 24 October 1996, Case C-72/95, *Kraaijeveld BV v. Gedeputeerde Staten van Zuid-Holland*, [1996] ECR I-5403. For a more recent consideration, see ECJ 10 July 2008, Case C-156/07, *Salvatore Aiello and Others v. Comune di Milano etc*, [2008] ECR I-5215, which held that where a Member States chooses to determine on a case-by-case basis which of the projects listed in Annex I or II must be subject to EIA, it must ensure that the criteria in Annex III are clearly referred to in the decision-making process, and that, if one is considered, they must all be considered.

66 J. De Mulder (2008) 'The Institutional Context for Transboundary Environmental Impact Assessment in Belgium: Multi-Level Setting – A Matter of Smooth Governance?', 26:4 *Impact Assessment and Project Appraisal*, 277.

67 ECJ 2 May 1996, Case C-133/94, *Commission v. Belgium*, [1996] ECR I-2323.

68 Ibid., at paragraph 53.

69 ECJ 9 November 2006, Case C-216/05, *Commission v. Ireland*, [2006] ECR I-10787.

70 ECJ 16 March 2006, Case C-332/04, *Commission v. Spain*, [2006] ECR I-40.

71 See references in note 34.

72 Koyano comments: 'An inquiry mechanism is extremely useful for environmental issues of which effective management is often impeded by conflicting views on scientific and technological matters between the Parties concerned'. See M. Koyano, note 13, at 301.

73 This indicates that other causes of action that may be available to States, such as any customary rules of international law relating to environmental harm, international cooperation or liability, and/or other treaty obligations, remain.

74 P. Sands (1995) *Principles of International Environmental Law I: Frameworks, Standards and Implementation*, Cambridge University Press, at 588.

75 W. Schrage, 'The Convention on Environmental Impact Assessment in a Transboundary Context', in J. Petts (ed) (1999) *Handbook of Environmental Impact Assessment*, vol. 2, Blackwell Science, at 88.

76 J. Ebbesson (1999) 'Innovative Elements and Expected Effectiveness of the 1991 EIA Convention', 19 *Environmental Impact Assessment Review*, 47, at 48.

77 J. Knox (2002) 'The Myth and Reality of Transboundary Environmental Impact Assessment', 96:2 *American Journal of International Law*, 291, at 301.

78 K. R. Gray (2000) 'International Environmental Impact Assessment: Potential for a Multilateral Environmental Agreement', 11:1 *Colorado Journal of International Environmental Law and Policy*, 83, at 101–102.

79 See J. Knox, note 77, at 303.

80 See J. Ebbesson, note 76, at 50.

81 See W. Schrage, note 75, at 96–97.

82 See ibid.

83 For the first review, see Decision III/1, Review of Implementation, found in *Convention on Environmental Impact Assessment in a Transboundary Context, Review of Implementation 2003* (ECE/MP.EIA/6, 30 August 2004), Annex I, Decision III/1, Review of Implementation, available at www.unece.org/env/documents/2004/eia/decision.III.1.e.pdf. Note that although Ireland had ratified the Convention on 25 July 2002 and the questionnaire was issued in late 2002 and following some amendments in mid-2003, Ireland did not respond. Bearing in mind the substance and timing of its dispute with the UK, which ironically was the lead country for the review, this is somewhat of a surprise. The relationship between EC law and international law is considered in more detail in the Implementation Report submitted by the EC to the UNECE in relation to the Aarhus Convention, which refers to the binding nature of treaties for EC institutions as well as Member States and the primacy of international treaties over EC secondary law. In relation to the latter, the Commission states: 'The primacy of international agreements concluded by the Community over provisions of secondary Community legislation also means that such provisions must, so far as is possible, be interpreted and applied in a manner that is consistent with those agreements'. The direct effect of treaties concluded with non-EC Member States is also considered, despite the absence of case law on the Aarhus Convention in particular. See United Nations Economic and Social Council, Economic Commission for Europe, *Meeting of the Parties to the Convention on Access to Information, Public Participation in Decision-Making and Access to Justice in Environmental Matters, Third Meeting Riga, 11–13 June 2008, Implementation Report Submitted by the European Community* (ECE/MP.PP/IR/2008/EC, 6 June 2008), at 2, available at www.unece.org/env/documents/2008/pp/mop3/ece_mp_pp_ir_2008_ec_e.pdf

84 UN Economic Commission for Europe, *Review of Implementation of the Espoo Convention* (United Nations, 2008), available at www.unece.org/env/documents/2008/eia/ece.mp.eia.11.pdf. The EC responded to this survey in part only, taking the view that the obligations contained within the Convention were for states, with its role merely to adopt framework legislation for its implementation. No mention

is made of the potential direct effect of the Espoo Convention or to the role of the Commission or ECJ in relation to it. See *European Community Response to the Questionnaire on Implementation of the Espoo Convention prepared pursuant to Decision II/11 of the Second Meeting of Parties to the Convention* (23 July 2003), available at www.unece.org/env/eia/documents/Review%202003-2005/Questionnaire%20-%202003%20%20EC.pdf; and see *Report of the European Community for 2003–2005 on the Implementation of the Espoo Convention on Environmental Impact Assessment in a Transboundary Context* (undated), available at www.unece.org/env/eia/documents/Review%202003-2005/Questionnaire%20-%202003-2005%20%20EU%20-%20annex.pdf. For the completed responses (with date received) by the UK (18 August 2006) and Ireland (7 February 2008) (the latter which, because of its lateness, was not taken into account in the ECE Report), see K. Chowns, *Questionnaire for the Report of the United Kingdom of Great Britain and Northern Ireland for 2003–2005 on the Implementation of the Espoo Convention on Environmental Impact Assessment in a Transboundary Context for the Period mid-2003 to end of 2005* (24 August 2006), available at www.unece.org/env/eia/documents/Review%202003-2005/Questionnaire%20-%202003-2005%20-%20United%20Kingdom.pdf; and see B. O'Neill, *Questionnaire for the Report of Ireland for 2003–2005 on the Implementation of the Espoo Convention on Environmental Impact Assessment in a Transboundary Context for the Period mid-2003 to end of 2005* (6 February 2008), available at www.unece.org/env/eia/documents/Review%202003-2005/Questionnaire%20-%202003-2005%20-%20Ireland.pdf

85 J. De Mulder, 'The Expansion of Environmental Assessment in International Law: The Protocol on Strategic Environmental Assessment to the Espoo Convention', 18 *Environmental Law and Management*, 269, at 270.

86 See European Commission, notes 57 and 58. On Table 16 of the report, Finland, Germany, the Netherlands, Sweden and the UK had all identified nuclear plants, uranium enrichment or nuclear re-processing as transboundary projects in their own countries.

87 This is reported on the Espoo website as a case study by C. Pulbere, *Applying the Espoo Convention on NPP Cernovoda Units 3–4, Panel Discussion on Energy Related Projects, Fourth Meeting of the Parties, Bucharest, Romania, 19–21 May 2008* (undated), available at www.unece.org/env/eia/documents/bucharest/Pulbere.pdf

88 It is also of note (although beyond the scope of this article to comment further) that the ICJ considered the environmental obligations of Slovakia and Hungary for the Danube in ICJ 25 September 1997, *Hungary v. Slovakia* (*Gabciovo-Nagymaros case*), [1997] ICJ Rep. 7, which in the Separate Opinion of VP Weeramantry included the concept of sustainable development and the principle of 'continuing EIA'. He viewed EIA obligations as built into the treaty under consideration, with his focus on the need for ongoing monitoring of environmental effects (at 111–113). For a summary, with reference to the relevant literature, see T. Stephens, note 5, at 173–187.

89 This provision would also have triggered an assessment for the MOX Plant. It reads: 'LIST OF ACTIVITIES 2. (a) Thermal power stations and other combustion installations with a heat output of 300 megawatts or more, and (b) Nuclear power stations and other nuclear reactors, including the dismantling or decommissioning

of such power stations or reactors 1/(except research installations for the production and conversion of fissionable and fertile materials, whose maximum power does not exceed 1 kilowatt continuous thermal load). 3. (a) Installations for the reprocessing of irradiated nuclear fuel; (b) Installations designed: – For the production or enrichment of nuclear fuel; – For the processing of irradiated nuclear fuel or high-level radioactive waste; – For the final disposal of irradiated nuclear fuel; – Solely for the final disposal of radioactive waste; or – Solely for the storage (planned for more than 10 years) of irradiated nuclear fuels or radioactive waste in a different site than the production site'. See Espoo Convention, note 11, Appendix 1.

90 It is not clear from the Ministry website if the process is yet complete. See Ministry of Environment, *Cernovoda Nuclear Power Plant Units 3 and 4* (undated), available at www.mmediu.ro/departament_mediu_en/cernavoda_power_station.htm

91 It is of course possible for more than one party to dispute the significance of the environmental effects, which is not apparently provided for in Appendix IV.

92 Decision IV/3 (Inquiry Procedure) (ECE/MP.EIA/10, 28 July 2008) was made by the MOP and follows the report of the first inquiry commission (note 93). It provides for Secretariat support if requested among other incidental matters.

93 See UN Economic Commission for Europe, Working Group on Environmental Impact Assessment, *Inquiry Procedure – Review of the First Inquiry Procedure, Note by the Secretariat, Meeting of the Parties to the Convention on Environmental Impact Assessment in a Transboundary Context, Tenth Meeting, Geneva, 21–23 May 2007*. See also *Report on the Espoo Inquiry Commission: Danube Delta, May 2007*, available at www.unece.org/env/eia/documents/WG10_may2007/Presentation_InqCie.pdf

94 Convention on Wetlands of International Importance, Especially as Waterfowl Habitat (Ramsar, 2 February 1971).

95 Convention Concerning the Protection of the World Cultural and Natural Heritage (Paris, 16 November 1972).

96 Sixteenth Session of General Conference of UNESCO, Programme on Man and the Biosphere (1970).

97 See M. Koyano, note 13, at 304–305. Koyano (at 305) credits its success with the fact that the parties were cooperative, NGOs and external experts were involved, the conclusion was unanimous and its work also covered mediation through the recommendations. See also W. Schrage, note 12, at 46–47.

98 The documentation submitted to Romania is available on the Ministry website. See Ministry of Environment, *Bâstroe Channel* (26 February 2009), available at www.mmediu.ro/departament_mediu_en/bastroe_channel.htm

99 Information on the latest situation can be found at UN Economic and Social Council, Economic Commission for Europe, *Meeting of the Parties to the Convention on Environmental Impact Assessment in a Transboundary Context, Implementation Committee, Seventeenth Session, Geneva, 14–18 September 2009, Annotated Provisional Agenda for the Seventeenth Session* (ECE/MP.EIA/IC/2009/3, 19 May 2009), available at www.unece.org/env/eia/implementation/implementation_committee_meetings.htm

100 UN Economic Commission for Europe, Working Group on Environmental Impact Assessment, Large-Scale Transboundary Projects – Application of the Convention to Complex Activities, Note by the Secretariat, (ECE/MP.EIA/WG.1/2009/4,

2 March 2009), available at www.unece.org/env/documents/2009/eia/wg.1/ece.mp.eia.wg.1.2009.4.e.pdf

101 This is of course also the subject of the Kiev Protocol to the Espoo Convention (not in force), which concerns SEA. See S. Marsden (2008) *Strategic Environmental Assessment in International and European Law*, Earthscan.

102 With respect to the UK and the future of its own nuclear power programme since the conclusion of the *MOX Plant* case, it is also of note that the Planning Act 2008 has introduced provisions for 'national policy statements' including for nuclear power, the assessment of which is to be a key part of the new procedures. See Planning Act 2008 (UK), Section 5. Potentially the application of the SEA Directive will also be triggered.

103 Decision II/4 (Review of Compliance) (ECE/MP.EIA/4, 7 August 2001), Annex 1V of MOP-2, revised as Decision III/2 (Review of Compliance) (ECE/MP.EIA/6, 13 September 2004), Annex I of MOP-3, which provides the structure and functions of the Implementation Committee and procedures for review of compliance. Paragraphs 14 and 15 of Decision III/2 deal with the relationship between dispute settlement and the Inquiry Procedure. The first states that the Implementation Committee NCP is 'a non-adversarial and assistance-oriented procedure', which is 'without prejudice to the settlement of disputes provisions in Article 15 of the Convention'. The second that 'Where a matter is being considered under an inquiry procedure', it cannot be referred to the Implementation Committee.

104 UN Economic Commission for Europe, *Meeting of the Parties to the Convention on Environmental Impact Assessment in a Transboundary Context: Review of the Work Done by the Working Group on Environmental Impact Assessment and Adoption of Decisions – Review of Compliance with the Convention, Report by the Implementation Committee, Fourth Meeting, Bucharest, 19–21 May 2008.* Note also T. Koivurova, 'The Convention on Environmental Impact Assessment in a Transboundary Context', in G. Ulfstein et al (eds) (2007) *Making Treaties Work: Human Rights, Environment and Arms Control*, Cambridge University Press, 218, at 229–234.

105 M. A. Fitzmaurice (2001) 'Environmental Dispute Avoidance and Settlement', in *International Protection of the Environment, Recueil de Cours*, vol 293, Hague Academy of International Law, 336.

106 Only four parties have to date done this: Austria, Bulgaria, Liechtenstein and the Netherlands. Other than Liechtenstein, these are all EC Member States which, in disputes with other EC Member States, would therefore fall foul of the ECJ's 2006 ruling.

107 Note the Opinion of the Environment Commissioner in relation to the Nord Stream Pipeline Project, which involves the construction of an offshore gas pipeline from Russia through the Baltic Sea to Germany. See S. Dimas, *Environmental Aspects of the 'Nordstream Pipeline', Hearing on 'Nordstream' by the Committee of Petitions, in association with the Committee on Foreign Affairs, and the Committee on Industry, Research and Energy of the European Parliament* (Speech/08/43, 29 January 2008), available at http://europa.eu/rapid/pressReleasesAction.do?reference=SPEECH/08/43. There are nine affected parties, and although Russia is only a signatory to the Espoo Convention and is not an EC Member State, it has agreed to be bound by the Espoo requirements. See also N. Grishin, *Environmental Impact*

Assessment of the Nord Stream Project Within the Framework of the Espoo Convention, Tenth Meeting of the EIA Working Group, UNECE, Geneva, 21–23 May 2007 (undated), available at www.unece.org/env/eia/documents/WG10_may 2007/Presentation_NordStream.pdf

108 For a summary of the correct procedure to follow, see N. Lavranos, 'MOX Plant Dispute – Court of Justice of the European Communities', note 2, at 463.

109 Note the discussion in N. Lavranos, 'The Epilogue in the MOX Plant Dispute', note 2, at 183–184, concerning the reluctance of the ECJ to give direct effect to provisions in international agreements including UNCLOS.

110 See ibid.

111 See UNECE, note 84.

112 See E. Albrecht, note 14, at 364.

6

Enforcing Non-discrimination in Transboundary Environmental Impact Assessment: Advantages for EU Citizens from the Transposition of the Espoo and Aarhus Conventions?

Simon Marsden

Introduction

The focus of this chapter is on the enforcement of public rights to information and participation in transboundary environmental impact assessment (TEIA),[1] with specific reference to the transposition of the United Nations Economic Commission for Europe (UNECE) Espoo[2] and Aarhus[3] Conventions in EU law.[4] The non-discrimination principle, which is recognized by the Aarhus Convention[5] and Treaty on European Union,[6] permits individuals and NGOs of one state to exercise these rights in another state in the context of a TEIA procedure operated by another. Yet it has recently been argued that TEIA is not as effective as domestic EIA because an underlying reason for the success of domestic EIA is the administrative law context for its enforcement;[7] in the international law sphere the absence of a comparable context suggests enforcement is less effective, with few opportunities for individual or NGO challenges.[8] Treaty non-compliance procedures, especially the Compliance Committee (CC) of the Aarhus Convention, go some way to providing a forum to raise these matters, although they are not designed as a means of individual dispute settlement as such.[9]

While TEIA may indeed be less effective than domestic EIA when contrasting international and municipal law alone,[10] when municipal law is harmonized in a regional context, in theory at least, the position is entirely different. This chapter therefore challenges the absence of an enforceable context to TEIA procedures by arguing that with respect to TEIA as implemented by EU law, there are several opportunities for individuals and NGOs to challenge procedural failings; when international law is incorporated into this regional context it is applied in a framework of enhanced judicial mechanisms for enforcement, in particular via national courts bound by EU law, the role of the European Commission as 'guardian of the [European] treaties', and ultimately the European Court of Justice (ECJ).[11] For Espoo and Aarhus Convention Parties which are EU Member States, the obligations of these international treaties have therefore been largely implemented by EU secondary law, with the EIA Directive[12], for example, having vertical direct effect on Member States. Justiciable individual rights against Member State governments have hence been created across the EU, which as a result of European citizenship means these rights can be exercised in any of the Member States regardless of which is their state of residence.[13] For provisions in these treaties that have not been transposed, they may arguably be claimed to have direct and indirect application and effect on the Member States,[14] although since these mainly concern dispute settlement procedures between Member States outside EU law and implementation of access to justice obligations in areas other than EIA (which has been transposed) this is considered elsewhere.[15]

The second part of the chapter explains what TEIA is with reference to the Espoo Convention, and considers the procedural role of the public in relation to the stages of assessment leading up to the decision-making process. The third part examines the changes made by the Aarhus Convention to TEIA procedure and the fourth part focuses on the underlying context of EU law, which has transposed most aspects of these treaties and may hence provide a medium for the public to challenge breaches of non-discriminatory rights. Reviews of implementation and compliance undertaken by the Espoo Implementation Committee (IC), the Aarhus CC and the European Commission (in respect of the implementing directives) are analysed in relation to each of these parts to contrast the position of the public and NGOs in EU Member States and other UNECE state Parties. Conclusions follow, making reference to further implementation of the Aarhus Convention, with implications outlined.

THE ESPOO CONVENTION

Objective and application

The objective of the Espoo Convention is to 'ensure environmentally sound and sustainable development' and to 'enhance international cooperation'.[16] It currently has 43 Parties including the EU, and it will be open to any UN state upon approval if the first amendment is ratified by sufficient current state Parties.[17] Under the Convention TEIA refers to a process applied by one or more states to evaluate the significant environmental effects from proposals within their own territory (e.g. a nuclear power facility, like the Sellafield 'MOX Plant' in the UK, which was not, however, assessed under the Convention[18]), or which physically cross borders (e.g. a gas pipeline, like the Baltic Sea 'Nord Stream Project' between Russia and Germany, which is currently being assessed and which may impact upon the territory of several states).

The process is designed to provide better information for decision-makers, as well as involve as many stakeholders in the process (including the public, NGOs and relevant government departments). The terminology 'Party of origin' and 'affected Party' is used under the Espoo Convention to differentiate the obligations and rights of each.[19] TEIA is based upon domestic EIA as applied within a state, sometimes conducted by national or state authorities, or by a mixture of both.[20] There may be more than one Party of origin and affected Party, and where proposals physically cross boundaries each state will be both a Party of origin and an affected Party, such as in the Nord Stream Project. Other international law may also have application.[21]

Proposals with a strategic dimension, commonly known as policies, plans and programme, may also be subject to TEIA; whether they are limited to the territory of a particular state or cross the boundaries of states, their assessment may expedite or avoid the need for assessment of specific projects at a later time.[22] Because of the improved benefits of Strategic Environmental Assessment (SEA) over project level EIA, such as determining priorities while all options are open and the avoidance of cumulative effects, it is an area of growing interest to many states.[23] Interest includes how best to implement requirements for public participation in this strategic context,[24] which includes transboundary strategic proposals.[25] While practice with project level TEIA is significant, there has been limited practice at the strategic level to date.[26] This may change following the likely coming into force of the SEA Protocol to the Espoo Convention, expected in 2010.[27]

Non-discriminatory procedures

Various definitions of terms and general procedural provisions are outlined in the Espoo Convention.[28] The procedure that applies in all instances for TEIA is very similar to that for domestic EIA, although there are some differences to provide for non-discriminatory rights. Schrage and Bonvoisin[29] identify the following: notification by the Party of origin to the affected Party[30] of the planned activity and likely significant effects (or the opportunity of the affected Party to request such a procedure[31]), response by the affected Party whether it wishes to be involved in the EIA procedure, sharing of information, preparation of documentation and distribution in the affected Party,[32] consultation between authorities and participation of the public in both origin and affected states,[33] decision-making taking into account documentation and comments,[34] sharing of information as to the decision and post project analysis.[35] Provisions for bilateral and multilateral cooperation are also set out.[36] The decision-making process must ultimately be based upon the information contained in the EIA documentation and the consultation undertaken, and reasons for the decision must be provided.[37] Although Ebbesson[38] and Knox[39] emphasize that many previous treaties have made provision for TEIA and that many of the core concepts of the Convention are drawn from these, Ebbesson indicates that because it operates to change the behaviour of 'non-state actors' as well as states, (such as individuals), formal compliance with the Convention is dependent upon the establishment of implementing provisions in domestic law.[40]

Implementation is also crucially dependent upon the principle of non-discrimination, which Gray emphasizes: 'The novel provisions of Espoo relate to public participation. The public in the affected area has a right to be informed of and to participate in the EIA procedure, even though the procedure takes place in another country.'[41] Knox identifies this as the 'principal innovation of the Convention',[42] with Ebbesson commenting that if the affected public is not given the opportunity to participate, 'the state in question does not comply with international law'.[43] Schrage also draws attention to this feature, commenting that the Convention 'reflects new trends in international environmental law, which allow all those who are likely to be affected by an environmental impact to be involved in the decision-making process'.[44] Ebbesson identifies two other innovative elements: it provides more specific procedures for notification and consultation regarding transboundary effects, and it advances the use and meaning of EIA in international law.[45]

Reviews of implementation and compliance

Implementation of the Espoo Convention has been reviewed twice by the IC, following detailed reports submitted to the Secretariat by the Parties to the Convention in response to a standard questionnaire.[46] The most recent was published in 2006 and was prepared for the period mid-2003 to the end of 2005, in which the relationship with EU law is only superficially discussed in relation to the summary information.[47] The report reveals a number of weaknesses in the implementation of the Convention, including deficiencies in information exchange, public participation and reporting obligations, and a lack of monitoring. There remains a continuing need for bilateral and multilateral agreements to identify contacts, and address differences in procedural steps,[48] timing of participation, the interpretation of certain terms, and the content of the EIA documentation. For Parties that are not also EU Member States or Parties to the Aarhus Convention, the problem of enforcing non-discriminatory access to justice was highlighted, as appeals by foreign citizens and residents were concluded to be 'very restricted' at this time.[49] Compliance issues in the Second Review of Implementation were discussed by the IC in 2009 where it was confirmed that Article 5 provides for specific consultations after completion of the EIA documentation, that information about the possibilities to appeal should be included by the Party of origin in the final decision, and that there is an obligation to inform the public concerned in the affected Party about the final decision.[50] The next report, for the period 2006 onwards, is currently being discussed and the Working Group on EIA has agreed the questions for the Parties.[51] The second amendment to the Convention will provide a legal obligation on Parties to report on their implementation, when this has been formally ratified and comes into force.[52]

Specific failures to comply with the Convention are dealt with by the IC,[53] the objective of which is to review compliance by the Parties with their obligations, in order to assist them fully to meet their commitments.[54] At the Fourth MOP a report on the activities of the IC was presented highlighting a number of topical issues, including those of a specific nature.[55] Public involvement was a matter discussed, and takes into account the guidance on public participation.[56] Case studies are included in the guidance, and indicate that 'equivalent' opportunities for public participation have been provided for those in affected states.[57] In one instance the proponent of the project advised the public in both the Party of origin and affected Party; the obligation overall however is on both Parties to ensure that this is carried out. Time limits are also considered, with one of the case studies highlighting how the competent authorities in the Party of origin provided more time for comment from the public in the affected Party than for its own public.[58] Specific matters are raised with Parties when a response to an implementation questionnaire indicates non-compliance. In a few instances, these matters have included failings to comply with provisions related to public participation, such as in Finland.[59] Over the last five years, the IC has been especially busy with evaluating Ukraine's compliance with the Convention, after this was raised by Romania in relation to the Bystroe Canal Project proposed for the Danube Delta, which is discussed in the next section.[60]

The Aarhus Convention

Objective and application

The objective of the Convention is to guarantee procedural rights in order to contribute to protection of the right of current and future generations to live in an environment adequate for health and well-being.[61] Since it applies generally to environmental decision-making procedures beyond EIA, it has had an even greater influence on the development of information provision, public participation and access to justice;[62] it has arguably also contributed to the development of democratic and constitutional principles in the Eastern European members of the UNECE,[63] and indeed beyond the nation state;[64] where democracy may not be attainable at the international level, it may also provide significant assistance to 'enhanced transparency, participation, more balanced voting procedures and effectiveness of decisions as possible ways of making decisions more legitimate'.[65] It has also served to link the environment and human rights in international law.[66] All of these are very significant claims indeed. The Convention currently also has 43 Parties and is already open to accession by any UN member with the approval of the parties;[67] even where it is not directly applicable, it has influenced state behaviour globally, including conditions attached to loans provided to developing states by international institutions.[68]

Procedures under the Espoo Convention are enhanced as a result of the Aarhus Convention, and there are several ways in which procedural failings with TEIA by governments that are bound by these treaties can be remedied by individuals, whether they affect access to information or public participation obligations. First, a member of the public can ask their own government to challenge a non-compliant state through the IC of the Espoo Convention. Second, individuals and NGOs can raise procedural failings directly through the unique CC of the Aarhus Convention. Third, the Aarhus Convention's access to justice obligations require Parties to provide opportunities for the public to take proceedings, so a member of the public must be allowed to access the courts of its own state or of another Party when it has breached its obligations. As seen below, for EU citizens these opportunities extend to complaints to the European Commission and action by the ECJ.

Non-discriminatory procedures

The Convention deals with access to and collection of environmental information,[69] public participation in decision-making,[70] and access to justice.[71] Environmental information must progressively be made available in electronic databases, which are easily accessible to the public through public telecommunications networks. Information in this form should include texts of legislation on or relating to the environment, and, 'as appropriate', policies, plans or programmes on or relating to the environment and environmental agreements. There are also measures for the dissemination of legislation and policy documents, and progress reports on their implementation, prepared at various levels of government.

The Convention distinguishes between three areas of environmental decision-making for which public participation is required: specific activities, plans, programmes and policies, and environmental legislation. Plans and programmes on the one hand and policies on the other are distinguished; the more specific the decision, the greater the procedural guarantees. The access to justice provisions deal with three situations: where the right of access to environmental information is impaired, where the right to public participation is impaired, and where acts and omissions of public authorities or private persons are in breach of the law. As noted, non-discrimination is a central part of the Aarhus Convention.[72]

Reviews of implementation and compliance

Reporting on the implementation of the Aarhus Convention is again a key part of the treaty institutional mechanisms. Questionnaires are also circulated to the state Parties, with a requirement to consult with the public when national reports are being prepared.[73] Unlike the Espoo IC, individuals and NGOs have the right to submit complaints about a state's compliance directly to the CC,[74] the latter regardless of demonstrating a specific interest or needing to prove they are affected by the decision.[75]

Both consultations with the public concerning national reports, public submission of complaints and the openness of proceedings before the CC make practical sense given that the substantive provisions of the Convention relate to the public rather than states. The relationship between European and international law is considered in more detail in the 2008 Implementation Report submitted by the EC to the UNECE in relation to the Aarhus Convention, which refers to the binding nature of treaties for EU institutions as well as Member States and the primacy of international treaties over EU secondary law.[76] In relation to the latter, the Commission states: 'The primacy of international agreements concluded by the Community over provisions of secondary Community legislation also means that such provisions must, so far as is possible, be interpreted and applied in a manner that is consistent with those agreements.'[77]

A total of 37 cases under the CC have been initiated by communications from the public and some have been discussed in the literature;[78] a few of these have involved national or European courts. Only one case relates to TEIA, not surprisingly the Danube Delta dispute discussed in relation to the Espoo IC above.[79] At the same time as the Espoo Inquiry Commission procedure was initiated in 2004, both Romania and a Ukrainian NGO, Ecopravo-Lviv, made a 'submission' (by a Party) and a 'communication' (by the public) to the CC in relation to Ukraine's decision to construct the navigation canal. Ecopravo-Lviv, as a member of the 'public' and 'public concerned' contended that Ukraine had breached Article 6 by failing to consult; Romania, as the affected Party, raised similar issues in relation to its own public. The CC agreed with both sets of contentions and recommended the matter be determined by the MOP so that Ukraine would fully comply with its obligations.[80] Despite the issuance of a caution, Ukraine would however appear to still have some work to do to fully comply.[81] While Romania is an EU Member State, the fact that the Ukraine is not means that the EU enforcement mechanisms discussed below are not applicable to it, a clear disadvantage to the Ukrainian NGO, which while representing the public in the Party of origin rather than the affected Party must nonetheless rely on the IC and CC (or use Ukrainian domestic means) to put pressure on the Ukraine through the MOP to ensure compliance.

Treaty implementation by EU law

EIA Directive

The EIA Directive as amended by the Public Participation Directive (below) implements almost all of the provisions of the Espoo and Aarhus Conventions regarding TEIA.[82] It is possible for some provisions of the EIA Directive to have vertical direct effect (by an individual against a Member State),[83] depending on whether the conditions as originally laid down have been complied with.[84] Although some of the case law in the national courts initially suggested that because the third of these conditions could not be satisfied, it was not possible for the EIA Directive (which allowed discretion) to be directly effective, other cases have since suggested otherwise, particularly because of the limits to the discretion exercised. Tromans indicates that the lack of clarity in ECJ rulings, despite reference to direct effect in the cases, has been a major problem, with judgments failing to examine the matter with the precision needed.[85] However, Harwood has since highlighted that the Wells decision[86] allowed the individual to require the UK to enforce the Directive and carry out an assessment.[87]

The amended EIA Directive[88] delegates to the Member State the 'detailed arrangements for implementing' all of the aspects of the transboundary consultation procedure,[89] which is triggered when a Member State is aware that the project is likely to have significant effects on the environment of another Member State, or if the latter requests its application.[90] Documentation must be forwarded to the affected Member State describing the project, possible transboundary impact, and on the nature of the decision to be taken,[91] giving the affected Member State a reasonable time to indicate whether it wishes to participate. If it does, the proponent Member State must send all the information gathered,[92] and information on the procedure, including the request for development consent. Both or all involved Member States (if more than one) must arrange for the information to be made available within reasonable time-frames to relevant authorities and the public concerned in the affected Member State. Those authorities and the public concerned must be given an opportunity before the granting of the development consent to forward their opinion.[93] Consultations within an agreed timeframe must be entered into regarding the potential transboundary effects and measures proposed to mitigate or prevent them.[94]

Transboundary matters have only been considered once by the ECJ, perhaps not surprisingly in Belgium, which recognizes three territory-related Regions under constitutional law (Flanders, Wallonia and Brussels) and allows implementation of TEIA between them as well as in the relationship with other states.[95] The Commission took infringement proceedings against Belgium, succeeding in various arguments including that the Flemish Region and the Region of the Capital City of Brussels had failed to transpose Article 7 and comply with requirements relating to works with a transboundary impact.[96] In rejecting the position put by the Kingdom of Belgium, the Court held:

> *The argument based on the geographical situation of the Region of the Capital City of Brussels is based on the assumption that only projects located in frontier regions are capable of affecting the environment in another Member State. Yet, as the Commission has rightly observed, that assumption is wrong since it ignores the possibility of air- or water-borne pollution.*[97]

Public Participation Directive

Transboundary procedure was also amended further under Directive 2003/35/EC (the Public Participation Directive),[98] with a number of additional significant amendments, many to comply with the Aarhus Convention.[99] As noted above, definitions of 'the public' and 'the public concerned' have been added,[100] the latter which includes not only the public affected or likely to be affected by the decision but also those having an interest in environmental decision-making, such as NGOs, who are explicitly referred to. The complete exemption from assessment of projects for national defence has been removed,[101] and information is now required on alternative assessment procedures to be made public when a Member State decides to exempt specific projects from the Directive.[102] Detailed provisions establish which information is to be made available, and that it must be released early in the process.[103] Early and effective opportunities for public participation with reasonable time-frames are required,[104] and the public must be informed of the main reasons for the decision and of the consultation process.[105]

The ECJ has decided on the question of how much flexibility is available to a Member State in deciding the 'detailed arrangements' governing public participation in the ELA Directive, including the question of fees payable, which would also apply in respect of the Article 7 transboundary requirements. The Court decided that the action of a Member State must not undermine the effectiveness of a directive in practice,[106] especially the role of the public, which was acknowledged in an earlier case.[107] However, in the more recent decision, the Court held that a Member State is free to impose a participation fee provided it does not constitute an obstacle to the exercise of participation rights.[108]

The Public Participation Directive has also inserted a new Article 10a into the EIA Directive to implement the third pillar of the Aarhus Convention in specific regard to EIA. It recognizes the significance of this area for environmental decision-making across the EU, and operates in the absence of the proposed Access to Justice Directive.[109] This provides for access for the public concerned to the courts or other independent and impartial bodies, to challenge 'the substantive or procedural legality' of decisions, acts or omissions,[110] which must be 'fair, equitable, timely and not prohibitively expensive';[111] the public concerned must, however, have a 'sufficient interest' or 'maintain the impairment of a right'.[112]

Member States retain the power to decide in accordance with their national legal system at what stage decisions, acts of omissions may be challenged,[113] and what constitutes a sufficient interest or impairment of a right;[114] this limits the potential for challenges, although 'wide access to justice' is specified as an objective in determining what a 'sufficient interest' or 'impairment of a right' means,[115] which restricts Member State discretion. Member States must also provide the public with practical information on access to administrative and judicial review procedures.[116] Article 10a has been analysed in detail, including its application in an Irish domestic context.[117] It is likely to result in increased litigation with the EIA Directive requiring interpretation under Article 6 of the Aarhus Convention regarding specific activities,[118] and also in accordance with the principles of effective judicial protection and consistent interpretation (indirect effect).[119]

Reviews of implementation and compliance

The European Commission is tasked with carrying out periodic reviews of specific directives,[120] and prepares annual reports on the application of EU law generally.[121] With respect to the early application of TEIA, and referring to the two first EIA Directive implementation reports, Woodliffe comments 'there was scanty evidence on the use of Article 7' but that 'several states without a formal provision did in fact notify neighbouring states of projects'.[122] The 2003 Commission Report comments on the variety of issues that have arisen in applying Article 7 across the EU, including those of a procedural nature, such as establishing points of contact and obtaining translation; ways of overcoming such issues are also discussed.[123] At the time, few bilateral agreements had been entered into between Member States concerning these issues, but this would seem to have developed since then.[124]

The most recent review of the EIA Directive was concluded in 2009.[125] The Report comments on the need to generate synergy between the EIA Directive and Espoo Convention, and the EIA Directive has been identified as a potential instrument for a simplification exercise concerning transboundary projects.[126] Specifically, the Report draws attention to the fact that despite increasing public participation in the decision-making process there is still no standard practice across the EU, with the meaning of early and effective participation tending to be interpreted as public consultation once the information has been collected;[127] timeframes for consultation also vary considerably, as do arrangements for informing and consulting the public and criteria for standing in a review procedure, which are determined by the Member States.[128] Difficulties regarding TEIA procedures point to the differences between national EIA procedures, suggesting the need for further harmonization; language barriers also continue to cause concern.[129] Projects that physically cross Member State jurisdictions are singled out, with risks of duplication, inconsistency and cumulative effects highlighted; the need to improve consultation arrangements including coordinated, joint or single procedures is therefore emphasized.[130]

Access to justice—the role of the Commission and ECJ

Further to the access to justice provisions operating at a national level explained earlier, EU citizens also have the right to pursue non-compliance matters via the European institutions, in particular the Commission and ECJ.[131] In a few instances individuals and NGOs have initiated actions against Member States in national courts which have later been referred to the ECJ for a preliminary ruling[132] which allows submissions to be made by individuals.[133] An action can be begun in the courts of any Member State without discrimination,[134] although they depend in part on national standing rules as discussed. EU citizens can also submit formal complaints in relation to Member State non-compliance to the Commission,[135] which must ensure legislation is applied in the Member States.[136] Although the Commission is required to respond, recent changes to practice, notably a pilot scheme allowing for direct communication between the Member State and the individual or NGO,[137] may have reduced the effectiveness of this.[138] If it finds a Member State is at fault, it will first consult with the Member State and may provide a reasoned opinion requiring compliance;[139] failure to comply may result in the Commission initiating action in the ECJ against the Member State.[140]

Since the ECJ must also ensure the law is applied,[141] if it agrees with the Commission an order may be made for compliance, failing which payment of a lump sum or penalty payments may be ordered.[142] In some circumstances, individuals may also be able to claim damages from a Member State whose breach of EU law causes them to suffer loss, regardless of which branch of government is responsible.[143] Direct access to the ECJ, however, is limited because directives are addressed to Member States and the Commission (or another Member State) is generally responsible for taking action for breaches;[144] for individuals and NGOs to institute proceedings against decisions there is an obligation to show that the matter is of 'direct and individual concern' to them, the restricted interpretation to date of which has presented an insurmountable hurdle to standing.[145] This is currently subject to a challenge by an environmental NGO to the CC of the Aarhus Convention, which argues that blocking access to justice for individuals and NGOs in environmental matters is an erroneous reading of Article 230(4) of the EC Treaty, which should be subject to interpretation in the light of Article 9 of the Aarhus Convention.[146]

Conclusion

The Espoo and Aarhus Conventions contain detailed specific and related provisions for involvement by individuals and NGOs in TEIA regardless of their state of residence; these extend to opportunities for the public to raise issues of non-compliance, particularly through the Aarhus CC. The transposition of these treaties in EU law, however, provides additional advantages for EU citizens, with the European treaties providing substantial institutional means of enforcement. These have been judicially recognized, developed and applied to enable individuals and NGOs to initiate and contribute to enforcement action against their own or other Member States. While implementation of these treaties by EU law can be concluded to be advantageous, difficulties however remain as indicated by the reviews of compliance and the literature discussed. Primary among these are that public participation requirements are interpreted differently in the Member States, access to justice depends on national standing requirements, and there continues to be restricted access to the ECJ. Given wider knowledge and understanding of these matters and willingness to overcome them, there is, however, room to be cautiously optimistic that further experience with harmonization in this area will benefit EU citizens further in participating in and contributing to TEIA procedures. An obvious implication is that if individuals and NGOs in EU Member States benefit from greater opportunities to enforce TEIA provisions, individuals and NGOs in non-EU Member States are effectively discriminated against in the exercise and enforcement of their respective rights, which would need to be addressed by the Espoo and Aarhus treaty compliance procedures, and in general governance matters between the state Parties.

Notes

This chapter was originally published as: Marsden, S. (2009) 'Enforcing Non-discrimination in Transboundary Environmental Impact Assessment: Advantages for EU Citizens from the Transposition of the Espoo and Aarhus Conventions?', *Journal for European Environmental and Planning Law*, vol 6, no 4, pp437–460.

1 See generally K. Bastmeijer and T. Koivurova (eds) (2008) Theory and Practice of Transboundary Environmental Impact Assessment; IAPA 2008 (26), (a special issue on TEIA) p.234.
2 Convention on Environmental Impact Assessment in a Transboundary Context (Espoo), 30 ILM (1991), 802.
3 Convention on Access to Information, Public Participation in Decision-Making and Access to Justice in Environmental Matters (Aarhus), 38 ILM (1998), 517.
4 References to 'EU law' assume the coming into force of the Reform Treaty, Lisbon, adopted 17 December 2007, following Irish ratification in October 2009. One of the implications is that EC law will finally be replaced by EU law and the Union will have a single legal personality; the TFEU will replace the TEC and the TEU will remain. References to provisions in the TEU and TEC are however used in this chapter pending these changes.
5 Article 3(9).
6 TEU, Maastricht, 7 February 1992, Articles 17–22; See also TEC Article 12.
7 C. M. Kersten, Rethinking Transboundary Environmental Impact Assessment, (2009) Yale JIL 2009 (34), pp.173, 175 et seq.

8 T. Stephens, (2009) International Courts and Environmental Protection,, pp.247, 253–256 et seq, discusses opportunities for public interest proceedings in relation to international courts and tribunals. In relation to the ICJ contentious matters Article 50 permits expert advice from individuals and perhaps NGOs although it has not been triggered; in relation to advisory proceedings submissions may be received from NGOs (Article 66(2)) although this has not been welcomed. The Permanent Court of Arbitration (PCA) procedural rules relating to the environment may allow NGOs as parties (Article I(I)), but they have not been used. UN Convention on the Law of the Sea (UNCLOS), Montego Bay 1982, in force 16 November 1994 proceedings may provide various opportunities for NGOs to participate within its specific context.

9 V. Koester, (2005) The Compliance Committee of the Aarhus Convention: An Overview of Procedures and Jurisprudence, Environmental Policy and Law (37) pp.83, 85 et seq; V. Koester, Review of Compliance under the Aarhus Convention: A Rather Unique Compliance Mechanism, JEEPL 2005 (14) p.31; J. Ebbesson, Public Participation, in: D. Bodan-sky, J. Brunnee and E. Hey (eds), The Oxford Handbook of International Environmental Law, 2007, p.681; M. Fitzmaurice, Environmental Justice through International Complaint Procedures? Comparing the Aarhus Convention and the North American Agreement on Environmental Cooperation, in: J. Ebbesson and P. Okowa (eds), Environmental Justice in Context, 2009, pp. 211, 213 et seq.

10 Kersten's comments (see note 7) are directed to the situation in the United States in particular.

11 Note that for disputes between European Member States themselves in relation to mixed environmental agreements (which include the Espoo and Aarhus Conventions), these now have to be determined by the ECJ following the 2006 ruling in the MOX Plant case; Case C-459/03, Commission v Ireland [2006] ECR I-4635. See 5. Marsden, MOX Plant and the Espoo Convention: Can Member State Disputes Concerning Mixed Environmental Agreements be Resolved Outside EC Law?, RECIEL 2009 (18), p.312.

12 Council Directive 85/337/EEC on the assessment of the effects of certain public and private projects on the environment, [1985] OJ L175/40 (as amended by Council Directives 97/11 [1997] OJ L73/5) and 2003/35 [2003] OJ L156/17.

13 See note 6.

14 Direct application in this instance refers to the obligations in the international treaty binding the EU (and hence Member State) in the absence of an implementing directive, which is argued by some to flow given the 'monist' nature of the EU legal system, which in public international law means there is no need for treaties to be implemented by domestic law; direct effect confers rights on individuals, again regardless of implementing law; see generally J. Wouters, A. Nollkaemper and E. de Wet (eds.), The Europeanisation of International Law: The Status of International Law in the EU and its Member States, 2008. Indirect effect, or the doctrine of 'consistent interpretation', refers to the need of Member State institutions to apply domestic law within the context of their treaty obligations. Indirect application refers to the need to take general rules of international law into account. See: G. Betlem and A. Nollkaemper, Giving Effect to Public International Law and European Community

Law before Domestic Courts. A Comparative Analysis of the Practice of Consistent Interpretation', EJIL 2003 (14), pp.569.

15 See 5. Marsden, Invoking Direct Application and Effect by the European Court of Justice: Implications for International Environmental Law, under review 16 Preamble.

17 First amendment, adopted 2001.

18 For reasons, see Marsden, note 11.

19 Article 1.

20 Bastmeijer and Koivurova, see note 1; Kersten, see note 7.

21 Article 2(10) states that 'the provisions of this Convention shall not prejudice any obligations of the Parties under international law with regard to activities having or likely to have a transboundary impact'. For examples including the role of customary international law, see N. Craik, The International Law of Environmental Impact Assessment, 2008; A. Gillespie, Environmental Impact Assessments in International Law, RECIEL 2008 (17), p.221.

22 T. Fischer, The Theory and Practice of Strategic Environmental Assessment: Towards a More Systematic Approach, 2007; 5. Marsden, Strategic Environmental Assessment in International and European Law, 2008; B. Sadler et al (eds), (2010) Handbook of Strategic Environmental Assessment f.

23 UNECE, Benefits and Costs of Transboundary EIA: Applying the Convention on EIA in a Transboundary Context, 2007. For similar information, with specific reference to the UNECE Protocol on Strategic Environmental Assessment to the Espoo Convention (Kiev, 21 May 2003), not in force, see UNECE, Applying the Protocol on SEA, 2007. See also 5. Marsden and de Mulder, SEA and Sustainability in Europe: How Bright is the Future?, RECIEL 2005 (14), p.50; and Regional Environmental Center for Central and Eastern Europe and United Nations Development Programmme, Benefits of an SEA, 2003.

24 S. Heiland, Requirements and Methods for Public Participation in SEA, in: M. Schmidt, E. Joao and E. Albrecht (eds), (2005) Implementing Strategic Environmental Assessment, p.421.

25 H. Meyer-Steinbrenner, Public Participation for SEA in a Transboundary Context, in: M. Schmidt et al, id. p.433; E. Albrecht, Transboundary Consultations in Strategic Environmental Assessment, IAPA 2008 (26) p.289.

26 N. Bonvoisin, Transboundary Issues in Strategic Environmental Assessment, in: B. Sadler et al (eds) see note 22; J. de Mulder, The Institutional Context for Transboundary Environmental Impact Assessment in Belgium: Multi Level Setting – A Matter of Smooth Governance?, IAPA 2008 (26), pp.282, 285–287 et seq.

27 UNECE, Report of the Working Group on EIA on its 12th Meeting, Geneva, 11–13 May 2009, 4 et seq, ECE/MP.EIA/WG.1/2009/2, 14 July 2009. Partly because the SEA Protocol is not yet in force, the transposition of the SEA provisions in the Aarhus Convention (which primarily concern public participation in relation to plans and programmes) are not considered in this chapter.

28 Article 1 defines 'proposed activity', 'environmental impact assessment', 'impact', 'transboundary impact', 'competent authority' and 'public'. Article 2 contains general provisions for the management of effects, legal measures required, anticipation and notification, consultation and participation, exemptions, and requirements for more stringent measures.

29 W. Schrage and N. Bonvoisin (2008) Transboundary Impact Assessment: Frameworks, Experiences and Challenges, IAPA (26), p.234.
30 Article 3.
31 The Inquiry Procedure was established under the Espoo Convention to decide effect significance by an independent panel when a potentially affected Party disputes the determination made by the Party of origin. It was used for the first time in the Danube Delta case involving Ukraine and Romania. See UNECE, Working Group on Environmental Impact Assessment Inquiry Procedure – Review of the First Inquiry Procedure, Note by the Secretariat, Meeting of the Parties to the Convention on Environmental Impact Assessment in a Transboundary Context, 10th Meeting, Geneva, 21–23 May 2007. For discussion, see M. Koyano, The Significance of the Convention on Environmental Impact Assessment in a Transboundary Context (Espoo Convention) in International Environmental Law: Examining the Implications of the Danube Delta Case, IAPA 2008 (26), pp.299–314 for analysis of whether European Member States can still use this procedure, see Marsden, note 11.
32 Article 4.
33 Article 5.
34 Article 6.
35 Article 7.
36 Article 8. For examples, see E. Albrecht (2008) Implementing the Espoo Convention in Transboundary EIA between Germany and Poland, EIA Review (28), p.359; R. Albergaria and T. Fidelis (2006) Transboundary EIA: Iberian Experiences, EIA Review (26), p.614.
37 Article 6(1) contains the substantive obligation to give 'due account' to the outcome of the EIA, and Article 6(2) is the requirement for the reasons and other considerations on which the decision is based.
38 J. Ebbesson (1999) Innovative Elements and Expected Effectiveness of the 1991 EIA Convention, EIA Review, (19), p.47.
39 J. Knox (2002) The Myth and Reality of Transboundary Environmental Impact Assessment, AJIL, (96), p.291.
40 Ebbesson, see note 38 at p.48.
41 K. R. Gray, International Environmental Impact Assessment: Potential for a Multilateral Environmental Agreement, Colorado J.Int.Env.L. and Pol. 2000 (n), pp.83, 101–102 et seq.
42 Knox, see note 39 at p.303.
43 Ebbesson, see note 38 at p.50.
44 W. Schrage (1999) The Convention on Environmental Impact Assessment in a Transboundary Context, in: J. Petts (ed) Handbook of Environmental Impact Assessment, vol. 2, pp.96–97.
45 Ebbesson, see note 38. On the latter point, see Craik, note 21 and Gillespie, note 21.
46 See Annex I, Decision III/I of the 3rd MOP, Review of Implementation, Cavtat, 1–4 June 2004, ECE/MP.EIA/6. For discussion of some of the issues encountered in the early years of the Convention coming into force, see M. Hilden and E. R. Furman, Assessment Across Borders – Stumbling Blocks and Options in the Practical Implementation of the Espoo Convention, EIA Review 2001 (21), 537.

47 UNECE, Review of Implementation of the Espoo Convention, 2008, ECE/MP.EIA/II, paragraph 15. The EC responded to this survey in part only, taking the view that the obligations contained within the Convention were for states with its role merely to adopt framework legislation for its implementation; see European Community Response to the Questionnaire on Implementation of the Espoo Convention prepared pursuant to Decision II/11 of the Second Meeting of Parties to the Convention, Sofia, 23 July 2003. The brief Report of the European Community for 2003–2005 on the Implementation of the Espoo Convention on Environmental Impact Assessment in a Transboundary Context can be found at www.unece.org/env/eia/documents/Review%202003—2005/Questionnaire°/o20-0/o202003-20050/o20-0/o2oEU0/o20-°/o20annex.pdf.

48 See UNECE, Guidance on Sub regional Cooperation, 2001, derived from Decision III/ 5 Strengthening Subregional Cooperation, 3rd MOP, Cavtat, 1–4 June 2004. This focuses on the application of the Convention in the Balkan and Black Sea regions, Central and Eastern Europe and South Eastern Europe. This report also contains case studies including a German-Polish study that indicates the most problematic aspect was that of public participation because there was no direct communication between the public of the affected Party and the Party of origin (paragraph 52).

49 Ibid. at paragraph II(d).

50 UNECE, MOP to the Convention on Environmental Impact Assessment in a Transboundary Context, Implementation Committee, Report of the Implementation Committee on its 16th Sess., held from 10–12 March 2009 in Berlin, ECE/MP.EIA/IC/2009/2, 9 April 2009, paragraphs. 25–27.

51 UNECE, Meeting of the Parties to the Convention on Environmental Impact Assessment in a Transboundary Context, Working Group on EIA, 12th Meeting, Geneva, 11–13 May 2009, Report of the Working Group on EIA on its 12th Meeting, ECE/MP.EIA/WG.1/2009/2, 14 July 2009, paragraphs 22–24 and Annex I.

52 The second amendment, to incorporate a new Article 14 bis, was adopted 2004.

53 Article 11(2) of the Convention is the trigger for the IC, including in the terms of reference of the MOP to require Parties to continuously review implementation in the ways stated.

54 Decision II/4 of the 2nd MOP, revised as Decision III/2 of the 3rd MOP, provides the structure and functions of the IC and procedures for review of compliance. Paragraph 14 of Decision III/2 states the IC is 'a non-adversarial and assistance-oriented procedure'. It was established by the MOP in February 2001, and comprises eight members, nominated by Parties that are, in turn, elected by the MOP.

55 Report by the Implementation Committee, 4th Meeting, Bucharest, 19–21 May 2008. Note also T. Koivurova, The Convention on Environmental Impact Assessment in a Transboundary Context, in G. Ulfstein et al, (eds) (2007) Making Treaties Work: Human Rights, Environment and Arms Control, pp.218, 229–234 et seq.

56 UNECE Convention on EIA in a Transboundary Context, Guidance on Public Participation in EIA in a Transboundary Context, ECE/MP.EIA/7, adopted by Decision III/8, 3rd MOP, Cavtat, 1–4 June 2004.

57 Ibid. at p.8.

58 Ibid. at p.17.
59 In the case NGOs were found to have been notified but not the public generally. This was explained by Finland by time restrictions and the isolation of the area where effects may occur. Finland was not found to be in breach. See Ministry of the Environment, Implementation Committee Letter on Compliance with the EIA Convention: Compliance with the EIA Convention's Articles on Public Participation in Projects where Finland has been an Affected Party, 28 October 2005. See www.unece.org/env/eia/documents/ImplementationCommittee/2004—2008/IC_Finland_compliance2005_response_from_ Finland.pdf.
60 UNECE, Meeting of the Parties to the Convention on Environmental Impact Assessment in a Transboundary Context, Implementation Committee, 17th Session, Geneva 14–18 September 2009, Follow-Up to Decision TV/2 Regarding Ukraine (Paragraphs 7–14), Independent Review of Ukraine's Legal, Administrative and other Measures to Implement the Provisions of the Convention, ECE/MP.EIA/IC/2009/5 2 July 2009. Koyano, see note 31, at 305, credits its success with the fact that the Parties were cooperative, NGOs and external experts were involved, the conclusion was unanimous, and its work also covered mediation through the recommendations. See also W. Schrage, The Convention on Environmental Impact Assessment in a Transboundary Context, in Bastmeijer and Koivurova, note 1, at 46–47.
61 Article 1.
62 P. Davies, Public Participation, the Aarhus Convention, and the European Community, in: D. M. Zillman, A. Lucas and G. W. Pring (eds) (2002) Human Rights in Natural Resource Development: Public Participation in the Sustainable Development of Mining and Energy Resources, p.155; M. Lee and C. Abbot (2003) The Usual Suspects? Public Participation under the Aarhus Convention Modern Law Review, (66), p.80.
63 J. Jendroska and S. Stec (2001) The Aarhus Convention: Towards a New Era in Environmental Democracy, Environmental Liability, p.140; M. Zschiesche, The Aarhus Convention – More Citizens' Participation by Setting out Environmental Standards?, ELNI Review 2002 (1), p.21; J. Wates, The Aarhus Convention: A Driving Force for Environmental Democracy, JEEPL 2005 (2), p.2.
64 E. Petkova and P. Veit (2000) Environmental Accountability beyond the Nation State: the Implications of the Aarhus Convention.
65 S. Andresen and E. Hey (2005) The Effectiveness and Legitimacy of International Environmental Institutions, International Environmental Agreements (5), p.211.
66 E. Morgera (2005) An Update on the Aarhus Convention and its Continued Global Relevance, RECIEL (14), p.138; S. T. McAllister (1998) Human Rights and the Environment: The Convention on Access to Information, Public Participation in Decision Making, and Access to Justice in Environmental Matters, Colorado Journal of International Environmental Law (9). p.187.
67 Article 19(3).
68 J. Razzaque, Participatory Rights in Natural Resource Management: The Role of Communities in South Asia, in: J. Ebbesson and P. Okowa (eds) (2009) Environmental Justice in Context, p.117.
69 Articles 4–5.

70 Articles 6–8.

71 Article 9. With respect to the role of NGOs in particular, see C. Ziehm (2005) Legal Standing for NGOs in Environmental Matters under the Aarhus Convention and under Community and National Law, JEEPL (4), p.287.

72 See note 5.

73 Article 15. See UNECE, MOP to the Convention on Access to Information, Public Participation in Decision-Making and Access to Justice in Environmental Matters, Working Group of the Parties to the Convention, 7th Meeting, Geneva, 2–4 May 2007, Item 9 of the provisional agenda, Reporting Requirements: Guidance on Reporting Requirements, ECE/MP.PP/WG.1/2007/L.4, 20 February 2007.

74 The IC is aware of the differences between public involvement in the CC and the IC and is keeping the matter under review in the light of future experience. See UNECE, MOP to the Convention on EIA in a Transboundary Context, 4th Meeting Bucharest, 19–21 May 2008, Item 8(b) of the Provisional Agenda, Review of the Work Done by the Working Group on EIA and Adoption of Decisions, Review of Compliance with the Convention, Report on the Activities of the Implementation Committee, ECE/MP.EIA/2008/5, 27 February 2008, paragraph 20.

75 Article 2(5). Proceedings of the CC are largely in public, and members of the CC are independent, serve in their personal capacity, and may also be nominated by environmental NGOs. A Guidance Document has been published setting out these and other procedural matters: Guidance Document on Aarhus Convention Compliance Mechanism, 2008.

76 See also J. Jendroska, Aarhus Convention and Community Law: The Interplay, JEEPL 200 5(2), p.12; R. Hallo, The Aarhus Convention in Operation: EEB Survey Initial Results, ELNI Review 2007 (2), p.2.

77 UNECE, MOP to the Convention on Access to Information, Public Participation in Decision-Making and Access to Justice in Environmental Matters, 3rd Meeting Riga, 11–13 June 2008 Item 6(a) of the provisional agenda Procedures and mechanisms facilitating the implementation of the Convention: Reports on implementation. Implementation Report Submitted by the European Community, ECE/MP.PP/IR/2008/EC, 6 June 2008, paragraph 2.

78 Fitzmaurice, see note 9, 213 et seq; Koester, see note 9 (Environmental Policy and Law), 33 et seq.

79 Communication Concerning Non-Compliance by Belarus with the UNECE Convention on Access to Information, Public Participation in Decision-Making and Access to Justice in Environmental Matters, ACCC/C/2009/37, submitted 14 March 2009, may also relate to TEIA because it makes reference to a hydro power plant on the Neman River which passes Belarus, Lithuania and Russia, but since the communicant has requested confidentiality there is no way of knowing this.

80 UNECE, MOP to the Convention on Access to Information, Public Participation in Decision-Making and Access to Justice in Environmental Matters, Compliance Committee Report on the 7th Meeting, Addendum, Findings and Recommendations with Regard to Compliance by Ukraine with the Obligations under the Aarhus Convention in the Case of Bystroe Deep-water Navigation Canal Construction; adopted by the Aarhus Conventions Compliance Committee on 18 February 2005, ECE/MP.PP/C.1/2005/2/Add.3, 14 March 2005.

81 UNECE, MOP to the Convention on Access to Information, Public Participation in Decision-Making and Access to Justice in Environmental Matters, 3rd Meeting, Riga, 11–13 June 2008, Item 6(b) of the provisional agenda. Procedures and mechanisms facilitating the implementation of the Convention: Compliance mechanism Report by the Compliance Committee, Addendum, Compliance by Ukraine with its Obligations under the Convention and its Implementation of Decision 11/ 5b of the MOP. See also Implementation Committee, see note 60.

82 As indicated, provision for the Inquiry Procedure in the Espoo Convention has not been transposed by the EIA Directive, although Article 7(1) of the EIA Directive allows an affected Member State to request the application of the transboundary consultation procedure if it is likely to be significantly affected. The implication is that where the Member States disagree, there is no independent scientific procedure established to resolve the matter, which would need determining by the ECJ in the absence of agreement.

83 Courts have been unwilling to allow for horizontal direct effect of directives, (by an individual against another individual or a developer), regardless of the implications of those affected, who may be individuals challenging development by state authorities.

84 The content of a provision must be 'clear and unconditional and not contingent on any discretionary implementing measure'. Case 44/84, Hurd [1986] ECR 29, paragraph 47.

85 S. Tromans and K. Fuller (2003) Environmental Impact Assessment: Law and Practice, at 36–39.

86 Case C-201/02, Delena Wells v. Secretary of State for Transport, Local Government and the Regions [2004] ECR I-723.

87 R. Harwood, EIA, Development Consent and Duties on the Member State, JEL 2004 (16), pp.261, 275 et seq.

88 See note 12.

89 Article 7(5).

90 Article 7(1).

91 Article 7(1) (a) and (b).

92 Article 2.

93 Article 7(3).

94 Article 7(4).

95 De Mulder, see note 26.

96 Case C-133/94, Commission v. Belgium [1996] ECR I-2323.

97 At paragraph 53.

98 See note 12.

99 Article 3. Note also Parliament and Council Directive 2003/4/EC on public access to environmental information and repealing Council Directive 90/313/EEC, OJ 2003L 41/26, which implements the first pillar of the Aarhus Convention generally.

100 Article 1(2).

101 Article 1(4).

102 Article 2(3).

103 Articles 6(2) and (3).

104 Articles 6(4) and (6).
105 Article 9(1).
106 Case C-216/05, Commission v. Ireland [2006] ECR I-10787.
107 Case 0332/04, Commission v. Spain [2004] ECR I-40.
108 Case C-216/05, see note 106.
109 Proposal for a Directive of the European Parliament and of the Council on access to justice in environmental matters, Brussels, 24.10.2003 COM (2003) 624 final, 2003/0246 (COD).
110 Article 10a[2].
111 Article 10a[6].
112 Article 10a[1].
113 Article 10a[3].
114 Article 10a[4].
115 Article 10a[4].
116 Article 10a[7].
117 A. Ryall (2009) Effective Judicial Protection and the Environmental Impact Assessment Directive in Ireland.
118 A. Ryall, EIA and Public Participation: Determining the Limits of Member State Discretion, JEL 2007 (19), pp.247, 256 et seq.
119 S. Drake (2005) Twenty Years after Von Colson: The Impact of 'Indirect Effect' on the Protection of the Individual's Community Rights, European Law Review (30), p.329.
120 For example, Article 11(3) of the EIA Directive requires a review of application and effectiveness every five years.
121 For discussion on the latest report which makes specific reference to the EIA Directive, see Commission of the European Communities, Commission Staff Working Document accompanying the 2 5th Annual Report on Monitoring the Application of Community Law (2007) Situation in the Different Sectors, Brussels, 18.11.2008 SEC (2008) 2854, paragraph 6.2.3. One of the further challenges identified in this report is to ensure that citizens and NGOs can seek reviews at the Member State level, which means full implementation of Article 10a (paragraph 6.2.3.4).
122 J. Woodliffe, Environmental Damage and Environmental Impact Assessment, in M. Bowman and A. Boyle (eds) (2002) Environmental Damage in International and Comparative Law: Problems of Definition and Valuation, pp.133–147.
123 Commission of the European Communities, Report from the Commission to the European Parliament and the Council on the Application and Effectiveness of the EIA Directive (Directive 85/337/EEC as amended by Directive 97/11/EC), How Successful are the Member States in Implementing the EIA Directive; section 4.5 deals with transboundary impacts, at 78–82.
124 Albrecht, see note 25, for example discusses the implementation of the Espoo Convention on TEIA between Germany and Poland, making explicit reference to the requirements of Article 7 of the EIA Directive.
125 Commission of the European Communities, Report from the Commission to the Council, the European Parliament, the European Economic and Social Committee

and the Committee of the Regions on the Application and Effectiveness of the EIA Directive (Directive 85/337/EEC, as amended by Directives 97/11/EC and 2003/35/EC), Brussels, 23.7.2009 COM (2009) 378 final. The review of the Public Participation Directive is due to be submitted in 2009.

126 Ibid. at p.3.

127 Ibid. at p.6.

128 Ibid. at p.7. See also Annex 2 (at p.13) entitled: 'Preliminary observations on the implementation of the new provisions introduced by Directive 2003/3 5/EC', which states in respect to Article 10(a) 'it is not possible to draw any conclusions at this stage on the implementation of these provisions. However, the Commission has already identified transposition problems in some MS and initiated infringement procedures, in accordance with Article 226 EC Treaty.'

129 Ibid. at p.7.

130 Ibid. at p.7.

131 Article 194 TEC also provides the right to petition the European Parliament concerning the application of EU law. The effectiveness of this provision is discussed by the Commission Staff Working Document, see note 121, paragraphs. 6.3.1.3 and 6.3.1.4 at p.142.

132 Case C-435/97, World Wildlife Fund (WWF) v. Autonome Provinz Bozen [1999] ECR I-5613; Case C-201/02, see note 86.

133 Article 234 TEC.

134 Articles 17–22 TEU.

135 For a brief summary of procedure, see L. Kramer, Regional Economic Integration Organizations – The European Union as an Example, in D. Bodansky, J. Brunnee and E. Hey (eds) (2007) The Oxford Handbook of International Environmental Law, pp.853, 868–870 et seq.

136 Article 211 TEC.

137 Launched in April 2008, the intention is to try and expedite the resolution of complaints, although the Commission retains the right to take infringement proceedings where necessary. For evolution of the approach of the Commission in response to complaints, see: Commission of the European Communities, Commission Communication Better Monitoring of the Application of Community Law, Brussels, 16.5.2003 COM (2002) 725 final/4, paragraph 3.2; Commission of the European Communities, Communication from the Commission A Europe of Results – Applying Community Law, Brussels, 5.9.2007 COM (2007) 502 final, paragraphs. 2.2 and 3.

138 P. Roller and L. Cashman, *Implementing EC Environmental Law: Compliance Promotion and Enforcement by the European Commission*, JEEPL 2009 (12), p.1; and L. Kramer (2009) *The Environmental Complaint in EU Law, JEEPL* (12), pp.13, 29 et seq.

139 Article 226(1) TEC.

140 Article 226(2) TEC.

141 Article 220 TEC.

142 Article 228 TEC. See Commission of the European Communities, Communication from the Commission Application of Article 228 of the EC Treaty, Brussels, SEC (2005) 1658.

143 Case C-6 /90 and C-9/90, Francovich and Bonifaci [1991] ECR 1-053 57.

144 L. Kramer, Environmental Justice in the European Court of Justice, in: J. Ebbesson and P. Okowa (eds) (2009) Environmental Justice in Context, pp.195, 199, 207–209 et seq.

145 Article 230(4) TEC. See Court of First Instance, Case T-585/93 Greenpeace and others v. Commission, ECR 1995, p. II-2205, and Court of Justice, Case C-321/95P, Greenpeace and others v. Commission, ECR 1998, p. I-1651. For discussion, see also Kramer, note 138. at 16; E. Biernat, The Locus Standi of Private Applicants under Article 230(4) EC and the Principle of Judicial Protection in the European Community, Jean Monnet Working Paper 12/03, NYU School of Law.

146 ClientEarth, Communication to the Aarhus Convention's Compliance Committee, ACCC/C/2008/32/EC. The detailed arguments of the communication and related documents can be found at www.unece.org/env/pp/compliance/Compliance%20 Committee/32TableEC.htm.

Part III

Convention Law and Practice

7

Transboundary EIA: Iberian Experiences

Rita Albergaria, Teresa Fidelis

EIA AND TRANSBOUNDARY ISSUES, AN INTRODUCTION

Environmental Impact Assessment (EIA) is one of the most critical environmental policy instruments. It emerged in the second half of the 20th century as a response to the growing need to assert environmental values in decision-making processes related to development projects. It took the form of a new analytical approach that makes it possible to evaluate and interpret, prior to decision-making, any potential for negative impacts (Clark and Canter, 1997; Partidário and Pinho, 2000). Public hearings, and related input, constitute integral parts of this evaluation. In principle, the EIA should lead to the discarding of environmentally unacceptable actions and the mitigation – until an acceptable level is reached – of the environmental effects of proposed activities (Sadler, 1996; Wood, 2003).

EIA is defined as 'the process of identifying, predicting, evaluating and mitigating biophysical, social and other pertinent effects of development proposals before basic decisions are made and commitments taken on'.[1] The main objectives of EIA are (Partidário and Jesus, 2003):

- To ensure that the environment is explicitly considered and incorporated into the decision-making process;
- To anticipate and avoid, minimize or compensate for substantial biophysical, social and other pertinent adverse effects;
- To protect the productivity and capacity of natural systems and of the ecological processes that sustain their functions;
- To promote sustainable development that optimizes the use of resources and management opportunities.

Considerations relating to environmental impacts across national borders were first highlighted in the Stockholm Declaration (1972) in which the signatory nations committed themselves to respecting other countries' environments. Some of these principles were further emphasized in the Bruntland Report (1987). This led to the process of drawing up various international agreements and conventions on transboundary impacts, many of which were strengthened as a result of the United Nations Conference on Environment and Development (1992).[2] One of the outcomes was the Espoo Convention on Transboundary Environmental Impact Assessment (TEIA) – signed by 29 countries and by the European Union (EU) (1991). The signing of the Convention integrated the strategic objective of promoting cooperation between countries in preventing or reducing negative transboundary impacts. The European Directive (97/11/CE) introduces inter-state consultations as mandatory. Together with the Espoo Convention, this represents a force for the harmonization of existing national EIA systems (Connelly, 1999; Hildén and Furman, 2001). Although the transboundary issues had already appeared in a number of European documents, the Espoo Convention constitutes a major multilateral treaty stipulating procedural rights and obligations of the parties regarding transboundary impacts and their consideration in the decision-making process (Petts, 1999). Compared to most international agreements on environmental protection, this convention is innovative in that it provides specific routines for notification and consultation. It refines the EIA concept at the level of international legislation and secures the recognition that successful implementation of the EIA in the international context does not depend solely on governments, but also on non-governmental actors (Ebbesson, 1999). Being a convention, it is natural that the terms are presented in a generalist manner. However, implementing it satisfactorily requires the clarification of a number of steps by means of national legislation or administrative measures.

The implementation of the Espoo Convention involves a complex process of technical and participatory assessment of transboundary impacts. The particular approach selected commits the country of origin to involving the public and officials of the affected country. The bi- and multi-lateral agreements between the parties – setting standards, contacts and timelines – act to reduce potential obstacles to the application of the EIA Convention, given that the fundamental issue in the EIA process is to avoid discrimination between affected border areas during the decision-making process (Boer, 1999).

The Iberian experience in applying the rulings of transboundary EIA procedures established by the Espoo Convention is fairly limited. It essentially amounts to four projects which have been subject to consultations between Portugal and Spain:[3]

- An Integrated Study of the Alqueva Project (1995);
- A Project to Link the Portuguese and Spanish Natural Gas Networks – Crossing the Minho River (1997);
- The Hydroelectric Prospects for the Sela Dam (1998);

- A Proposal to Define the Scope of Linking the Castela (Spain) Highway to the IP5 (Portugal) (2003/2004).

In all these cases, the institutional relationship between these neighbouring countries seems to have been formalized merely through the exchange of documents between the Ministries of the Environment and through the diplomatic channels of the Ministry of Foreign Affairs, as required by the European Community (EC) Directive. There is, however, little evidence of any input sent from Spain for inclusion in the final version of the environmental impact statements (EISs).

It was the 'hydroelectric prospects' for the Sela and Alqueva dams that provided the greatest stimulus for consultations with Spanish officials and public (Serrano, 2004). This was largely related to the implied economic scenarios and environmental impacts as well as to their importance for national and peninsular development, especially in the case of Alqueva. The limited discussion on TEIA procedures and the need to consider alternative options, in order to prepare adequate environmental impact statements, justify the relevance of this study. The scarcity of research into how collaboration between entities (on both sides of the border) should proceed, how public input should be integrated and how environmental impacts on both territories should be considered also adds importance to this study.

The fulfillment of adopted national and international legislation on TEIA brings strategic advantages for both countries, as is shown in the decision-making processes associated with development projects with transboundary environmental impacts. In this context, the chapter aims to critically analyse legal and procedural features of bilateral cooperation through the comparison of two case studies related to water management projects (Sela and Alqueva dams). The chapter highlights procedural weaknesses and puts forward a 'Good Practice Model' for cooperation between Portugal and Spain during TEIA processes.

THE ESPOO CONVENTION

The Espoo Convention was promoted by the United Nations Economic Council on Europe (UN-ECE) and signed in Finland by 29 countries and the EU – including Portugal (April 6, 2000) and Spain (September 1, 1997). In terms of EIA in a transborder context (Petts, 1999), the Convention:

- stipulates the obligations of the parties to assess environmental impacts at an early planning stage;
- lays out means and procedures for avoiding, controlling or reducing any negative effect on the environment, especially those that are transborder in nature;

- clarifies that the impact assessment process must be carried out by the party responsible for causing said impacts.

Ten years after the initial UN initiative, the Convention had been signed by nearly 35 countries and the entire EU had ratified the Espoo Convention. However, there has been little experience in applying it. Most of the signatory countries referred to the Convention once or twice but few informed their neighbours of potential impacts in any systematic way. Only the Netherlands really made use of the Convention in carrying out approximately 10 TEIA studies.

The phases to be followed under the Espoo Convention include a series of steps as described in Figure 7.1. Once the transboundary nature of a proposed project has been defined – by the originating or affected country – there must be an initial exchange of contacts and documents (Article 3). The originating party must provide the affected party (officials and public) with all of the EIA documentation: EIS; non-technical summary; other studies (Article 4). The most important stage is that of the 'consultations': during which the countries discuss

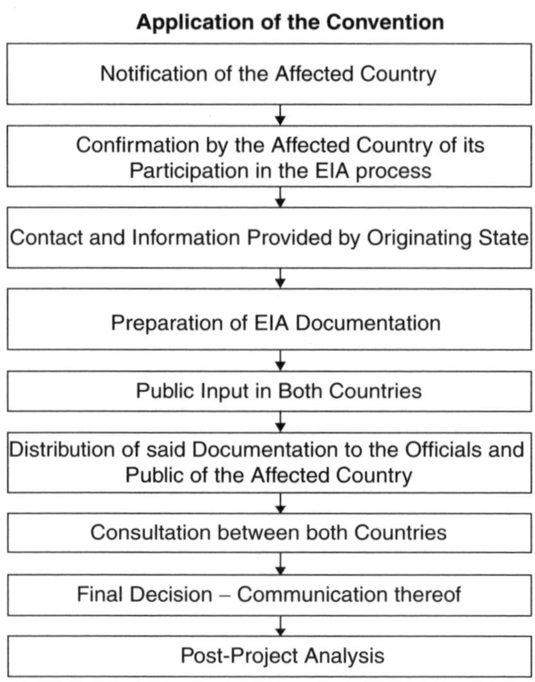

Figure 7.1 *Espoo Convention implementation strategy.*

Source: Convention on Environmental Impact Assessment in a Transboundary Context, United Nations Economic Council on Europe, 2003, www.unece.org/env/eia

together the significance of the impacts (Article 5). In this phase, the affected country has the chance to express its disagreement with specific aspects of the project that could affect it negatively. These consultations may take the form of negotiations but with no obligation to reach a common solution. The Convention is clear on this point: the final decision rests with the originating country.

One of the biggest challenges in applying the Convention is that of 'public input': both parties have the responsibility of ensuring that there is public input from the potentially affected areas (Article 3.8). The Espoo Convention requires that public opinions and objections be included in the final decision (Article 6.1). In practice, applying the articles of the Espoo Convention may create difficulties that are explicit, notably in:

- The allotment of tasks and costs:
- which entities should be involved?
- how are contacts made and how are joint efforts defined?
- which methods of impact assessment and public input should be adopted?
- who pays for translations and public hearings?
- Evaluation of the significance of the impacts:
- how to deal with the subjectivity of value judgment of the countries involved?

To overcome these difficulties, the Espoo Convention requires the adoption of particular measures (Tesli and Husby, 1999; Ebbesson, 1999; Boer, 1999; Hildén and Furman, 2001). One of these refers to the establishment of bilateral agreements between the parties in order to determine a uniform process with the definition of stages, intervening agents and methods for impact assessment and public input. A further measure refers to the importance of cooperation between regional governments and those agents directly responsible for the affected territory and population. Finally, another refers to the inclusion of the opinions of the public and/or any directly affected entities at an earlier stage of the EIA process.

EC DIRECTIVES AND NATIONAL EIA LEGISLATION

At the EC level, transboundary factors surface in Directive 85/337/CEE (Article 7): 'Where a Member State is aware that a project is likely to have significant effects on the environment in another Member State or where a Member State likely to be significantly affected so requests, the Member State in whose territory the project is intended to be carried out shall forward the information gathered pursuant to Article 5 to the other Member State at the same time as it makes it available to its own nationals. Such information shall serve as a basis for any consultations necessary in the framework of the bilateral relations between two Member States on a reciprocal and equivalent basis'. However, the rulings of the

Espoo Convention are only incorporated with the publication of Directive 97/11/EC, which introduces the following obligations in its Article 13: 'The Member State on whose territory the project is expected to be carried out must send information on the project to those Member States to be affected by transborder impacts as soon as possible and no later than when informing its own citizens'. In the event that the affected state wishes to participate in the EIA process, the Directive requires it to make information on the project available to relevant public officials and to the public. The Member States shall consult, reciprocally, on the potential transborder impacts. Portugal and Spain incorporated the EC directives into their own legal systems. The main measures established in the national legislation of these two countries include the aspects described below.

In Portugal, the content of the legislation (Decree-Law 69/2000, 3 May, with the changes introduced by the Decree-Law 197/2005, 8 November) states that the EIA authority (Environmental Institute) shall take steps such that the Ministry of Foreign Affairs and the governmental officials of the affected country shall receive a description of the project, accompanied by preliminary information on the nature of the decision. The affected country shall have 30 days to declare its interest in participating in the consultation process. In the case of an affirmative response, all those components subject to public dissemination should be made freely available. These are: the EIS, the Non-technical Summary, the Public Hearing Report, the Final Report of the Evaluation Commission, the Declaration of Environmental Impact and the Authorizing Decision. The conclusions from the public input of the affected country shall be considered by the Evaluation Commission in its final report.

In Spain, the content of legislation (Law 6/2001 8 May, which has introduced changes to the Royal Decree-Law 1302/1986, 28 June) states that whenever a project may have significant environmental repercussions for another EU member state, the procedures governed by the Espoo Convention shall be followed. It also states that the environmental agency participating in the project EIA shall contact the affected country through the Ministry of Foreign Affairs. The Autonomous Communities legislate the EIA through specific decrees (Galicia: Decree 442/190, 13 September; and Extremadura: Decree 45/1991, 16 April), but they do not consider the transboundary question, which belongs to the Spanish Central Administration.

The case studies

The sharing of water resources has been one of the most delicate issues in the relationship between Portugal and Spain, because 40 per cent of the water resources in Portugal come from Spain. Mutual acknowledgement of the need to solve the shared water transboundary problems has opened the way for a growing cooperation between the Portuguese and Spanish authorities, making this subject an interesting and useful field for research. Of the list of projects submitted to

EIA procedures and involving transboundary environmental impacts, as referred to in Section 1, only two, the Sela and Alqueva dams, were related to the management of water resources. This fact, their Iberian strategic relevance and their comprehensiveness and procedural relevance for EIA transboundary purposes have substantiated their selection for this research.

The Sela dam

Portugal and Spain signed the Luso–Spanish International Convention in 1968 at a time when the building of a dam on the shared basin of the Minho River was being planned. Construction contracts for the dam were awarded to the following companies: EDP (Portugal) and Unión Fenosa (Spain). The EIA process consisted of an EIS (prepared by the twin-users EDP/Unión Fenosa), EIS Technical Assessment (carried out by a bilateral Luso–Spanish commission), and of the public contributions from both countries. Figure 7.2 shows the EIA procedure and its transboundary stage. The Final Report was unfavourable to the project due to the perceived negative impacts. The Bilateral Commission, whose composition is described, was of the opinion that the area subject to the study would be particularly affected in terms of nature conservation and cultural heritage. The initial cooperation between Portugal and Spain, which included the

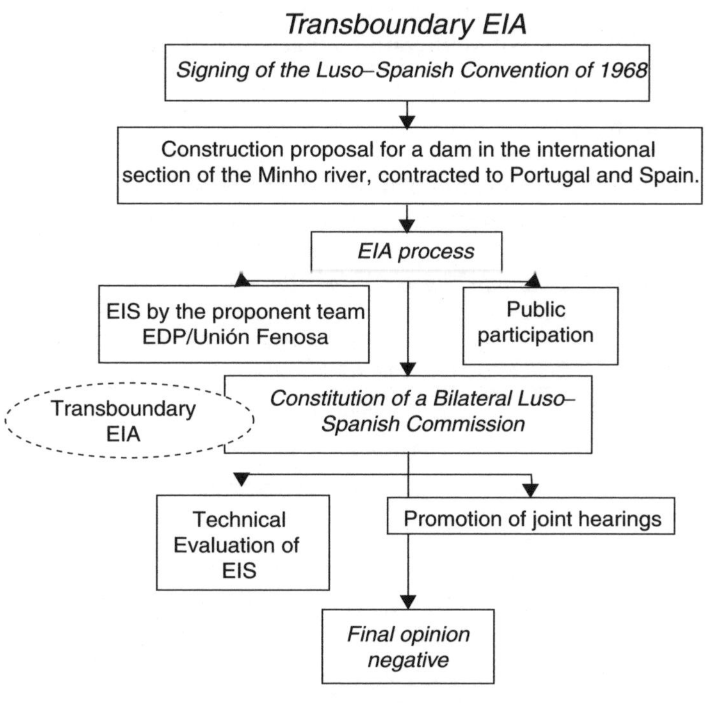

Figure 7.2 *EIA procedure for the Sela dam and its transboundary stage.*

Table 7.1 *Organizations involved in the Bilateral Working Group*

Portugal	*Spain*
Instituto Nacional da Água (INAG) [National Water Institute] Direção Regional do Ambiente e Ordenamento do Território do Norte (DRAOT-N) [Regional Department of the Environment and Land Use for the North] Instituto de Promoção Ambiental (IPAMB) [Institute of Environmental Promotion] Instituto de Conservação da Natureza (ICN) [Institute for the Conservation of Nature]	Dirección General Obras Hidráulicas y Calidad de las Aguas (DGOHCA) [Central Department of Water Works and Water Quality] Dirección General de Calidad y Evaluación Ambiental (DGQEA) [Central Department of Environmental Quality and Assessment] Confederación Hidrográfica del Norte (CHN) [Water Consortium for the North]

Source: Bilateral Working Group Report Luso–Spanish Commission to Regulate the Use and Development of International Rivers in Border Areas (1998).

involvement of the contractor companies, in the assessment of the environmental impacts of the Sela dam resulted in the joint and unified development of the EIS (Madrid, 1992). In order to monitor and validate the entire EIA process, a Bilateral Working Group was set up under the aegis of the Luso–Spanish Convention (Madrid, 1996). Table 7.1 shows the organizations that constituted the Bilateral Luso–Spanish Commission. Table 7.2 shows details of its scope and Table 7.3 shows examples of deliberations that resulted from the joint meetings.

The Alqueva dam

In 1968, the Luso–Spanish International Convention assigned the hydraulic development of the international basin of the Guadiana River, where the Alqueva

Table 7.2 *Actions taken by the Bilateral Working Group*

Tuy, 1997	*Comparison of standard hierarchical procedures in Portugal and Spain*
Valença do Minho, 1997	Visits in situ
	Analysis of EIS contents
	Request for additional information
Madrid, 1997	Visits in situ
	Schedule of public hearings
	Analysis of additional documentation
Melgaço, 1998	Determination of the organization of public hearings
Lisboa, 1998	Final non-compliance report

Source: Bilateral Working Group Report Luso–Spanish Commission to Regulate the Use and Development of International Rivers in Border Areas (1998).

Table 7.3 *Examples of deliberations resulting from the joint meetings*

Meeting in Valenfa, Portugal, 07/05/1997
"(...)The Portuguese party said that compensation on the Spanish side included not only lands but also commerce and small industry. These criteria would be applied differently in Portugal, as the compensation would concern only the land that was expropriated for the construction and exploitation of the dam (...)"
"(...) As regards the consultation of the general public, the Portuguese party made a brief statement on the way things were supposed to evolve in Portugal, and only after that was the same thing done on the Spanish side (...)"
Meeting in Madrid, Spain, 14/11/1997
• "(...) As regards the measures that are to be taken in terms of public consultation, the measures taken by the two countries were re-evaluated and, after reflecting on the issue, the two delegations agreed to: • Begin the consultation on the 12th January 1998, if all participants are available, and continue until mid-February • Make arrangements for simultaneous information and consultation dates in both countries, whenever possible • Make the EIA on the Portuguese side available in all City Halls and distribute the Non-Technical Summary to all parishes in the surrounding area • Conduct two public hearings in Melgaço and Monção, with the presence of a Spanish delegation • Create a leaflet in Portuguese and Galician to publicise the public consultation and to then conduct hearings • Arrange a preparatory meeting of the bilateral group with the concessionaires, on the 28th (...)"

Source: Bilateral Working Group Report Luso–Spanish Commission to Regulate the Use and Development of International Rivers in Border Areas (1998).

dam was due to be built, to Portugal. In 1995, after years of indecision involving a series of advances and reversals in project evaluation, the Sociedade de Estudos de Impactes Ambientais (SEIA) [the Environmental Impacts Study Company] carried out the Integrated Environmental Impact Study (IEIS) for Alqueva. The EIA for Alqueva, described in Figure 7.3, integrated the IEIS, the technical assessment of the IEIS (Commissão de Avaliação Nacional – National Assessment Commission), the public consultation (hearings in Portugal and Spain) and a 'Technical Report' produced by the Ministry of Public Works, Transportation and the Environment (Department of Environmental Policy) in Spain. This last document conditions the positive final report to the presentation of an environmental management programme with an eye towards the programming and allocation of resources in order to provide ways of minimizing and compensating for negative impacts. For operational purposes, two entities were created:

- EDIA (Empresa de Infra-Estruturas de Alqueva) [The Alqueva Infrastructures Company]

- CAIA (Comissão de Acompanhamento Ambiental das Infra-Estruturas de Alqueva) [The Alqueva Infrastructure Environmental Monitoring Company].

With the aim of complying with the programme for minimizing impacts on the land lying on the left bank of the Guadiana (the Spanish region), three joint Luso–Spanish groups were established. One coordinating group was in charge of choosing the members of the working groups, defining methodologies and approving and monitoring the programme for the implementation of minimization measures. Another working group, focusing on environmental issues was in charge of locating and detailing the most significant environmental impacts (as already identified in the IEIS, SEIA, 1995) and defining the corresponding minimization measures. Finally, another group, was established to deal with heritage issues. It was empowered to engage in systematic prospecting in the area subject to intervention and to develop an overall heritage reference framework as well as to take appropriate 'rescue' actions. The composition of these groups is described in Tables 7.4–7.6. Some examples of actions decided on in the joint meetings are described in Table 7.7.

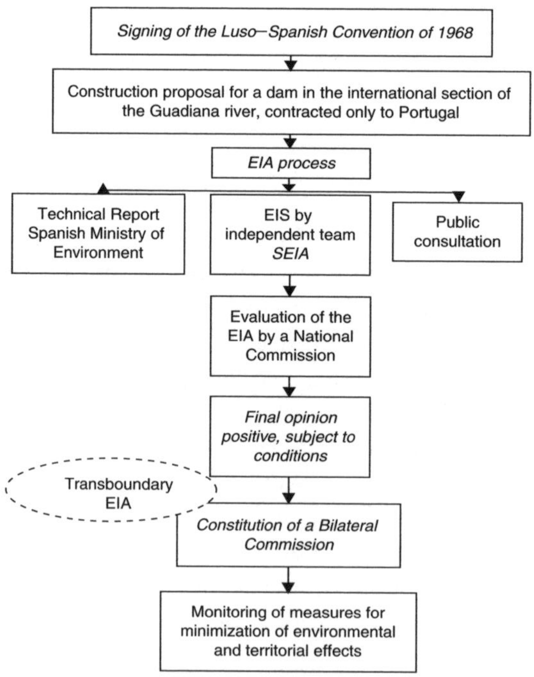

Figure 7.3 *Flowchart that schematizes the EIA procedure of the Alqueva dam and its transboundary stage.*

Table 7.4 *Composition of the Steering Group*

Portuguese representatives	*Spanish representatives*
Instituto Nacional da Água (INAG) [National Water Institute]	General Department of the Environment Dirección General de Calidad y Evaluación Ambiental (DGQEA) [Central Department of Environmental Quality and Assessment]
Instituto Português de Arqueologia (IPA) [Portuguese Institute of Archaeology]	Dirección General Obras Hidráulicas Y Calidad de las Aguas (DGOHCA) [Central Department of Water Works and Water Quality]
Empresa de Infra-Estruturas de Alqueva (EDIA) [The Alqueva Infrastructures Company]	Junta de Extremadura [The Extremadura Council]
Direcção Regional do Ambiente e Ordenamento do Território do Alentejo (DRAOT-Alentejo) [Regional Department of the Environment and Land Use for the Alentejo]	

Source: Minutes of the Meetings of the Programme to Minimize Impacts on the Environment and Patrimony for the Territories of Badajoz, Olivença, Alconchel, Cheles and Villanueva del Fresno (1997–2001)

Table 7.5 *Environmental Working Group*

Portuguese representatives	*Spanish representatives*
Direcção Regional do Ambiente e Ordenamento do Território do Alentejo (DRAOT-Alentejo) [Regional Department of the Environment and Land Use for the Alentejo]	Department of the Environment
Instituto de Conservação da Natureza (ICN) [Institute for the Conservation of Nature]	Junta de Extremadura [The Extremadura Council]
Empresa de Infra-Estruturas do Alqueva (EDIA) [The Alqueva Infrastructures Company]	University of Badajoz
–	Dirección General de Calidad y Evaluación Ambiental (DGQEA) [Central Department of Environmental Quality and Assessment]

Source: Minutes of the Meetings of the Program to Minimize Impacts on the Environment and Patrimony for the Territories of Badajoz, Olivença, Alconchel, Cheles and Villanueva del Fresno (1997–2001).

CRITICAL ANALYSIS OF THE CASE STUDIES

The conclusion to be drawn from the two case studies points to potentials and weaknesses in the assessment procedures for projects that have impacts in Portugal and Spain. In both cases, the transboundary process followed, but went little beyond, those obligations stipulated by the first European Directive. The

Table 7.6 *Heritage Working Group*

Portuguese representatives	*Spanish representatives*
Instituto Português de Arqueologia (IPA) [Portuguese Institute of Archaeology]	Consejería de Cultura e Patrimonio de la Junta de Extremadura [Culture and Patrimony Advisory Group of the Extremadura Council]
Empresa de Infra-Estruturas de Alqueva (EDIA) [The Alqueva Infrastructures Company]	Junta de Extremadura [The Extremadura Council]

Source: Minutes of the Meetings of the Program to Minimize Impacts on the Environment and Patrimony for the Territories of Badajoz, Olivença, Alconchel, Cheles and Villanueva del Fresno (1997–2001).

Table 7.7 *Examples of deliberations resulting from the joint meetings*

Meeting in Alqueva, Portugal, 25/05/1999

"(...) The Portuguese party was in charge of sending their Spanish counterparts the following documents: Environmental Management Programme; Business Plan (...)"

"(...) The Reference Terms for Cultural Heritage will be developed by the Spanish party; (...) the Spanish party will provide the list of the competent entities according to the subject that is to be consulted (...)"

"(...) The EDIA will send the Reference Terms of the works on the Natural Heritage to all participants through the President of the Coordinating Group, Eng. Pedro Serra, (...)"

Meeting in Alqueva, Portugal, 2000

10 March – The minutes were sent to the Spanish party to be analysed.

05 April – The president of the Portuguese delegation on the Luso-Spanish Commission received the minutes of the meeting that took place in Mérida, in February.

14 April – The Direccion General de Patrimonio Cultural received the originals of the "Technical Evaluation of the Proposals" documents for signature.

Meeting in Alqueva, Portugal, 2001

8 February – The INAG has received the documentation it has requested from Spain on the previous item and the Spanish party calls for a meeting of the Environmental Working Group.

27 March – The INAG has received and sent to the Director General de Patrimonio Cultural de la Junta de Extremadura, a copy in Spanish of the Report of Archaeological Prospecting carried out in the territories affected by the Alqueva Backwater located in Badajoz, Olivença, Cheles, Alconchel and Villanueva del Fresno, conducted by the company ERA – Arqueologia, Lda. On the same date, the scheduling of a work meeting was suggested, to be held on the 17th April, to analyse the same report.

17 April – Meeting of the Working Group of Cultural Heritage, in Badajoz, to analyse the report on archaeological prospecting in the territories affected by the Alqueva Backwater, located on the left bank of the Guadiana River, with the following agenda:

- Identification of the minimization actions that are to be carried out
- Definition of a methodology to concretize those actions.

Source: Minutes of the Meetings of the Programme to Minimize Impacts on the Environment and Patrimony for the Territories of Badajoz, Olivença, Alconchel, Cheles and Villanueva del Fresno (1997–2001).

Portuguese Ministry of the Environment communicated to its Spanish equivalent, via the Ministry of Foreign Relations or 'External Affairs', the contents of the EIS procedures carried out on its territory.

In the case of Sela, cooperation between the countries showed a positive level of development over time. The organized and systematic performance of a Bilateral Commission allowed entities from both countries to jointly evaluate the feasibility of the project. The Portuguese authorities were also guided by the Convention on the Use of International Luso–Spanish Rivers Basins (1968) given that the Minho River basin belongs to both countries. Public participation also brought in interested parties from both countries and took account of all the requirements imposed by the legislation. It was, however, diligent action by affected towns and non-governmental environmental organizations that instigated the involvement of the relevant institutions in the shutting down of the dam building process, thus avoiding any negative consequences for the region.

In Alqueva, the meeting between officials and technicians from both countries, which was designed to jointly identify impacts and propose means of mitigation, was only held after the completion of the (conditionally favourable) final report. This took place under the aegis of the Joint Commission and was designed to evaluate the means of minimizing environmental and heritage impacts on Spanish territory. Collaboration between the two countries for the purposes of elaborating the preliminary impact assessment was limited to a mere formal exchange, through diplomatic channels, of documents between the two Ministries of the Environment.

A comparison of practices for the implementation of the transboundary legislation in force (The Espoo Convention and the European Directive of 1997) in both case studies shows that the EIA of the Sela dam was exemplary in fulfilling many of the requisite conditions. In this case, the importance give to the evaluation of the EIA process by the Bilateral Commission must be called into question given that this resulted in a decision to reject dam construction despite the qualified approval contained in the evaluation report.

The transboundary cooperation at EIA level involves significantly more than the mere formal exchange of documentation and agreements on a diplomatic level. The need for such cooperation was established by the signing of the Espoo International Convention, which was also adopted by Portugal and Spain. This is the first multilateral treaty to impose a set of rights and obligations on all parties, in the case of projects with transboundary impacts. Its major innovation has to do with the fact that the originating country is obliged to involve both authorities and the affected population in their own EIA process. It proposes agreements between the central and regional administrations of the countries involved with the creation of joint bodies and taking advantage of pre-existing institutions. It attaches great importance to transboundary contacts at a regional level, encouraging the parties to establish agreements to ensure its full implementation within a limited geographical area.

From the detailed analysis of the EIA of the two case studies, it is possible to summarize the following good practices, which indicate that the countries have already acknowledged the benefits of transboundary cooperation at EIA level:

1(a) Sela Dam

- The evaluation of EIS and promotion of public participation was undertaken by a bilateral team which also determined the final decision (constituted in the context of the Luso–Spanish Convention of 1968). Luso–Spanish cooperation existed before the decision was taken. This allowed for the technical conformity of the EIS to be evaluated by both countries and for the public participation phase to include joint discussions on project impacts involving both Portuguese and Spanish citizens. In this case study, the TEIS was not limited to the mere exchange of EIA documentation given that institutional relations were complemented by proactive cooperation.
- Impact identification was carried out by the Luso–Spanish working groups before approval of EIS content. The bilateral team analysed the EIA documentation and formed Luso–Spanish working groups to identify impacts and carry out in situ investigations into the degree of conformity of the contents of the EIS.

1(b) Alqueva Dam

- There was an exchange of inter-ministerial documentation that followed the transboundary procedure stipulated in the 1985 EC Directive. This allowed the affected country to analyse the EIA documentation and evaluate the importance of any impacts on its territory.
- The IEIS was devised by a team set up by the European Commission (SEIA – Sociedade de Estudos de Impactes Ambiental). Good practice dictates that it should not be the scheme proposer that carries out the environmental viability study for the project. Given that the proposer has a vested interest in the project, their opinion must certainly be taken into account. In this case, the EIS was carried out by a team independent of the proposing party.
- There was bilateral cooperation for the monitoring of the mitigation measures, in the form of mixed working groups.

Good practice weaknesses were found in the following areas:

1(a) Sela Dam

- Non-involvement, at an early stage, of regional/local entities in the EIA procedures. Luso–Spanish cooperation did take place at the highest level but only involved central administrative bodies. Those interested parties who directly administer the project area were not consulted.

- Public participation promoted only in Portugal (origin country): Spanish representatives were present. Members of the Spanish public were obliged to go to Portugal in order to take part in public hearings on an Iberian project. There should have been separate hearings held in each territory.
- Carrying out of the EIS by a proposing team (EDP/UF). It is not considered good practice to have the environmental viability study carried out by the project proposer.
- Public consultation after the EIS, raising opinions that blocked the process. The opinions of local bodies and civil society (city halls, producers' associations, nature conversation associations, business people and others) should have been taken into account before this stage. If such opinion gathering takes place in parallel it may lead to differences critical to the closure of the EIA process.

1(b) Alqueva Dam

- Absence of bilateral cooperation in the evaluation of transboundary impacts. This happened only after the final opinion was settled. There was no opportunity to consider alternatives together with adequate minimization measures for the affected areas. In such an important project, the post-project phase needs to be able to relay on strategic planning in order to guarantee success.
- Absence of working groups (at a regional level) to carry out research studies leading to decision-making proposals. At an earlier stage, local bodies involved in development planning for the area and affected members of the public should be brought into the process, and their opinions should be incorporated into the EIS.

Good practice model for TEIA (Portugal/Spain)

Critical analysis of the two case studies allowed us to synthesize potentials and weaknesses in TEIA procedures, as evidenced in these two cases that involve Portugal and Spain. The features highlighted in section 2 led to the proposal of a Good Practice Model designed to improve effective implementation of the Espoo Convention in similar future cases.

The Good Practice Model, schematized in Figure 7.4, integrates, in a single procedure, the bilateral consultation process for impact assessment between two countries: Portugal and Spain. It aims to comply, objectively, with the current requirements of the Espoo Convention and EC Directive requirements on TEIA. It also includes refinements to the strategies used in the case studies. Thus, for conducting TEIA processes, we propose to formalize a 'Permanent Bilateral Body' that would work to facilitate regional Iberian Peninsula relations and to ensure greater efficiency in the implementation of EIA procedures. As for the remaining entities we suggest that, for each level of decision-making, responsibilities, and related deadlines, regarding TEIA processes should be clearly defined. The 'National Government', which initiates contacts with the neighbouring country, will, naturally, issue the final decision, taking into consideration the most recent EIA process report, national interests, international relations and good neighbour policy. Subsequently, the EIS is carried out, shared with specific national bodies[4] and the proposing party and then submitted to regional governments.[5] The 'Regional Body', concurrently with contact being established with the Spanish counterpart team,[6] undertakes consultations for initial assessments of the baseline status and potential impacts of the project:

1. Preliminary consultations should be wide open[7] and a scoping phase should be mandatory. The participation of local entities,[8] at an early stage, will mean increased efficiency and transparency, avoiding, thus, parallel and late-delivery reports that often block the process.
2. The initial assessment must be carried out by a 'Luso–Spanish Working Group' (with note-takers) that will issue a report. This report and the results of the consultations will be integrated into the EIS.[9]

The regional-level mission ends with the evaluation of the EIS and issuance of a new report. The National Government is responsible for the final decision, following analysis of the regional report. It takes into consideration the main points of the project to be evaluated, eliminating bureaucracy wherever possible. The proposal for inter-departmental evaluations with similar weight is based on the conviction that to increase efficacy in decision-making it is necessary to bring together a determinant range of knowledge including environmental, societal and economic perspectives. The final decision is to be made by the Ministry of the Environment of the originating country and shall be binding.

The main features of the Good Practice Model relate to the following issues:

1. Selection of Projects and Notification – As established in the Espoo Convention (Article 2 and 3), and in the specific case of hydroelectric projects, the country of origin is in charge of defining the transboundary nature of the project (as listed in Appendix 1 of the Espoo Convention for projects that cause significant impacts on another country). However, whenever the affected country becomes aware of a project with impacts on its territory, it should notify the country of origin, even if it is uncertain of its significance (Article 2 Espoo Convention). This first stage represents the starting point for discussion between proponent/country of origin/affected

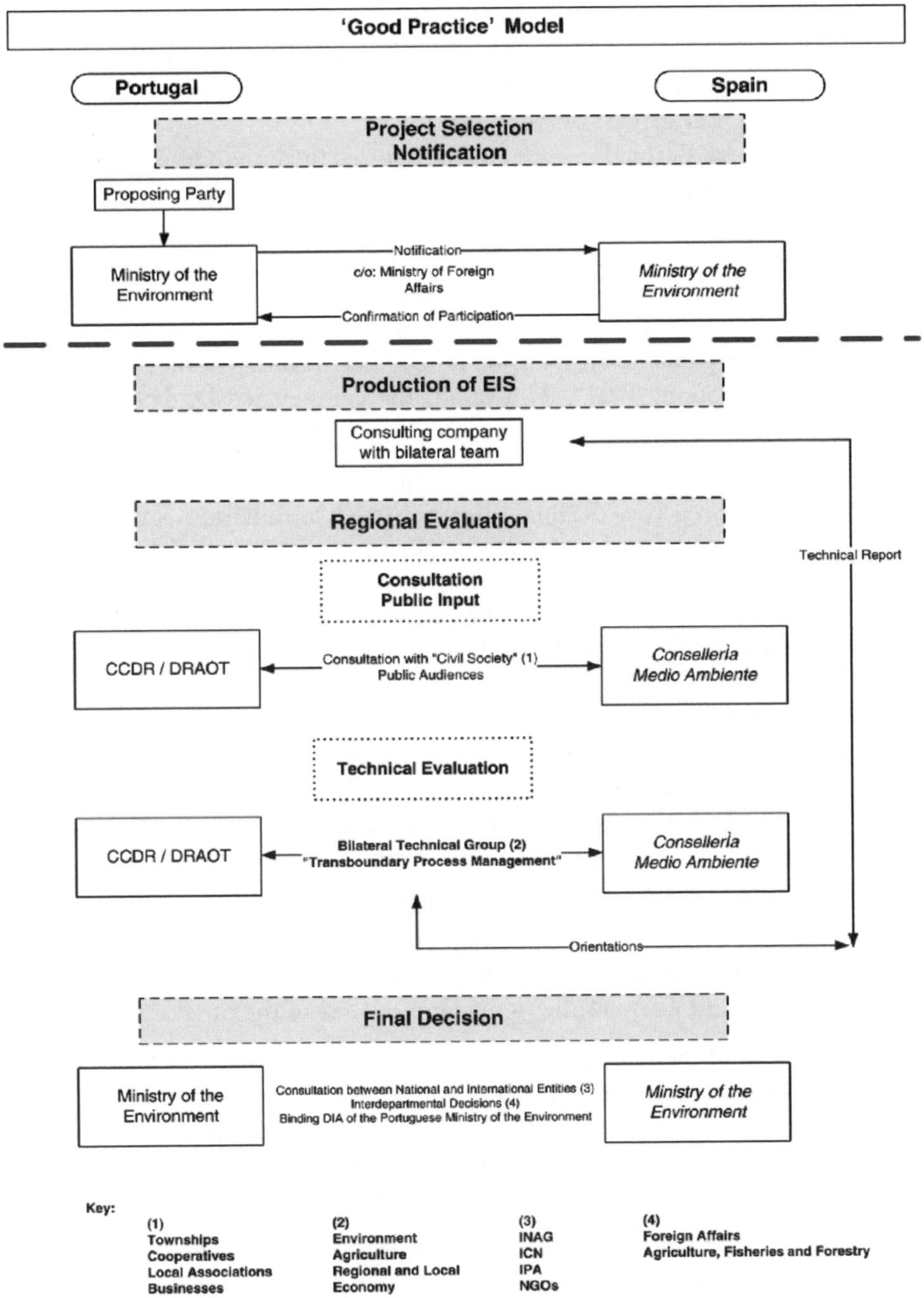

Figure 7.4 *Good Practice Model.*

country, through the Ministries of Environment and Foreign Affairs (Article 33 DL 69/2000; Article 6 Ley 6/2001). The parties should exchange prior information on the project and potential significant impacts on the affected area (Article 3 of Espoo Convention).

2. Initial Regional Evaluation – Having recognized the importance of evaluating impacts on the affected area of the project, the regional entities should carry out the procedures of initial technical evaluation, EIS evaluation and public participation. After communication has taken place between the aforementioned central entities, each country sends the above information on the project to the regional entities, thus initiating the process of evaluation of impacts at a regional level (Decision II/8, second meeting of the Parties of the Espoo Convention, 2001). As a result, the competent EIA departments at regional level[10] should:

   - consult entities directly affected by the project or who may have relevant opinions on it (Ley 6/2001 –'initiation and consultation' stage);
   - make an initial evaluation of the situation and potential impacts. The successful examples of the bilateral working groups, created to make the transboundary evaluation of the Sela (before the decision) and Alqueva (after the final opinion) dams, should be taken into account in future cases. The need for similar working groups and the appointment of a person to be in charge at the highest regional level – a 'Transboundary Processes Manager' (who would have relevant experience in integrated management of similar processes) should be systematically considered. There should be periodical meetings between the regional administrations, presided over by the head of the delegation from the territory in which they are conducted. Working languages should be Portuguese and Spanish.

3. Formulation of the EIS – The company/institution that wins the public tender for carrying out of the EIS should proceed to the full evaluation – independent and free – of the significant impacts of the project:

   - for both of the territories
   - for the various factors: environment/economy/society.
   - Its technical team, preferably Luso–Spanish, should analyse each of the items included in the technical specifications, using the same evaluation criteria on both sides of the border.

4. Public Participation and Evaluation of the EIS – The final evaluation of the EIS and the promotion of public participation is the final task of the Luso–Spanish working group at regional level. According to the explicit indications of the Espoo Convention (Article 3 and 4), the public participation phase must be promoted by both parties and take place before the decision-making process. Both countries should reach an agreement on tasks and the shared costs implied for this stage:

- Joint hearings promoted in both countries
- Publicity in the press, radios and public places
- Formulation of the 'Public Consultation Report.

5. Evaluation of the EIS – Both parties should be obliged to make a bilateral consultation on the significance of impacts (Article 4 and 5 Espoo Convention), and therefore the regional bilateral team should confirm the credibility and technical accurateness of the EIS, ensuring that it includes the opinions of those entities consulted. It should also evaluate the descriptors of the most significant effects, and present alternative proposals and minimization measures.
6. Final Decision – The process goes back to the central authorities (Environment Institute) in the form of a technical document, which must include all documents resulting from the public consultation phase. In this stage, the regional results and the EIS will be evaluated. The degree to which the facts conform to national and international legislation will be verified, as will the issue of national interest and/or security, according to the state of relations between the states. Those aspects relating to the balance between development and environmental protection should be taken into account at a national level (Ministry of Environment), as should any issues concerning good neighbourly relations and as raised by the Ministry of Foreign Affairs. The final decision should include the results of the bilateral evaluation and of the joint public participation, and is announced by the Ministry of Environment of the country of origin. It should be legally binding (Article 6 of Espoo Convention).
7. Post-evaluation – This is the stage of implementation of the mitigation measures proposed in the EIS, with the purpose of avoiding, minimizing or compensating negative impacts and augmenting positive impacts. In this stage, joint, systematic/periodical activities should be conducted to monitor the environmental and social–economic variables, in space and time (the joint leadership of this stage is not compulsory, but recommended) (Article 7 Espoo Convention).

CONCLUSIONS

In the context of water resources cooperation between Portugal and Spain, which includes the sharing of five hydrological basins covering about 45 per cent of the Iberian landmass, two case studies in the field of hydroelectricity were selected for evaluation – the Sela and Alqueva dams. These are both transboundary projects, because they are located on the international sections of the Minho and Guadiana rivers, respectively. In the case of Sela, the evaluation of the EIS by a Bilateral Commission was based on the Iberian partnership that promoted the project (EDP + Unión Fenosa). In Alqueva, the proponent is exclusively Portuguese. However, both these projects cause transboundary impacts in Spain and, therefore, should have been submitted to a transboundary EIA.

The key regulation covering this field, TEIA, is the Espoo Convention. This document, signed at an international level, goes far beyond EC Directives as it implies the organization of a complex procedure of notification and bilateral consultation on the significance of transboundary impacts. The highly innovative Espoo Convention proposes agreements between the central and regional administrations of the countries involved, as well as the objective definition of contact points between existing national institutions. It recommends the joint and early promotion of public participation, and it intends that the final decision includes the comments of both the general public and the authorities of the affected country.

The empirical study of the two cases and its comparison with current transboundary legislation, revealed a gap between the practices adopted by the EIA in Iberia and the formal requirements adopted at the Espoo Convention. This study revealed limitations but also some potentials. The limitations of transboundary cooperation as regards EIA in the two case studies included:

- The exchange of inter-ministerial documentation where the joint evaluation of transboundary impacts is not proactive. If it had been, the final resolution would have been more participatory, accurate and fair.
- The absence of consultation of relevant entities, especially at a regional/local level and the general public that is affected by the project (Luso–Spanish scoping).
- The carrying out of the EIS by the proponents themselves and not by independent teams, preferably Luso–Spanish in make-up.
- The tendency to overestimate national interests, good neighbourly relations and the filtration of regional/local needs.

The potentials of transboundary cooperation evident in the EIA of the two case studies included the constitution of a bilateral team (Portugal/Spain) with representation of regional and local entities (concerned with water, environment, town planning, nature conservation and environmental promotion). This team planned tasks, shared responsibilities, met periodically, performed ground research, conducted joint hearings, all with the purpose of evaluating the EIS and reflecting on the significance of the projects' impacts by considering the affected area in both countries.

In order to consolidate the results of the critical and comparative analysis – case studies vs. transboundary legislation – the relevant good practice was systematized in one single procedure for the evaluation of transboundary projects, taking the Espoo Convention fully into account. The proposed Model of Good Practices provides objectivity, efficiency and transparency to the TEIA procedure. It proposes a development that is in accordance with the current legislation and ensures the right level of bilateral evaluation (Portugal/Spain). It is also accurate, impartial and equitable. This implies, in turn, that better project alternatives are designed, adequate measures for impacts minimization are proposed and the actual construction embodies maximum sustainability. In outline, the main steps include the need to:

- Institutionalize a 'Luso–Spanish Working Group' for the EIA process within the Iberian Peninsula;
- Institutionalize the 'regional level of cooperation for transboundary EIA';
- Formally choose someone responsible for managing the transboundary processes who has recognized competence to articulate and create consensus for extremely complex processes;
- Institutionalize a deadline for public participation, namely from local, regional bodies and the general public.

Despite its objectiveness, the operationalization of the proposed procedures may come across a number of obstacles. These obstacles relate to increased complexity of the EIA process, to new stages and new participants. The transboundary nature of the EIA means focusing on a wider impact area and, as a result, the need to take a higher number of descriptors into account. Resultant increases in bureaucracy, cost and time taken for the decision process will be unavoidable difficulties when these have not been previously distributed correctly by the parties involved. However, the increasing importance of transboundary issues in EIA requires a correspondingly increased effort to be made. Pilot EIA procedures using the proposed model could help to improve current practices and consequently, improve Luso–Spanish relations in the context of the EIA.

Acknowledgements

The authors are grateful to the officials from central EIA departments of Portugal and Spain for sharing their experience and allowing the collection of information on the case studies, to the Coordinators of the Master Programme on Environmental Policies and Management for the financial support to present a draft version of this paper on the Conference 'International Experience and Perspectives in SEA', International Association for Impact Assessment, University of Agriculture, Prague, Czech Republic, and to Tim Oswald for his comments and recommendations to improve our text.

References

Bilateral Working Group Report Luso–Spanish Commission to Regulate the Use and Development of International Rivers in Border Areas, 1998. Assessment of Environmental Impact of Hydroelectric Use at Sela on the Minho River. Lisbon.

Boer J. (1999) Bilateral agreements for the application of the UN-ECE convention on EIA in a transboundary context. Environ Impact Asses Rev (19):85–98.

Clark R, and Canter L. (1997) Environment policy and NEPA, past, present and future. Boca, FL: St. Lucie Press. Connelly R. (1999) The UN convention on EIA in a transboundary context: an historical perspective. Environ Impact Asses Rev (19):37–46.

Ebbesson J. (1999) Innovative elements and expected effectiveness of the 1991 EIA convention. Environ Impact Asses Rev (19):47–55.

Hildén M, and Furman E. (2001) Assessment across borders. Stumbling blocks and options in the practical implementation of the Espoo convention. Environ Impact Asses Rev (21):537–51.

Minutes of the Meetings of the Program to Minimize Impacts on the Environment and Patrimony for the Territories of Badajoz, Olivença, Alconchel, Cheles and Villanueva del Fresno. 1997–2001. CAIA – The Alqueva Infrastructure Environmental Monitoring Company.

Partidário, M., and Jesus, J. (2003) Rudiments of Environmental Impact Assessment. Universidade Aberta.

Partidário M, and Pinho P. (2000) Handbook for the new environmental impact assessment legislation. IPAMB.

Petts J. (1999) Handbook of environmental impact assessment: vol. 1. Process, methods and potential; vol. 2. EIA in practice: impacts and limitations. Blackwell Science.

Sadler B. (1996) Environment assessment in a changing world: evaluating practice to improve performance. Ottawa, ON: Canadian Environment Assessment Agency.

Serrano, A. (2004) The Convention on Environmental Impact Assessments in a Transboundary Context (The Espoo Convention) and the Protocol to the Convention on Strategic Environmental Assessment. Mestrado em Gestão e Políticas Ambientais: Relações Internacionais do Ambiente. Universidade de Évora.

Tesli A. and Husby S. (1999) EIA in a transboundary context: principles and challenges for a coordinated Nordic application of the Espoo convention. Environ Impact Asses Rev (19):57–84.

Wood C. (2003) Environment impact assessment: a comparative review. 2nd ed. Upper Saddler River, NJ: Prentice Hall.

Notes

This chapter was originally published as: Albergaria, R. and Fidelis, T. (2006) 'Transboundary EIA: Iberian Experiences' *Environmental Impact Assessment Review*, vol 26, pp614–632.

1 IAIA – International Association for Impact Assessment.

2 According to Principle 19 of the International Conference on Environment and Development, Rio de Janeiro: 'States shall provide prior and timely notification and relevant information to potentially affected States on activities that may have a significant adverse transboundary environmental effect and shall consult with those States at an early stage and in good faith'.

3 'Impactos Transfronteiriços', 2004, Instituto do Ambiente. ['Transboundary Impacts', Environment Institute.]

4 Ministries of the Environment – Portugal/Spain.
5 Comissão de Coordenação e Desenvolvimento Regional (CCDR) [Regional Commission for Coordination and Development].
6 Consellería de Medio Ambiente.
7 Involving local and regional entities, as well as, those relevant at the national level.
8 Townships, cooperatives, producers, businesses, NGOs, etc.
9 Performed by the independent team.
10 Comissões de Coordenação e Desenvolvimento Regional of the Portuguese regional administrations and Consellarias de Medio Ambiente of the Spanish autonomous communities.

8

Transboundary Environmental Assessment in the Arctic

Timo Koivurova

The Arctic is a region of vast natural resources and a relatively clean environment compared with most areas of the world. For better or for worse, it is often compared to its southern counterpart, the Antarctic. Both do have extreme climatic conditions, receiving less radiation from the sun than other parts of the globe, and the ecosystems have had to adapt to very cold and dark environments with short and intense growing seasons. In such conditions, the ecosystems are simple, containing only a few key species, and are thus more vulnerable to human-induced pollution than those of more temperate areas. The comparison between the circumpolar areas may, however, be misconceived given that the two poles show more differences than similarities: the Arctic consists of ocean surrounded by continents, whereas the Antarctic is a continent surrounded by ocean; the Antarctic has no permanent human habitation, while the Arctic is inhabited by indigenous peoples and other local communities,[1] and the sovereignty is effectively frozen in the Antarctic whereas much of the Arctic is under the sovereignty of eight states (see Koivurova, 2005; Rothwell, 1996, for comparison). While the Antarctic Treaty System (ATS) provides important checks on the economic development of the region, the Arctic will in all likelihood face environmental pressures from various economic activities, in particular increased fishing, maritime navigation and hydrocarbon exploitation.

The poles are also different with respect to their environmental assessment (EA)[2] regulations, even though both poles have their own environmental impact assessment (EIA) regulations. In ATS, EIA has figured prominently from quite early on, culminating with the moratorium on mining and setting out detailed legally binding EIA rules over other activities in the region, as enshrined in Article 8 and Annex I in the 1991 Protocol to the Antarctic Treaty of 1959 (see Bastmeijer and Roura, 2008, for a recent in-depth study of the Antarctic EIA system). Also, the Arctic has its own set of regulations covering EA, Guidelines for EIA, adopted as part of the Arctic Environmental Protection Strategy (AEPS) for co-operation between the eight Arctic states in 1997. Yet, these guidelines are legally non-binding, with the result that the document has not been incorporated in national legislation nor administrative guidelines of the eight Arctic states, and set out only general guidance for states on how to implement EA in their Arctic regions.

The main intent of this chapter is to examine the extent of applicable international treaties and other regulations that obligate the concerned states to perform transboundary EA (TEA) in the Arctic, and whether these have induced any state practice. One real-life case study is chosen to demonstrate what a quality TEA can mean in the Arctic context.

Before these main tasks, it is crucial to introduce the very particular administrative reality that prevails in the Arctic. Because much of the Arctic is part of the sovereign territory of the eight Arctic states, it is primarily the national (and sub-national) authorities that apply EA in the northernmost areas of these states, and thus it is important to examine the basic administrative structures within which EA takes place. Another important development to be studied is the Arctic-wide operation between the eight Arctic states that commenced in 1991, and was revised slightly with the establishment of the Arctic Council in 1996 (transition from AEPS to Arctic Council took place 1996–1998). It is useful to introduce this unique inter-governmental process and examine in particular what kind of work has been done in the fields of EA and TEA.

Introduction to the Arctic governance framework

Before commencing the study of the prevailing governance system in the Arctic, it is pertinent to try to define it. The question of defining the southernmost boundary of the Arctic is complicated since several different criteria can be presented as to where this boundary should be drawn. Possible natural boundaries are, for instance, the tree line (i.e. the northernmost boundary where trees grow) or the 10°C isotherm (i.e. the southernmost location where the mean temperature of the warmest month of the year is below 10°C). In Arctic-wide co-operation, the Arctic Circle itself has been used as a criterion for membership, with only those states that possess areas of territorial sovereignty above the Arctic Circle invited to participate in co-operation. It thus includes the Nordic countries (Norway, Sweden, Finland, Denmark (Greenland) and Iceland), Canada, Russia and the US (through Alaska), and will be used in this article as a point of departure.

The first stage of the Arctic-wide co-operation started with the 1991 AEPS. In the strategy, six priority environmental problems facing the Arctic were first identified (persistent organic contaminants, radioactivity, heavy metals, noise, acidification and oil pollution), together with international environmental protection treaties that apply in the region and, finally, specific actions to counter these threats were laid out. As part of the environmental protection action by the eight Arctic states, four environmental protection working-groups were established:

- Conservation of Arctic Flora and Fauna (CAFF);
- Protection of the Arctic Marine Environment (PAME);
- Emergency Prevention, Preparedness and Response (EPPR); and
- The Arctic Monitoring and Assessment Programme (AMAP).

In addition, a task force on sustainable development and utilization, under which the EIA Guidelines were negotiated, was established after the 1993 Nuuk ministerial meeting. Three ministerial meetings (after the signing of the declaration and the strategy) were held in this first phase of Arctic co-operation, generally referred to as AEPS co-operation.

The establishment of the Arctic Council in 1996 broadened the mandate of the co-operation to all common issues facing the Arctic (excluding matters related to military security), especially those relating to environmental protection and sustainable development; the four environmental protection working groups of the strategy were integrated into the structure of the Council, and one new working-group was established (Sustainable Development Working Group (SDWG)). In the absence of a permanent secretariat, the work of the Arctic Council is heavily influenced by the priorities that the chair-states lay out for their two-year chair period, at the end of which a ministerial meeting is organized. Senior Arctic Officials (SAO), a group of high-level officials, guides the work of the Council between the ministerial meetings. The Arctic Council has also adopted new programmes related to environmental protection, such as the Arctic Council Action Plan to Eliminate Pollution in the Arctic (ACAP), which was recently turned into a sixth working group, and the Arctic Climate Impact Assessment (ACIA).

What is unique in the Arctic Council is the role it has given to the region's indigenous peoples, as they are normally accorded the status of NGOs in different inter-governmental organizations and forums. They are defined as permanent participants, a distinct category of membership between members proper and observers, whom the Arctic Council member states must consult prior to any consensus decision-making. The group of observers is large, and consists of inter-governmental organizations and NGOs as well as states active in the Arctic region (see Koivurova and VanderZwaag, 2007, for a recent comprehensive analysis).

Within this approximate region, all the levels of law – international, European and national (and sub-national) – come into play, as much of the region falls under the sovereignty and sovereign rights of the eight states. Taking a closer look, we can identify three federal states (Russia, Canada and the US), with varying kinds of mandates accorded to their sub-administrative units (the state of

Alaska, the northern territories of Canada and the varying kinds of subjects of the Russian Federation), wherein indigenous peoples have been given different powers and rights. Finland, Sweden and Denmark are Member States of the European Union (EU), but since the Danish protectorate, Greenland, chose to exit from the then European Economic Community (and is thus not part of the European Community (EC) and EU), it possesses extensive autonomous powers in the form of a home-rule government. The European Free Trade Association (EFTA) states Iceland and Norway are bound by much of European law as parties to the European Economic Area (EEA) Agreement (Agreement on the European, 1992), with the exception of the Svalbard Islands, which were excluded from the EEA agreement by a special protocol (Agreement on the European, 1992)[3] due to its unique status established through an international treaty in 1920 (Svalbard Treaty, 1920).

All of the land area – continents as well as islands – is firmly under the sovereignty of the Arctic states, and much of the Arctic waters now falls under their exclusive maritime jurisdiction. The core of the Arctic Ocean remains part of the high seas as well as some holes enclosed by the exclusive economic zones of Arctic coastal states. The deep seabed is governed by the International Sea-Bed Authority, although some of the Arctic states are developing submissions to the Commission on the Limits of Continental Shelf to extend their continental shelf to even the deep sea-bed ridges of the Arctic Ocean floor.[4] There are also ongoing and potential disputes over the location of some of the maritime borders, especially those between Canada and the US in the Beaufort Sea, and Russia and Norway in the Barents Sea, although in general it can be noted that the Arctic states have been able to resolve their maritime boundary disputes peacefully, via negotiations, conciliation and judicial procedures.[5]

The Arctic EA system seems most complex when national EA procedures are examined. Since these are extremely complicated systems, here only a brief mention can be made. The state of Alaska applies the federal National Environmental Policy Act (NEPA) EA procedures in certain cases, but does not have its own EA procedures (Manager of the Alaska Operations Office (Oil and Gas sector), part of the Pacific Northwest region of the US Environmental Protection Agency, personal communication email, 14 April 2008; Ferester, 1992, for the general overview of federal-state EA system) and Russia's regional EAs are much influenced by federal EA regulations.[6] Canadian provinces and territories have quite extensive autonomous powers in many fields of policy, including environmental policy; the northern territories of Canada have their own kinds of EA procedures, which place more emphasis on the rights and interests of indigenous peoples.[7] Of the Arctic states, only Finland and Sweden are required to implement the EU's EIA and SEA Directives as Member States of the EU. However, because of the EEA agreement between EFTA and EU, EFTA states Iceland and Norway are required to implement the Directives as well. An exception is the Svalbard Islands, which was excluded from the EEA agreement and now has an EA procedure of its own, enacted by Norway (Act of 15 June 2001, especially chapter VII). Since Greenland is not part of the EU and has extensive self-government, it has a right to specify its own EIA rules.[8]

Even though there are some pieces of EA legislation that apply particularly to Arctic conditions, such as the EA procedures applicable to the Svalbard Islands and the Nunavut territory in Canada, it is mostly the case that the EA rules in international, European and national law do not take into account the very specific Arctic circumstances. With the exception of Iceland, the capitals of the Arctic states are far away from the states' Arctic territory, and it is thus no wonder that the Arctic perspective does not figure in their EA procedures.

THE DEVELOPMENT OF ENVIRONMENTAL ASSESSMENT AND TRANSBOUNDARY ENVIRONMENTAL ASSESSMENT FROM THE ARCTIC PERSPECTIVE

The end of the Cold War was at least a partial factor explaining the emergence of two important international processes that had a distinct impact on Arctic EA and TEA: the signing of the AEPS and the conclusion of the United Nations Economic Commission for Europe (UNECE) Convention on Environmental Impact Assessment in a Transboundary Context. Both were signed in 1991 and in Finland, in Rovaniemi and Espoo, respectively. Both also started to function from 1991 onwards. The AEPS, an Arctic-wide co-operation agreement, commenced with working-group activities, whereas the Espoo Convention, even though it entered into force as late as in 1997, started immediately with meetings of the signatories.

The AEPS, which was signed on 14 June 1991, contained relevant guidance in the field of EA in general but nothing of TEA, even though all the eight Arctic states that were members of the AEPS had adopted the Espoo Convention on 25 February 1991, and most of them had signed it on 26 February 1991;[9] some of them had accepted it as binding fairly soon after its signature.[10] There were also high expectations immediately after the signing of the convention that the US and the Russian Federation would ratify it within a short period of time. The report from the second meeting of the signatories states: 'The delegations of ... the Russian Federation and the United States of America ... expected that their countries would be able to ratify the Convention in the course of 1993' (Report of the second meeting, 1992). Hence, the convention did contain a potential at the time to become a pan-Arctic legally binding framework for transboundary EIA (TEIA).

This manifested itself especially during the years 1993 to 1996, when the AEPS co-operation placed growing emphasis on the Convention. The Espoo Convention was expressly promoted in the 1993 ministerial meeting in Nuuk, Greenland, where the ministerial declaration states:

> *Therefore, we shall maintain, as appropriate, or put into place as quickly as possible, an internationally transparent domestic process for the environmental impact assessment of proposed activities that are likely to have a significant adverse impact on the Arctic environment and are subject to decisions by competent national authorities. To this end we support the implementation of the provisions of the Convention on Environmental Impact Assessment in a Transboundary Context. (Nuuk Declaration, 1993: paragraph 8)*

Paragraph 9 of the Nuuk Declaration is also relevant: 'We underline the importance of prior and timely notification and consultation regarding activities that may have significant adverse transboundary environmental effects.' The 1996 Inuvik Declaration considered the Espoo Convention much more briefly: 'reaffirming the support of our countries … on Environmental Impact Assessment in a Transboundary Context' (Inuvik Declaration, 1996: Preface). In the 1997 Alta Declaration, the Arctic states agreed to apply the 1997 EIA Guidelines, which contain a separate chapter on transboundary impacts that specifically mentions the Espoo Convention.[11]

Yet, the promise of the Espoo Convention becoming a pan-Arctic TEA regime has still to materialize. The ministerial meetings after the Inuvik did not pay any further attention to the Convention, and when the Espoo Convention entered into force on 10 September 1997, Iceland, the Russian Federation and the US had not accepted the Convention as binding – a situation that has persisted to this day. The reason for this in the USA has probably been the focus on the environmental side agreement – the North American Agreement on Environmental Cooperation (1993) – to the North American Free Trade Agreement, under which an attempt has been made to formulate TEA rules (Craik, 2008). For Iceland, the reasons are said to include the improbability of TEA procedures because of its location and the fact that priorities need to be established for a country with such a small civil service, given that international activities require investment in time, attention and money (Official from the Ministry of the Environment of Iceland, personal communication, email, 3 April 2008).[12] In addition, it must be remembered that via the EEA Agreement, Iceland is legally obligated to perform a TEA procedure towards other EEA agreement parties.[13]

THE ESPOO CONVENTION AS A BASIS FOR ARCTIC TRANSBOUNDARY ENVIRONMENTAL ASSESSMENT

Even though three of the Arctic states have not become parties proper to the Espoo Convention, it is still this Convention that is and will be the primary standard for TEA in the Arctic. This is due to many factors. Even though not all states are parties to it, increasingly the Convention is used as a global standard for how to implement the requirements of customary law principle of no-harm, which is legally binding on all states of the world (Bastmeijer and Koivurova, 2008). In many instances, states apply the Convention even though they are not legally obligated to do so. A good example of this is Finland's recent notification to the Russian Federation on the basis of the Espoo Convention of a planned mining project in Sokli – which is located above the Arctic Circle, 12km from the Russian border – even though Russia is not a party to the Convention. [14] It is also good to keep in mind that Iceland, Russia and the US are signatories to the Convention and thus required not to frustrate the object and purpose of the treaty as stipulated in the customary law of treaties[15] and can be expected to become parties to the Convention at a later stage.

Since the Espoo Convention contains political commitment only from the Espoo Convention parties where evaluation of the effects of strategic level policies and plans is concerned (Article 2(7)), the parties decided to develop a special protocol on SEA. This was signed on 21 May 2003 by 35 states and the EC, with the Arctic signatories, including Norway, Finland, Sweden and Denmark. The protocol focuses on creating national SEA procedure but also stipulates rules by which transboundary SEA is to be organized in certain cases of transboundary environmental effect.[16] This Protocol was largely inspired by the SEA Directive of the EC, which also contains provision on transboundary SEA (TSEA) (Directive 2001/42/EC, 2001, article 7). The TSEA procedure has far less potential in the Arctic since four of the Arctic states have not even signed the Protocol, and the Protocol has not entered into force as yet.

Even though the Espoo Convention clearly established the primary legal basis for an Arctic TEA, it is not the only treaty providing for communication – or even full-blown TEA – procedure between states in cases where potential adverse impact from a proposed activity or plan at the other side of the border likely threatens the environment of the potentially affected state (see Koivurova, 2002: 181–286 for a detailed assessment). There are, in effect, many treaties in force between the eight Arctic states that provide for a TEA type of procedure. There are many applicable treaties and other regulations between the Nordic states[17] and between the US and Canada (thus also covering the Alaska–Yukon border)[18] and one between Canada and Denmark (Greenland) (Agreement between Denmark and Canada, 1983). There are also treaties that apply throughout most of the region and contain a general level TEA procedure (United Nations Convention on the Law of the Sea (UNCLOS), 1982, Article 206; Convention on Biological Diversity (CBD), 1992, article 14).[19] The primary function of all these treaties and other regulations is to substitute for the Espoo Convention in those cases where it does not apply. Yet, there are also borders which are not covered by any TEA type of treaty, such as those between the USA and Russia, and Russia and its Nordic neighbours. However, the other UNECE treaty, the Convention on the Transboundary Effects of Industrial Accidents

(1992) does provide a TEIA procedure between Russia and its Nordic neighbours in situations where it applies.[20]

As briefly mentioned above, the Arctic states were able to adopt the Guidelines for Environmental Impact Assessment in the Arctic, a Finnish initiative, in 1994. At the time, there were great prospects of having the Espoo Convention to become a pan-Arctic convention, which partly inspired the idea of developing these guidelines, given that the Espoo Convention does not only regulate TEIA but sets out certain minimum requirements for national EIAs. In the preamble, the legal nature of these EIA Guidelines is clarified:

> *These guidelines are not intended to replace existing procedures adopted by international, national or provincial laws, land claim agreements, regulations or guidelines. As they do not recommend any particular procedure for EIA, these guidelines are applicable across jurisdictional boundaries and in different EIA processes. They aim at providing suggestions and examples of good practice to enhance the quality of EIAs and the harmonization of EIA in different parts of the Arctic. (AEPS, 1997)*

The guidelines provide important guidance as to how EA should be conducted to give due consideration for the special conditions in the Arctic. The drafting of the instrument was prompted by the realization that the Arctic states share many challenges in applying EA in their Arctic areas. For example, the participation of the public in EA is constrained by the region's small population, which includes many indigenous peoples; the long distances and the limited number of cities and towns also affect how public participation is organized. Moreover, although environmental conditions vary in different parts of the Arctic, EA must address the similarities in the region's ecosystems and the challenge of integrating indigenous peoples and their traditional knowledge into the decision making processes.

Chapter 11 of the guidelines provides useful recommendations for the Arctic states on how to organize their TEA procedures. As all the Arctic states are signatories to the Espoo Convention, the guidelines are meant to adjust the requirements of the Convention to the Arctic. Above all, the guidelines urge that all activities assessed according to the national EA legislation should be screened also from the viewpoint of whether transboundary impacts are likely (AEPS, 1997, chapter 11, paragraph 8). Thus, all activities to which a national EA procedure is applied should be screened in view of likely transboundary impacts in the Arctic context. In addition, lower thresholds may be needed for those activities listed in the Espoo Convention if proposed to operate in Arctic conditions (AEPS, 1997, chapter 11, paragraph 8).

According to the guidelines, the origin state should initiate the transboundary EA procedure in a very early phase of its national EA procedure. The guidelines document recommends that, already in the scoping phase of the national EA procedure, potential transboundary impacts should be identified and methods to be used for assessing them should be agreed upon between the concerned states; joint steering groups are recommended to perform these tasks (AEPS, 1997, chapter 11, paragraph 4). The guidelines also urge cooperation in the implementation of the TEA procedures taking place in the Arctic (AEPS, 1997, chapter 11, paragraphs. 7 and 8).

The Espoo Convention provides for a basic right for all those private legal subjects of the affected state located in the area likely to be affected to participate in the TEIA procedure, just as the private legal subjects of the origin state may also participate. The guidelines go further and urge the Arctic states to be as inclusive as possible when organizing a TEA procedure: 'Communities in the area of anticipated impacts should be given an opportunity to participate, irrespective of their location relative to the border' (AEPS, 1997, chapter 11, paragraph 10). In the Arctic context, these communities normally are indigenous peoples, as referred to in chapter 11 (AEPS, 1997).[21] The guidelines document also emphasizes that even though the activities may be far away from the border, transboundary impacts may occur anyway, especially with respect to large-scale activities such as oil and gas activities (AEPS, 1997, chapter 11, paragraph 9).

Unfortunately, even though the Arctic states agreed to apply these EIA guidelines, in practice, according to the ministerial declaration from the Alta meeting, no real follow-up mechanism was established to oversee how these guidelines were in effect implemented. In the research conducted by the Arctic Centre's Northern Institute for Environmental and Minority Law for the Finnish Ministry for the Environment (MoE), it was found that only a few people from the stakeholders in the Arctic – environmental NGOs, indigenous peoples' organizations, companies, administrative agencies – even knew that the guidelines exist, let alone that they would have influenced any Arctic EAs (Koivurova, 2008: 151–174).

TRANSBOUNDARY ENVIRONMENTAL ASSESSMENT APPLICATION IN THE ARCTIC

TEA is becoming an increasingly commonplace phenomenon in state practice. This is because, although the Espoo Convention entered into force as recently as 1997, it is only now that states are starting to make greater use of it. A good illustration of this is Finland, which during 2007 received four notifications on the basis of Article 3 of the Espoo Convention.[22]

All in all, even though estimates are increasing that the Arctic's vast natural resources will be exploited, it is also true that this has not taken place as yet on any large scale. In many places of the region, more important is the functioning of the national or sub-national EAs since the international borders are distant from places of operation. This estimate was verified by the enquiries with the points of contact to the Espoo Convention.[23] On the basis of these answers, it can be concluded that relatively few TEIAs and SEAs are ongoing in the Arctic region.

Thus, with the exception of the core Barents region (northernmost parts of Scandinavia and northwest Russian) – where international borders are close to each other – and the border between long-term co-operating partners US's Alaska and Canada's Yukon, it seems there is little activity in the field of TEA. There are four pending Arctic TEAs and EIAs where Finland has commenced the procedure on the basis of either the Espoo Convention or the SEA Directive with its neighbouring countries of Norway, Sweden and Russia.[24] There is the potential construction of a natural gas pipeline being built from Alaska to Yukon and further to Alberta but this still awaits the decision by the competent authority.[25] Interestingly, an SEA is being undertaken for the development of Iceland's Dreki (Dragon) maritime area for gas exploitation, even though this SEA is, at least for the time being, confined to national context. Yet, since the area is part of a jointly divided maritime zone between Iceland and Norway's Jan Mayen Island, it is expected that a TSEA or TEIA (at a later stage) procedure will take place.[26]

In the not so distant future, the international borders of the Arctic states will also approach each other in the Arctic Ocean basin, given that Russia has made a large submission covering almost half of the basin as part of its continental shelf. Other Arctic states are now developing their submissions, and it seems that in many places these will overlap or be otherwise very close to each other, resulting in the possibility for future disagreements and TEA procedures. Here it needs to be kept in mind that those parts or pockets that remain part of the deep sea-bed are governed by the International Sea-Bed Authority and its EA rules (see Le Gurun, 2008: 221–264, for a recent study).

Case study: Arctic transboundary environmental impact assessment

There is one pending case in the Finnish Arctic where the problems and possibilities of the Guidelines for EIA in the Arctic manifest themselves in the context of Arctic TEIA. Finnish and Norwegian authorities have developed a plan to create a road connection from south of Lake Inari to the village of Nellim in Finland to Nyrud in Norway (see Ruokanen, 2006: 25–28, for a more detailed overview of the case with illustrations). The project is part of an overall road plan of 204km to connect the Finnish town of Ivalo with the Norwegian town of Kirkenes. The project on the Finnish side (from Nellim to the Norwegian border) is connected with plans in Norway to construct a road from Nyrud to the Finnish border. The new road connection would shorten the distance from Ivalo to Kirkenes by 32km.

The part of the road under consideration (from Nellim to the Norwegian border) on the Finnish side of the border is approximately 30km long. The project proponent, the Finnish Road Administration, asked the Lapland Regional Environmental Centre (LREC) on 29 September 2005 whether EIA for the road construction was needed. LREC proceeded to submit its views on the question to the Finnish Ministry for the Environment. As the road construction project involves a short distance, EIA is not mandatory according to the Finnish EIA legislation. However, according to Article 4.2 of the EIA Act, if the project is of such a nature that it is likely to cause significant adverse environmental impacts, the ministry may decide that the project must undergo EIA. According to the EIA Decree, in such cases it is pertinent to study the project's characteristics, location and the nature of its impacts.

According to the LREC, neither the scope of the project nor its other characteristics are such that they are likely to cause significant adverse environmental effects. Moreover, there are only few residential areas that would be impacted by the construction. Nevertheless, LREC is of the view that the area under consideration possesses particular characteristics: it consists mostly of wilderness, and human interference with nature is at present very limited in the area. The road would also be located close to wilderness and nature protection areas. Importantly, LREC opines that the regeneration rate of the area as a northern ecosystem is poor; that is, it consists of vulnerable ecosystems, whereby the project would irreversibly alter nature in the area. The fact that the road would create a new international road link, with increased traffic, also suggests likely significant environmental impacts. LREC has also taken the view that even though there are not many inhabitants in the area, it is significant that many of these are indigenous Sami and especially Skolt Sami. For these reasons, LREC proposed to the ministry that the project was likely to have significant adverse environmental consequences and that an EIA should be undertaken, together with a co-ordinated application of the Espoo Convention with Norway (LREC, 2006).

The proposal by the LREC was followed by negotiations at the Ministry of the Environment, where representatives from the road management authority, as well as the municipality of Inari, were present and stated that in their opinion EIA was not necessary for completion of the project (Finnish Ministry of the Environment, 2006: 1). Yet the ministry decided, on the basis of reasoning similar to that espoused by LREC, that the project must undergo EIA (Finnish Ministry of the Environment, 2006: 1). Finland will thus carry out the EIA as a joint implementation project under the Espoo Convention with Norway and, in all likelihood, will also try to involve the Russian Federation, even though Russia is not a party to the Espoo Convention (responsible official in the Ministry of the Environment, personal communication, email, 2 June 2006, on file with the author). For some time now, Finnish policy has been to treat Russia as if it were a party to the Espoo Convention in order to induce it to ratify the convention.

The official responsible for making the proposal to the ministry, and whose reasoning was accepted by the ministry, seemed to apply the guidelines when making her decision. She identified exactly the kind of considerations that are highlighted in the guidelines as a basis for her proposal; for example, the presence of indigenous peoples in the impact area of the proposed activity, cumulative impacts, the vulnerability of the Arctic environment and the possibility of transboundary impacts. When asked in an interview whether she was in fact applying the guidelines, she stated that this was not the case. She was aware that such an instrument had been created but she had not made use of it in making her decision (responsible official from the Lapland Regional Environment Centre (LREC), personal communication, telephone interview, 17 May 2006).

This case is a good illustration of the problems and potential of the Arctic EIA Guidelines. The main problem is, of course, that the persons responsible for Arctic EIA are rarely aware of the document, but even when they are, they do not consciously apply it. Yet, the case also shows the relevance of the instrument. The responsible official clearly conducted a very well-reasoned and thorough study, on the basis of which she submitted that the EIA Act should be applied to the project. Even though she was not aware of it, she was making choices in implementing an Arctic EIA that are recommended by the EIA Guidelines (responsible official from the LREC, personal communication, telephone interview, 17 May 2006). Evidently, in all EA procedures, including Arctic ones, there is a great deal of room for the officials in charge to make choices as to what considerations are given weight when making decisions. In some cases, these lead to poor quality EAs and disregard for the special characteristics of the Arctic but in some instances, as in this case, Arctic considerations are given special weight. The value of the EIA guidelines, if they were known and made use of, would be to guide the responsible officials in making exactly the choices that are most pertinent for the special conditions in the Arctic.

Conclusion

As the 2004 Arctic Climate Impact Assessment (ACIA, 2004)[27] and more recent studies and assessments have shown, the Arctic can be seen as the barometer of climate change. We can already witness the consequences of climate change in the region simply because snow and ice react very swiftly to climate change and these consequences will be more intense in the Arctic. Together with accelerating economic globalization, this will mean vast transformation for the Arctic region. The previously inaccessible region, and its vast natural resources, will be open for exploitation.

Estimates abound that much of the world's undiscovered hydrocarbon reserves lie in the Arctic, resulting in increased pressure to exploit the Arctic's offshore hydrocarbon resources. The development trajectory is real and it is underscored in GLOBIO Report of the United Nations Environment Programme, which observes:

> *In the last part of the 20th century, the Arctic has been increasingly exposed to industrial exploration and exploitation as well as tourism. The growth in oil, gas and mineral extraction, transportation networks and non-indigenous settlements are increasingly affecting wildlife and the welfare of indigenous people across the Arctic ... A 2050 scenario was made using reduced, stable, or increased rates of infrastructure growth as compared to the growth between 1940–1990. The scenario revealed that at even stable growth rates of industrial development, 50–80% of the Arctic may reach critical levels of anthropogenic disturbance in 2050, rendering most of these areas incompatible with traditional lifestyles of many subsistence-based indigenous communities. (UNEP, 2001: 2)*

In principle, most areas of the region are covered already by national and sub-national EIAs and SEAs, which can at least study the potential consequences for the vulnerable Arctic environment. TEA agreements also cover most relations between the Arctic states although some gaps remain. It would be important for Iceland, the Russian Federation and the US to become parties to the Espoo Convention since this would further clarify the basic steps that need to be taken in order to also examine the cross-border impacts of planned activities and plans.

What would be needed, however, are clearer legally binding rules on how to perform EA and TEA in the unique Arctic conditions, where the environment is particularly vulnerable to human-induced pollution. Before such ambitious regulations could be in place, it would be of utmost importance to make use of the EIA guidelines (which apply to project level EIAs and also SEAs), both in conducting national EAs and TEAs in the region. All officials who execute or co-ordinate the making of a TEA or EA have some room for discretion as to how to take the particular circumstances of the Arctic into account, as was clearly shown in the case study above, and that is well captured in the EIA guidelines. If taken to active use, and perhaps revised and made more concrete – an action that the present chair of the Arctic Council Norway identified in its plans for its chair-period 2006–2008 – the EIA guidelines could be a very important regulatory tool to meet the vast challenges posed to the region by the increasing exploitation of its natural resources.[28]

Two issues stand out for making TEA effective in the Arctic region. The role of the working-groups of the Arctic Council is of vast importance, given that national line-agencies and region's indigenous peoples work in these bodies to promote environmental protection and sustainable development in the region. The officials and representatives in these working groups share information on the economic development in their Arctic regions, and thus also become knowledgeable of planned economic activities possibly posing transboundary threats. Some of these working-groups also conduct scientific assessments, which raise awareness of the areas planned for economic development. One such recent assessment is the Oil and Gas Assessment by the AMAP working group, which also contains important information of the planned hydrocarbon activities (see e.g. AMAP, 2008). All this sharing of information is likely to lead to enhanced awareness of the scope of the probable environmental consequences of development plans, thus including also transboundary impacts.

As pointed out in Chapter 11 of the EIA guidelines, the role of indigenous peoples' organizations is of huge importance in promoting an effective TEA. Especially the transnational organizations of indigenous peoples, such as the Saami Council and Inuit Circumpolar Council, which play an increasingly important role in supervising and reacting to the transboundary concerns of indigenous peoples.[29] These organizations have become influential largely because of their unique status in the Arctic Council as permanent participants, thereby having access to various information sources as to what happens in the Arctic. Of great importance is also their role in making sure that indigenous peoples' traditional knowledge is taken into account in various scientific assessments, the most influential to date being the Council's ACIA. With these capacities, the Arctic indigenous peoples' organizations can promote public participation and in general enhance the quality of TEAs in the Arctic.

Notes

This chapter was originally published as: Koivurova, T. (2008) 'Transboundary Environmental Assessment in the Arctic', *Impact Assessment and Project Appraisal*, vol 26, no 4, pp265–275.

1 The 2004 Arctic Human Development Report (AHDR) by the Arctic Council applies the narrower definition of the Arctic, yielding a population of 4 million people. Furthermore, the report highlights that it is extremely difficult to assess how many of these people are of indigenous origin, given the differing definitions adopted in census statistics in the Arctic countries (AHDR, 2004: 27–41).

2 In this chapter, the term 'EA' will refer to both environmental impact assessment (EIA), which applies to projects, and strategic environmental assessment (SEA), which refers to assessment of plans, projects and policies. This also applies in the transboundary context: TEA is therefore composed of project level TEIA and strategic level TEA.

3 A special protocol was adopted as part of the EEA Agreement to the effect that Norway may decide whether to apply the EEA Agreement to Svalbard or not (Protocol 40). Norway decided to exclude the Svalbard Islands.

4 See the submissions to the Commission on the Limits of Continental Shelf by the Russian Federation in 2001 and Norway in 2006, and reactions to these by other

states, at the Commission's website at www.un.org/Depts/los/clcs_new/clcs_home.htm, last accessed 14 April 2008.

5 For an analysis, see Koivurova (2002: 56–64). The most recent was the conclusion of an agreement between the Government of the Kingdom of Norway on the one hand, and the Government of the Kingdom of Denmark together with the Home Rule Government of Greenland on the other hand, concerning the delimitation of the continental shelf and the fishery zones in the area between Greenland and Svalbard, Copenhagen, 20 February 2006 (signed 20 February 2006, entry into force 2 June 2006), United Nations Treaty Series 42887.

6 For interesting inter-disciplinary perspectives on the EA from the Russian Arctic point of view, containing also articles on the federal EA procedure and how it plays out in various Arctic regions of Russia, see the Special Issue on the Oil and Gas Industry, Local Communities and the State (guest editors Emma Wilson and Florian Stammler), *Sibirica*, 5(2), Autumn 2006. Available at www.berghahnbooksonline.com/journals/sib/index.php?pg=toc5-2, last accessed 14 April 2008.

7 For the three northern territories, the EIA system is based on relevant land claim and the federal Canadian Environmental Assessment Act (CEAA): the Yukon Environmental and Socio-Economic Assessment Act (S.C. 2003, c. 7), at www.canlii.org/ca/sta/y-2.2/, last accessed 10 April 2008; for the Mackenzie Valley in the Northwest Territories, see http://laws.justice.gc.ca/en/showdoc/cs/M-0.2//20080410/en?command=HOME&caller=SI&search_type=all&shorttitle=mackenzie&day=10&month=4&year=2008&search_domain=c s&showall=L&statuteyear=all&lengthannual=50&length=50, last accessed 10 April 2008; EIA for Nunavut is established by Article 12 of the land claim agreement between the Inuit of the Nunavut Settlement Area and Her Majesty the Queen in Right of Canada, at www.ainc-inac.gc.ca/pr/agr/pdf/nunav_e.pdf, last accessed 10 April 2008. See also Green and Binder (1995: 343–345). For a recent analysis of the role of indigenous peoples in EAs, see O'Faircheallaigh (2007).

8 Information about home rule and the applicable environmental legislation in Greenland (not including EA), at www.nanoq.gl, last accessed 14 April 2008. According to the head of department in spatial planning in Nuuk Greenland, Greenland has a right to specify their EIA rules (not yet authority to specify SEA rules), and this will take place within a year or so (email communication on 21 April 2008).

9 Of the Arctic states, Norway signed it on 25 February and the Russian Federation on 6 June 1991. See www.unece.org/env/eia/convratif.html, last accessed 14 April 2008.

10 The first of the Arctic states to ratify the Espoo Convention was Sweden, which did so on 24 January 1992. Thereafter, Norway accepted the Convention as binding (23 June 1993), followed by Finland (10 August 1995), Denmark (14 March 1997), the EC (24 June 1997) and Canada (13 May 1998).

11 Chapter 11, 'Transboundary impacts', contains the following reference (pp40–41): 'The UNECE Convention on EIA in a Transboundary Context, the Espoo Convention (1991, entered into force in 1997), provides a comprehensive framework for dealing with activities likely to have significant adverse transboundary impacts.'

12 The author has not been able to find reasons for why Russia has not become a party to the Espoo Convention.

13 Accordingly, Article 19 of the Icelandic EIA Act states: 'Should a project be deemed likely to have significant environmental impact in another state of the European Economic Area, the National Planning Agency shall provide this state with a description of the project together with available information on its conceivable cross-border effects. The National Planning Agency may require the developer to compile information on the potential effects in the state in question in the language of that state. When it is deemed likely that a project carried out in Iceland may have significant effects on the environment in another state of the European Economic Area, such a state shall be given the opportunity to express itself on the issue.' See www.skipulag.is/focal/webguard.nsf/Attachment/MÁU%20lögin%20á%20ensku/$file/MÁU%20lögin%20á%20ensku.pdf, last accessed 14 April 2008.

14 See the environmental assessment programme of the project developed in Finnish language, at www.ymparisto.fi/download.asp?contentid=83123&lan=FI, last accessed 14 April 2008. The notification was made on 8 April 2008 (*Lapin Kansa*, 16 April 2008: 8).

15 According to Article 18 of the Vienna Convention on the Law of Treaties, and the identical norm in the customary law of treaties: 'A State is obliged to refrain from acts which would defeat the object and purpose of a treaty when: (a) it has signed the treaty or has exchanged instruments constituting the treaty subject to ratification, acceptance or approval, until it shall have made its intention clear not to become a party to the treaty; or (b) it has expressed its consent to be bound by the treaty, pending the entry into force of the treaty and provided that such entry into force is not unduly delayed.' Reprinted at 8 International Legal Materials 679 (1969).

16 In the SEA protocol the transboundary article is in Article 10; see www.unece.org/env/eia/documents/legaltexts/protocolenglish.pdf, last accessed 14 April 2008.

17 In addition to the above-mentioned EIA and SEA directives, there are, for example: the 1974 Nordic Environment Protection Convention (NEPC), the English text can be found from 3 International Legal Materials 591 (1974); the 1976 Guidelines for Communication Between Finland, Norway, Sweden and Denmark on Security Issues Related to the Nuclear Installations Constructed Near the Border, Finnish Treaty Series 19/1977 www.finlex.fi/fi/sopimukset/sopsteksti/1977/19770019 (Finnish language); the 1992 Convention for the Protection of the Marine Environment of the North-East Atlantic, which can be found from 32 ILM 1069 (1993).

18 The following apply between the US and Canada: the 1975 Agreement between the US and Canada Relating to the Exchange of Information on Weather Modification Activities, the text that is reproduced in 14 ILM 589 (1975); the 1987 Agreement between the Government of Canada and the Government of the United States of America on the Conservation of the Porcupine Caribou Herd, which can be downloaded at http://arcticcircle.uconn.edu/ANWR/anwrint-agreement.html, last accessed 14 April 2008; the 1991 Agreement between the Government of the United States of America and the Government of Canada on Air Quality, which is reproduced in 30 ILM 676 (1991). There is also pending work to conclude a TEIA treaty under the auspices of the North American Free Trade Agreement, see more closely Craik (2008).

19 All other Arctic states apart from the US are parties to both of these treaties.

20 The applicability of the Espoo Convention derives from its definition of 'hazardous activity' as 'any activity in which one or more hazardous substances are present or may be present in quantities at or in excess of the threshold quantities listed in Annex I to the Convention and which is capable of causing transboundary effects', which encompasses most large-scale industrial activities. However, there is large list of exclusions from the scope of the Convention.

21 Paragraph 10 reads: 'The Inuit Circumpolar Conference, the Sami Council and the Indigenous Peoples Secretariat are accredited non-governmental organizations on the Arctic Council, and which are active in several arctic countries. They may thus provide useful links to the public on both sides of the border.'

22 Finland received a notification from Estonia concerning: plans by the developer OÜ Nelja Energia to construct five offshore wind farms near Hiiumaa (an island located in Western Estonia), as well as a power transmission line; and concerning the commencement of an EIA pertaining to an energy complex and a detailed zoning plan for the complex. Finland received a notification from Sweden about a windmill park being planned by Finngrunden Offshore AB and its sister company WPD Scandinavia AB in the area of Finngrunden in Selkämeri (Baltic Sea). Finland, among others, received a notification from Lithuania of the commencement of an EIA for the construction of a new nuclear power plant. For a more detailed treatment, see Finland's country report 2007 in the *Yearbook of International Environmental Law.*

23 These points of contact are mentioned also for those states that are not (yet) parties to the Espoo Convention in the Espoo Convention website, at www.unece.org/env/eia/points_of_contact.htm, last accessed 14 April 2008. The points of contact were contacted via email and telephone. The idea was only to gain an indication whether TEAs are rare or more frequent phenomena and to look into the current TEAs, not to study the past procedures. No answer could be obtained from Canada or the Russian Federation so indeed the enquiry can only give a sense of the number of TEAs pending in the Arctic.

24 The most current TEA procedures are taking place in the North Calotte area. In addition to the Nellim road project transboundary EIA (the case study below), which is in its starting phase, there is a plan to build a phosphate ore mine in Sokli, located in the northeast of Finland, very close to the Russian border. As noted above, Finland notified Russia of this project explicitly on the basis of the Espoo Convention even though Russia is not a party to the convention. There are also two zoning plans to which Finland applies the SEA Directive and involves its neighbouring countries, for the Teno River zoning plans and for the mountain–Lapland regional plan. In Kolari and Pajala, there is a plan for a joint ironore mine, with TEA to be organized on both sides of the border (email communications from the officer from the LREC, Finland, on 19 March 2008, and from the officer of Swedish Environmental Protection Agency on 26 March 2008.)

25 The pipeline project proposed by TransCanada is yet to be approved by the Alaskan legislature. In the past, there have been transboundary procedures related to transboundary effects of mining between Alaska and Yukon (email communication from the manager of the Alaska Operations Office (Oil and Gas sector), part of the Pacific Northwest region of the US Environmental Protection Agency 14 April 2008). For further information about this Alaska highway pipeline project, see www.transcanada.com/company/alaska_highway_pipeline_project.html.

26 The Dragi SEA is interesting as it aims to apply all applicable international treaties in a very comprehensive manner; see http://eng.idnadarraduneyti.is/media/Rafraen_afgreidsla/Report_on_oil_exploration-KK.pdf, last accessed 14 April 2008. Information obtained from the officer of the National Planning Agency of Iceland on 25 March 2008 (email communication). There is also a pending EIA for a coal mining project in Svalbard, although not involving transboundary procedures (email response from officer of the Norwegian Ministry of the Environment, 12 March 2008). Iron-ore mine will be re-opened in Kirkenes, Norway, within a year. The site is located very close to the Russian border, but it is not yet clear whether and in what way Russia will be involved in the EIA procedure.

27 The ACIA Scientific Report was published in 2005 and contains a vast amount of information regarding climate change and its impacts, including future prediction with respect to both the Arctic and the global context (ACIA, 2005).

28 In fact, Norway, the current chair of the Arctic Council (2006–2008), has identified the Arctic EIA Guidelines as one focal area in its Programme for the Norwegian Chairmanship: 'It would be useful to establish a set of operational guidelines for assessing the impact of projects, plans and programmes in the Arctic. If this is done, they should be based on experience and evaluation of national and international guidelines for impact assessments.' The programme can be downloaded from the Arctic Council website at http://arctic-council.org/article/2007/11/norwegian_programme, last accessed 14 April 2008.

29 The Saami Council has, for instance, been a very strong actor in trying to persuade the Finnish Forestry Board (*Metsähallitus*) to stop logging in traditional Saami reindeer herding areas (Saami Council website: www.saamicouncil.net/?deptid=3754, last accessed 14 April 2008).

References

ACIA, Arctic Climate Impact Assessment (2004) *Arctic Climate Impact Assessment: Impacts of a Warming Arctic.* Cambridge University Press. Available at http://amap.no/acia, last accessed 14 April 2008.

ACIA, Arctic Climate Impact Assessment Secretariat and Cooperative Institute for Arctic Research (2005) *Arctic Climate Impact Assessment.* Cambridge University Press. Available at http://amap.no/acia/, last accessed 14 April 2008.

Act of 15 June 2001 No. 79 relating to the protection of the environment in Svalbard, Norway. Available at www.regjeringen.no/en/doc/Laws/Acts/Svalbard-Environmental-Protection-Act.html?id=173945, last accessed 14 April 2008.

AEPS, Arctic Environment Protection Strategy (1997) *Guidelines for Environmental Impact Assessment (EIA) in the Arctic: Sustainable Development and Utilization.* Finland: Finnish MoE. Available at http://arcticcentre.ulapland.fi/aria/, last accessed 14 April 2008.

Agreement between Denmark and Canada for Cooperation Relating to the Marine Environment (1983). 23 ILM 269 (1984). Agreement on the European Economic Area, Oporto, 2 May 1992. Official journal L 001, 3 January 1994.

AHDR, Arctic Human Development Report (2004) *Arctic Human Development Report.* Akureyri: Stefansson Arctic Institute. Available at www.svs.is/AHDR/AHDR%20 chapters/Chapters%20PDF.htm, last accessed 14 April 2008.

AMAP, Arctic Monitoring and Assessment Programme (2008) *Arctic Oil and Gas 2007.* The Oil and Gas Assessment Overview Report. Available at http://amap.no/oga/, last accessed 14 April 2008.

Bastmeijer, K. and R. Roura (2008) Environmental impact assessment in Antarctica. In *Theory and Practise of Transboundary Environmental Impact Assessment*, K Bastmeijer and T Koivurova (eds), pp.175–220. Leiden: Martinus Nijhoff.

Bastmeijer, K. and T. Koivurova (2008) Conclusions: globalisation of transboundary environmental impact assessment. In *Theory and Practise of Transboundary Environmental Impact Assessment*, eds. K. Bastmeijer and T. Koivurova, pp.347–389. Leiden: Martinus Nijhoff.

Convention on Biological Diversity, 5 June 1992, 31 ILM 818 (1992).

Convention on the Transboundary Effects of Industrial Accidents, Helsinki, 17 March 1992. 31 ILM 1330 (1992).

Craik, N. (2008) Transboundary environmental impact assessment in North America: Obstacles and opportunities. In *Theory and Practise of Transboundary Environmental Impact Assessment*, eds. K. Bastmeijer and T. Koivurova, pp. 93–118. Leiden: Martinus Nijhoff.

Directive 2001/42/EC of the European Parliament and of the Council of 27 June 2001 on the assessment of the effects of certain plans and programmes on the environment. Availableathttp://eur-lex.europa.eu/LexUriServ/LexUriServ.do?uri=CELEX:32001L0042: EN:NOT, last accessed 14 April 2008.

Ferester, P. M. (1992) Revitalizing the National Environmental Policy Act: substantive law adaptations from NEPA's progeny. *Harvard Environmental Law Review*, 16, pp.207–209.

Finland – country report 2007 forthcoming. *Yearbook of International Environmental Law.* Oxford: Oxford University Press.

Finnish Ministry of the Environment, 27 June 2006. Decision on application of the EIA Act, on file with the author.

Green, N. and R. M. Binder (1995). Environmental impact assessment under the Western Arctic (Inuvialuit) land claim. In *Integrating People and Wildlife for a Sustainable Future*, Proceedings of the first International Wildlife Management Congress, eds. J. A. Bisonette and P. R. Krausman, pp.343– 345. Bethesda, MD: The Wildlife Society.

Inuvik Declaration 1996. Declaration at the Arctic Council. Available at http://arcticcouncil.org/filearchive/The%20Inuvik%20Declaration.pdf, last accessed 14 April 2008.

Koivurova, T. (2002). *Environmental Impact Assessment in the Arctic: a Study of International Legal Norms.* Aldershot: Ashgate.

Koivurova, T. (2005) Environmental protection in the Arctic and the Antarctic: can the polar regimes learn from each other? *International Journal of Legal Information*, **33**(2), pp.204–218.

Koivurova, T. and D. VanderZwaag (2007) The Arctic Council at 10 years: retrospect and prospects. *UBC Law Review*, 40(1), pp.128–159.

Le Gurun, G. (2008) Environmental impact assessment and the International Sea Bed Authority. *Theory and Practise of Transboundary Environmental Impact Assessment*, eds. K. Bastmeijer and T. Koivurova, pp.221–264. Leiden: Martinus Nijhoff.

LREC, Lapland Regional Environment Centre 19 April 2006. Submission by the Lapland Regional Environment Centre to the Finnish Ministry of the Environment, on file with the author. North American Agreement on Environmental Cooperation 1993. 32 ILM. 1480 (1993).

Nuuk Declaration 1993. Declaration at the Arctic Council. Available at http://arcticcouncil.org/filearchive/The%20Nuuk%20Declaration.pdf, last accessed 14 April 2008.

O'Faircheallaigh, C. (2007) Environmental agreements, EIA follow-up and aboriginal participation in environmental management: the Canadian experience. *Environmental Impact Assessment Review*, 27, pp.319–342.

Report of the second meeting of the signatories to the Espoo Convention 17 December 1992. ENVWA/WG.3/4.

Rothwell, D. R. (1996) *The Polar Regions and the Development of International Law*. Cambridge: Cambridge University Press.

Ruokanen, L. (2006) Road connection between Norway and Finland: Nellim – Pasvik. *Environmental Impact Assessment in the North Calotte Transboundary Areas*, North Calotte Council publication series, Report No. 69, pp.25–28.

Svalbard Treaty (Treaty concerning Spitsbergen), Paris, 9 February 1920. The Treaty is reproduced in the *American Journal of International Law*, 18, 1924, Suppl. 1924, pp.199–208.

UNEP, United Nations Environment Programme 2001. *GLOBIO: Global Methodology for Mapping Human Impacts on the Biosphere, the Arctic 2050 Scenario*. Available at www.globio.info/region/polar/#arctic, last accessed 14 April 2008.

United Nations Convention on the Law of the Sea, Montego Bay, 10 December 1982. 21 ILM 1261 (1982).

9

Transboundary Environmental Impact Assessment in the Case of the Baltic Sea Gas Pipeline

Timo Koivurova and Ismo Pölönen*

Introduction

Most modern international treaties related to environmental protection have to come to grips with changing circumstances. This adaptive capacity is often built into a regime from the outset by first concluding a framework convention with broad and general obligations that are later complemented by protocols or by inserting flexible amendment/adjustment mechanisms into the treaty. Another modern development is the conclusion of treaties that create a plenary organ which is given powers to establish subsidiary bodies that elaborate the convention regime.[1] This approach contrasts starkly with some of the older international environmental treaties, which established no institutional mechanisms to develop the regime.[2]

The 1991 Espoo Convention on Environmental Impact Assessment in a Transboundary Context (the Espoo Convention) is clearly a modern dynamic international treaty regime, designed to adapt to changing and more complex circumstances.[3] The Espoo Convention was negotiated under the auspices of the United Nations Economic Commission for Europe (UNECE) and signed in Espoo, Finland, in 1991.[4] Following six meetings of the signatories, the Espoo Convention entered into force in 1997 and the first meeting of the parties (MOP) took place in Oslo in 1998. Three MOPs have been held since, at three-year intervals: in 2001 in Sofia, Bulgaria; in 2004 in Cavtat, Croatia; and in 2008 in Bucharest, Romania. At present, there are 42 parties to the Espoo Convention. The Strategic Environmental Assessment (SEA) Protocol to the Espoo Convention was signed in 2003 as part of the 'Environment for Europe' process and has been signed by 37 states as well as by the European Community (EC).[5]

Before moving to the more recent challenges for the Espoo Convention regime, it is important to review the basic content of the Espoo Convention. The Espoo Convention regulates situations where a significant adverse transboundary impact is likely to be caused to a state's environment by a proposed activity in another contracting state (the origin state). The Convention tries to manage these situations by requiring the parties to cooperate with each other before the activity is undertaken. In order for this procedure to function effectively, the Espoo Convention requires the states' parties to establish national environmental impact assessment (EIA) procedures that allow for the integration of foreign impacts and foreign actors.[6]

The origin state is first required to notify the potentially affected state of the likely significant adverse transboundary impact and to provide basic information regarding the proposed activity. The affected state must next confirm that it wants to participate in the procedure.[7] The origin state is then obligated to study the transboundary impacts together with the affected state and allow the public[8] of that state to participate in the process on the same terms as its own public would be entitled to.[9] After the EIA, the affected state has an opportunity through consultations with the origin state to comment on the proposed activity and its likely impacts; the public of the affected state is entitled to provide its comments on the proposed activity on the same terms as apply to the public of the origin state.[10] The final decision taken on the proposed activity in the origin state must take due account of the comments from the potentially affected state and its public and must be delivered to the affected state.[11] The states' parties are not required to determine whether the impacts studied ultimately materialize, as post-project analysis is optional.[12]

One of the most recent challenges to the Espoo Convention has arisen from its being applied to an ever-increasing number of activities, currently some 30 to 50 new ones per year, and to increasingly complex projects as well.[13] The parties to the Espoo Convention have confronted situations that were not originally envisaged or that in other ways pose challenges to the application of the regime. A note prepared by the Espoo Convention Secretariat[14] lists the following examples of 'complex activities':

(a) Pipelines, roads or other linear infrastructure projects that are part of an energy or transport network crossing several Parties;
(b) Nuclear power plants and related installations, such as waste storage facilities, which generate concerns in several Parties or which might have long-range impacts;
(c) Large industrial installations located on the banks of water bodies shared by several Parties;
(d) Large energy projects that could affect sub-regional policies.

Clearly, one of the reasons why the Espoo Convention Secretariat prepared the note (discussion paper) is the challenge faced in organizing the transboundary EIA (TEIA) for the proposed Baltic Sea Gas Pipeline (BSGP),[15] which would cross the jurisdictions of five Baltic Sea littoral states (Denmark, Finland, Germany, the Russian Federation and Sweden) and potentially affect the entire Baltic Sea.[16] Recently, the Working Group on EIA – one of the sub-bodies of the Espoo Convention – decided to establish a task force to discuss how to apply the Espoo Convention to such large-scale, complex activities.[17]

Now, as the BSGP TEIA process approaches its conclusion, it is useful to examine how the states concerned were able to apply the Espoo Convention to an activity whose complexity was not envisaged when it was negotiated. It is also important to examine whether the BSGP TEIA could have been handled better, given that the note prepared by the Espoo Convention Secretariat advances a different view on how the Espoo Convention should be applied to such large-scale multi-jurisdictional projects. That views on the issue vary naturally prompts questions on whether lessons can be learned from the BSGP TEIA for similar types of activities in the future, a discussion to which this chapter will also contribute. The Espoo Convention Secretariat has drawn its own conclusions and recommended in its discussion paper that the parties consider future guidance to overcome some of the challenges posed in the case of particularly complex activities.[18]

The focus of this chapter is to examine carefully how the TEIA was organized in the planned BSGP. It is only with such an in-depth study that it is possible to gain knowledge as to why the BSGP TEIA was designed in a certain manner and whether it was done in a legally appropriate, practically effective and even commendable manner. The chapter proceeds as follows. It first takes up the important question of why the BSGP presents a challenge for the Espoo Convention. This will show what the contact points of the Espoo Convention had to confront when they started to design the TEIA. This is followed by an examination of how the TEIA in effect proceeded, which is needed in order to understand how the system ultimately performed in practice. Because the TEIA had to be linked to the five national EIA procedures of the four origin states and the Russian Federation, the next section studies how this linking was accomplished in the case of one of these systems, the Finnish EIA procedure. It is useful to take the Finnish case as an illustrative example, given that the Finnish EIA authorities – both those supervising the national EIA and those overseeing the international IEIA – faced the most challenges in organizing the EIA procedures, as will be shown below.

Even though an argument will be made that the TEIA was organized in a responsible manner by the states concerned, it is also clear that many difficulties were encountered; these manifested themselves in the very critical comments by the affected states (and others) on the impact assessment studies made by the project proponent, Nord Stream AG.[19] Before proceeding as outlined, however, it will first be useful to say a few words about the BSGP project.

The BSGP has its origins back in the 1990s, when feasibility studies of pipeline routings were performed by North Transgas, whose major shareholders were the Russian Federation energy company Gazprom and the Finnish company Fortum Oil and Gas. Various routing options were considered that included different combinations of onshore and offshore segments. The project was subsequently taken over by Gazprom – due to a change of overall company strategy at Fortum – which in 2004 contracted with PeterGaz to produce a conceptual design for a system of two parallel offshore pipelines connecting the Russian Federation to Germany and extending further to the Netherlands and the United Kingdom. In late 2005, an agreement was signed between Gazprom, BASF AG and E.ON AG to form Nord Stream AG, which has now taken over to develop and implement the offshore project in the Baltic Sea. The company is registered in Switzerland.

The project involves a natural gas pipeline transmission system (two parallel pipelines) from the Russian Federation to Germany with connections to onshore systems in the Russian Federation and Germany. As planned, the pipeline passes through the Exclusive Economic Zone (EEZ) of five countries: Denmark, Finland, Germany, the Russian Federation and Sweden, and the territorial waters of the Russian Federation and Germany (and recently, due to re-routing, Denmark). At full capacity, it will supply 55 billion cubic metres per year to consumers in northwest Europe, provided it passes the EIA and meets the permitting requirements of the five coastal states.

The pipeline will run from Portovaya Bay near Vyborg on the Russian Federation's coast of the Gulf of Finland, to Greifswalder Bodden in Germany. Plans originally called for a service platform to be built in the Swedish EEZ, east or northeast of Gotland, but these have now been abandoned. The company intends to have both pipelines in use by about 2012. The BSGP is the longest offshore pipeline ever planned. Similarly, the TEIA procedure that was applied to it was the most challenging TEIA procedure ever conducted by the international community; the procedure involved five origin and nine affected states.[20]

From the beginning, there were various concerns about the environmental impacts of the pipelines. For instance, what is the effect of seabed interventions – when the pipelines are laid down to the seafloor – on the resuspension of sediments, and in particular on organic particles that can travel long distances via sea currents? The commissioning and de-commissioning of the pipelines were thought to be harmful to the fish stocks and therefore to fisheries in the Baltic Sea. The planned pipeline route also traversed areas that were identified as key wintering and staging sites for a large variety of waterfowl and harbour porpoise populations, with obvious problems, especially when these were already protected under Natura 2000 sites. Detonations of wartime munitions dumped onto the seabed were another concern, given that these had the potential of causing dispersal of seabed sediments and thus increase the environmental load in the already vulnerable Baltic Sea. Of particular concern were sites of dumped chemical weapons, which, when the seabed was disturbed, also could disperse chemicals into the upper water column and accumulate in marine populations.

Transboundary EIA in the case of the BSGP – a challenge to the Espoo Convention

As the BSGP passes through the jurisdiction of four contracting parties (Denmark, Finland, Germany and Sweden)[21] to the Espoo Convention, it was clear that the Espoo Convention had to be applied to the project.[22] The scope of the Espoo Convention covers large-diameter gas pipelines that are likely to cause significant adverse transboundary impacts.[23] However, the BSGP is also to be located in marine areas under the jurisdiction of the Russian Federation, causing an obvious problem for the smooth application of the Convention: the Russian Federation has signed the Espoo Convention but has not ratified it, meaning that it is not legally bound by the instrument.[24]

If the Espoo Convention does not cover the whole planned pipeline route, one obvious question that can be raised is whether there are international conventions that would apply to all the Baltic Sea littoral states and include provisions on TEIA. There are, in effect, two such conventions: the United Nations Law of the Sea Convention (UNCLOS) includes two articles covering TEIA, but these are too general in nature to be used for organizing TEIA.[25] However, the LOS Convention refers to a competent international organization that could play a useful role in TEIA. One such body is the Helsinki Commission (HELCOM), established by the 1992 Baltic Sea Convention, which has all the coastal states of the Baltic Sea as its members.[26] The Baltic Sea Convention contains a specific provision on TEIA in its Article 7.[27] Yet, as is readily apparent from the wording of the Article, the Baltic Sea Convention refers to other international treaties and supranational regulations when it comes to governing TEIA.[28] Hence, even though UNCLOS and the Baltic Sea Convention are generally applicable to the BSGP, they do not contain specific obligations that are suitable for organizing transboundary EIA. This task is, however, specifically handled by the Espoo Convention and the EIA Directive – instruments of international and EC law that apply to all parts of the BSGP other than the section in the Russian Federation.[29]

The problem of the Espoo Convention and the EIA Directive not being applicable to the Russian Federation was resolved in April 2006, at a meeting of the countries under whose jurisdiction the pipeline will traverse, in which they decided how to apply the Espoo Convention to the BSGP TEIA. The Russian

Federation agreed to apply the Espoo Convention to the extent permitted by its own national legislation.[30] Indeed, it is difficult to imagine that the Russian Federation would have consented to having European Union (EU) internal rules – the EIA Directive – govern the EIA. In addition, because the Espoo Convention was negotiated under the auspices of the UNECE, of which North American, European, Central Asian states and the Russian Federation are members, the forum seems best suited to governing TEIA involving EU Member States and the Russian Federation. Furthermore, even though it is not a party to the Espoo Convention, the Russian Federation is still a signatory.[31]

It was clear from the very beginning that organizing a TEIA for the BSGP would be very challenging. The typical situation which the Espoo Convention was originally negotiated to address is one where a proposed activity in state A, located close to an international border between states A and B, is likely to cause a significant adverse transboundary impact on the environment of state B. These situations are characterized for the most part by state A benefiting economically from the proposed activity and state B suffering the environmental consequences, thus creating, at least potentially, an adversarial relationship between the two neighbouring states. Accordingly, the Espoo Convention includes provisions that aim to help the parties find a solution enabling them to organize the TEIA in a mutually beneficial manner.[32] In these typical situations, state A integrates the transboundary impacts and foreign actors into its national EIA, the goal being to examine the entire area of likely impact.

Contrast these typical situations with that of the BSGP. A private company (Nord Stream AG) with Swiss headquarters proposes to build a gas pipeline traversing the maritime areas under the jurisdiction of five states, with the activity potentially causing environmental consequences for all the littoral states of the Baltic Sea. The benefits of the proposed activity would accrue to its shareholders and would serve to diversify the natural gas supply of Western Europe. This does not mean that the Espoo Convention does not apply to the BSGP. As mentioned above, the scope of the Espoo Convention covers large-diameter gas pipelines that are likely to cause a significant adverse transboundary impact. And, according to Article 1, from its inception the Espoo Convention was designed to accommodate activities with a plurality of origin and affected states.[33] Yet, what exactly the Espoo Convention requires from the four Espoo contracting states and the Russian Federation in the case of the BSGP is not altogether clear. The party or parties of origin are obligated to notify the other state(s) if a significant adverse transboundary impact is likely to ensue from the proposed activity within its jurisdiction; the presumption here is that under normal circumstances the activities listed in Appendix I trigger the duty for the origin state to commence a TEIA by notifying the affected state(s). However, the Espoo Convention leaves this decision ultimately to each and every origin state. Hence, one of the origin states might decide to notify one or more of other littoral states on the basis of Article 3, while others may take a different course, perhaps even concluding that no littoral state will suffer a significant adverse transboundary impact. Even though this was never a likely scenario in the case of a project as large and complicated as the BSGP, it is useful to keep this in mind in order to understand the innovative solutions that the concerned states came up with during the EIA procedure.

How the transboundary EIA was organized in the case of the BSGP

Arrangements at the transboundary level

The basic features of the BSGP TEIA were decided in a meeting on 19 April 2006 between the contact points from all the states under whose jurisdiction the planned pipeline would fall (Denmark, Finland, Germany, the Russian Federation and Sweden).[34] The officials decided that there were three categories of concerned states:

1. Four origin states and the Russian Federation, whose jurisdiction the planned pipeline will traverse[35] and whose national EIA and permitting procedures the company must undergo.
2. The nine Baltic Sea coastal states are deemed to be affected states.
3. Four sole affected states (Estonia, Latvia, Lithuania and Poland), whose jurisdiction the pipeline will not traverse and whose national EIA procedures and permitting procedures will not be triggered.

It bears emphasizing here how important this decision is from the legal perspective. By accepting the position of origin states, the four states and the Russian Federation – to the extent permitted by its legislation – committed themselves legally to all the obligations in the Espoo Convention vis-à-vis all nine affected states. Hence, each one of the four origin states and the Russian Federation opened its national EIA procedure to the eight other states and their public,[36] vastly increasing the level of complexity of organizing the transboundary procedure.

What might have been missing, then, was an EIA procedure that could examine the proposed activity in its entirety, not only each sector of the pipeline individually. This prompted another innovative move on the part of the concerned states: they decided to establish international Espoo contact-point meetings (international coordination meetings) for the duration of the BSGP TEIA. The affected states and the company attended most of these meetings; 16 have been held to date.[37] The main task of these meetings has been to ensure that the company makes an environmental impact statement (EIS) for the whole

BSGP (hereinafter the Espoo Report), in addition to assessing the individual sectors on the basis of national EIA legislation. The international coordination meetings took it upon themselves to oversee that the minimum technical requirements of the Espoo Convention were met. These meetings also served other important purposes:

1. They performed the immensely necessary task of coordinating the way the five national EIA procedures functioned. Also important was the direct communication between the contact points and the company to ensure that the Espoo Report, which the company was preparing, met the technical requirements of the Espoo Convention.
2. One of the main challenges faced by the international coordination meetings related to how the sets of data from the four origin states and the Russian Federation could be harmonized. If national EIA systems are very close to each other, it is possible to conduct a joint EIA, for instance, by establishing a steering committee. In the present case, such a committee would have the task of overseeing how the company studied each sector of the pipeline and the overall impacts of the pipeline. Yet, in the case of the BSGP, there was one non-party to the Espoo Convention, making a joint EIA practically impossible. The international coordination meetings were, however, able to push the company to compile the Espoo Report on the impacts of the entire pipeline system, an approach that achieved the dual goal of having the company produce data sets both for the entire pipeline and for each sector.
3. Another challenge related to how public participation was to be organized, given that each of the four origin states and the Russian Federation had committed itself to involving the public of the affected states on an equal footing with its own, as stipulated in Article 2(6) of the Espoo Convention. The most important function played by the international coordination meetings was that they provided a forum that enabled the concerned states to agree that the international scoping document produced by the company, and the final Espoo Report, would be publicly displayed and opened for comment at the same time in all the nine countries. The various assessment documents the company prepared – culminating in the final Espoo Report – were either delivered via contact points to become part of the material in national EIA procedures in the four origin states and the Russian Federation, or forwarded to the contact points in the sole affected states, most of whom placed them on public display and made them available for comment. The four documents prepared by the company were translated into all nine languages; the two most important documents, the initial scoping document and the final Espoo Report, were also translated into English. One result of the international coordination was that some of the countries where national EIA does not contain a scoping stage nevertheless publicly displayed the international scoping document and organized opportunities to comment on it.[38]

The international coordination meetings were limited to the nine affected states and, by invitation, to the company. This caused some concern about the transparency of such meetings, which was conveyed in a note prepared by the Espoo Convention Secretariat in the part urging future guidance:

> *Increase the transparency of preparatory meetings between focal points of concerned Parties, e.g., by making meeting reports available to the public. This would allow all stakeholders to know how the EIA procedure for complex activities would be coordinated and, possibly, how to make comments.*[39]

This is a commendable approach indeed. It would further legitimize the way in which projects on the scale of the BSGP are coordinated by increasing the transparency of international coordination meetings, yet at the same time keep them confined to a manageable number of authorities from the affected states.

Given the complexity of the EIA system thus created, it is useful to once more summarize the main elements of the procedure. First, it is imperative to differentiate between the two levels of EIA, which ultimately proceeded hand in hand: (a) The company prepared the Espoo Report on the entire BSGP project, with the assessments making up the Report mainly supervised by the international coordination meetings of the Espoo contact points from each of the nine affected states; and (b) The five national EIA procedures were conducted whereby the respective sectors of the pipeline were assessed on the basis of national EIA legislation and the quality of the assessments was assured under the rules of each national system. The way in which these two levels were coordinated was and still is of vast importance.

Synchronizing the functioning of the two levels of EIA were the national Espoo contact points. They were in charge of making sure that the international assessment of the entire project fulfilled the minimum technical requirements of the Espoo Convention, supplying the assessment documents of this international EIA procedure as part of the scoping and assessment information of the respective national EIA procedures, and notifying the other affected states of the comments made in the respective national EIA procedures. Overall, the system proceeded in the following way:

1. The company prepared an international-level scoping document at the end of 2006 (Project Information Document),[40] identifying which issues and impacts should be examined in the assessments. This was delivered to all the countries of origin and the Russian Federation, as well as to the four sole affected states, and translated into all the region's languages. In order for all nine affected states and their public to participate in one or more of the national EIA procedures, the countries of origin and the Russian Federation sent a notification to all nine affected states. The five national EIA procedures

were commenced at the same time so that the international scoping document could be included as material in those states where the national EIA contains a scoping stage with public participation.

2. The company prepared a mid-term study at the end of 2007 (Project Information: Status of Nord Stream Pipeline Route in the Baltic Sea),[41] explaining what kind of choices it had made with regard to the issues identified as meriting study. This was put on public display and opened for comment in all nine affected states.
3. The company prepared a specific routing document at the end of 2008 (the Status of the Nord Stream Pipeline Route in Denmark and in Germany)[42] concerning only one specific sector of the BSGP. This was again sent to all nine affected states, but was subjected to detailed scrutiny in Denmark and Germany.
4. The company, after approval from an international coordination meeting, submitted the Espoo Report, which was put on public display and opened for comment at approximately the same time in all nine affected states (at the beginning of March 2009). The completion of this document triggered all five national EIA procedures, as all of them contain a stage where the public and officials can comment on the final EIS. From the perspective of the national EIA procedure, the Espoo Report became part of the overall assessment material produced under the national EIA procedure. Hence, in principle, all nine affected states and their public could, by participating in one of the national EIA procedures, receive and comment on (a) information on the entire project and (b) the specific information produced on the sector under the jurisdiction of that state.

The transboundary EIA as linked to the Finnish EIA procedure

As a party of origin, Finland notified the eight other affected states of the BSGP project in November 2006. In the notification letter,[43] Finland requested that the affected countries indicate whether they intended to participate in the EIA procedure and provide comments concerning the scope of the assessment of the environmental impacts on their EEZ and territory. In addition, the affected countries were asked to submit the comments received from their public. The notification contained a scoping document prepared by the company entitled 'Project Information Document'. Concurrently, Denmark, Germany, the Russian Federation and Sweden notified Finland of the project and offered Finland the opportunity to participate in their EIA procedures.[44] The four parties of origin and the Russian Federation had agreed to send out identical notification letters to all affected parties.

At the same time, the national scoping document (assessment program[45]) required by the Finnish EIA legislation was submitted by the company to the Uusimaa Regional Environmental Centre (UREC), which acts as the liaison authority in the case. The liaison authority, a distinguishing feature of the Finnish EIA system compared to those in other jurisdictions, coordinates the EIA process from the very beginning and plays a central role in ensuring the quality of impact studies and public participation in the EIA. The crucial role of the liaison authority in the Finnish EIA system has been confirmed by court rulings. For example, the Finnish Supreme Administrative Court has given significant weight to the opinion of the liaison authority when reviewing the adequacy of environmental impact statements.[46]

The project information document and the assessment programme were put on public display and made available for comment in Finland from 27 November 2006 to 26 January 2007. A total of 50 comments were received from the public authorities and the public dealing with the project plan as a whole and, in particular, with the sector set to traverse the Finnish EEZ. After the commenting period, the liaison authority submitted its statement on the assessment program along with a summary of other statements and opinions.[47] In its statement, the liaison authority pointed out several salient considerations when preparing the EIS under the Finnish EIA legislation. It also took the responses received from Denmark, Estonia, Germany, Latvia, Lithuania, Poland and Sweden into consideration when preparing the statement.[48]

UREC's statement – together with the summary of other comments on the scoping document and the Ministry of the Environment's (MoE) own opinion – was forwarded by the MoE to Denmark, Germany, the Russian Federation and Sweden in the form of an additional answer to the notification.[49] In this answer, the MoE took up issues that had been raised in the comments nationally and suggested what the parties of origin and the Russian Federation should take up in the negotiations during the assessment process.[50] Finland's response to the Russian Federation's notification contained special comments on the assessment in the eastern part of the Gulf of Finland.[51]

There was fairly heavy criticism from the MoE in its responses to the other origin states and the Russian Federation. The assessment program was said to be too general in nature, the timetable for implementing it extremely tight, and the evaluation lacking in detail. In addition, Finland argued that the methods used and the bases of knowledge were not presented in nearly enough detail and that some research findings relied on in the document were out of date and might thus be inaccurate. Finland also urged the company to examine other alternatives for the routing of the pipeline and pointed out some important impacts needing further study.[52] The Finnish notification contained an annex prepared by the Finnish Institute of Marine Research that would serve as a monitoring program.[53]

It is good to keep in mind that at this early stage of the national EIA procedures – especially those that contain a scoping phase, such as Finland's – the responsible authorities could have asked the company to evaluate other alternatives for implementing the section of the BSGP that was under the jurisdiction of each coastal state. This applies in particular to Finland, given that the liaison authority, UREC, was obligated by law to request the company to analyse all routing alternatives, including land-based ones. However, UREC did not explicitly require such alternatives to be studied and presented in the EIS. Instead, it called for more detailed explanations of why the land-based alternatives were excluded from the EIA. This lack of explicit scoping advice to the company on the part of UREC had a distinct effect on the final EIS review, as the Finnish EIA legislation is built on the idea that the alternatives for the proposed activity should be outlined already in the scoping phase of the assessment.[54] Indeed, after this initial decision by UREC not to ask for a more extensive study of alternatives in the scoping phase, it becomes difficult to argue subsequently in the EIS phase that the assessment is insufficient when it comes to land-based alternatives.

The second document produced by the company – Project Information. Status of the Nord Stream Pipeline Route in the Baltic Sea (October 2007)[55] – described the on-going research and current situation on the basis of comments received. Finland displayed this document publicly and the MoE received 24 comments from both the public and the authorities. Again, the MoE sent a notification letter on 18 January 2008, to the other affected states, airing many of its concerns on how the EIA were carried out.[56]

Finland stated in its notification that from the beginning of the TEIA procedure, it had had concerns about how the company studies alternatives. Two specific points related to the routing near the Russian Federation's island of Gogland and the routing in the Gulf of Finland. The company had informed Finland on 13 December 2007 – after consultation with the Russian Federation's authorities – that it did not intend to study the alternative route south of Gogland.[57] According to the Finnish view, the decision by the company cannot be justified, especially without any supporting information.[58]

This view is well reasoned from the viewpoint of the Espoo Convention's objective of preventing and minimizing negative environmental impacts. The purpose of the Espoo Convention is to prompt the states concerned to find the best possible alternative from the environmental perspective by identifying various alternatives for implementing the proposed activity. However, the Espoo Convention leaves it to the country of origin to determine which alternatives within its jurisdiction are to be examined. Consideration of reasonable alternatives, including the no-action alternative, is stipulated in Article 4(1) and Appendix II of the Espoo Convention with the qualification "where appropriate". The implementation of this requirement seems to be thus largely left to the contracting states to decide. It can reasonably be argued that the Espoo Convention does not require examination of alternatives beyond the jurisdiction of the country of origin.[59] Even though Finland can try to influence the Russian Federation and the company to examine other alternatives, in the end it will be the Russian Federation (in preparing its national EIS) and the company (in preparing the final Espoo Report) that determine, within the jurisdiction of the Russian Federation, which alternatives are studied.

The problem in the Gulf of Finland was that the proposed route follows very closely the outermost sections of the Finnish EEZ without any sound justification from the perspective of environmental protection. In fact, according to the status report, the company had performed desk studies of the route which indicated that the sea bed was deeper and more even on the Estonian side of the Gulf – in Estonia's EEZ – and thus would seem to point to more favourable pipe-laying conditions.[60] However, the developer's application for survey permits was rejected by the Estonian Ministry of Foreign Affairs, and it thus was impossible in practice for the company to study the alternatives on the Estonian side of the Gulf.[61]

The final EIS – the Espoo Report – was to be completed by the company in January 2009. However, in their meeting on 15–16 December 2008, the Espoo contact points indicated that this timeline would not be met because there were still too many gaps and shortcomings of knowledge in the document.[62] Finally, in their 16th meeting, on 13 February 2009, the representatives of the nine Baltic Sea littoral states agreed that the Espoo Report as it stands fulfils the minimum technical requirements imposed by the Espoo Convention.

COMMENTS ON THE ESPOO REPORT

The final Espoo Report is clearly an improvement over the previous versions. One good example is how the report addresses the concerns Finland had about the way the mid-term report by the company (October 2007) dealt with the alternative southern routing via the island of Gogland, discussed above. In the final report, this request was taken into consideration. The southern alternative was taken up, even though the investigations were not as thorough as those dealing with the route north of Gogland.[63] This indicates that the active role of the contact points of the concerned states had clear impacts on the content of the EIA report.

Even though the contact points accepted that the company had now fulfilled the minimum technical requirements of the Espoo Convention with its Espoo Report, they also stressed that the quality of the information, the viability of the conclusions and the possible need for further studies would be examined via the national EIA procedures. This phase started at the beginning of March 2009 in all the origin states and the Russian Federation. Finland reported that, in keeping with its national EIA legislation, the EIS would be put on public display and opened to all affected states and their public for comment. Several public hearings would be organized as well.[64]

After the completion of the Espoo Report, the international and national aspects of the EIA split into two from the Finnish perspective. The MoE was in charge of commenting on the Report, which it did on 8 June 2009. UREC – as liaison authority in the Finnish national EIA procedure – is obligated by the national EIA legislation to give an authoritative statement on the quality of the assessment material submitted by the company. This took place on 2 July 2009.[65] The statement was attached to the final permit application. UREC focused its comments on the specific assessments related to the environmental impacts of the pipeline crossing the Finnish EEZ; the MoE's comments embrace the entire Espoo Report and thus all transboundary impacts likely to cause harm to the Finnish environment, livelihoods, and the Baltic Sea environment in general.

In its comments to other points of contact and the Russian Federation on 8 June 2009,[66] the MoE – on the basis of comments received from the public (16), authorities and research institutes (70) – expressed its concerns as to the quality of the Espoo Report in some aspects, as well as possible ways to remedy some of the shortcomings. The MoE requested more specific information from two of Finland's neighbouring states, as well as from Denmark and the company.

The most problematic sector of the pipeline from the Finnish viewpoint is that which falls under the Russian Federation's jurisdiction, given that, as a non-party, the Russian Federation has only agreed to apply the Espoo Convention to the extent permitted by its national legislation. While recognizing this, Finland still requested that the Russian Federation provide specific assessment information, which will be made available to Finns, and reserve an opportunity for consultation as stipulated in Article 5 of the Espoo Convention. More specifically, Finland was still dissatisfied with the way the Espoo Report describes the alternative route south of Gogland, even though the company did carry out additional studies on the matter. Finland asked for 'a more thorough description of the basis on which the south of Gogland alternative was developed and a more detailed comparison of the two routes',[67] and wants to discuss this issue in the consultations.

There are concerns in Finland that the pipeline construction in the Russian Federation and Swedish sectors would cause adverse impacts on Finland. According to the MoE, general information has been provided on this in the Report, but there are no specific data on transboundary impacts on Finland, and the MoE has now requested these from Sweden and the Russian Federation.[68] It has also asked the company to provide more specific data on the dispersal of sediments from the Russian Federation's and Swedish sectors of the planned pipeline, given that 'cumulative transboundary impacts from constructing two pipelines with repeated seabed interventions should be addressed more specifically'.[69]

There are several shortcomings that Finland would like Nord Stream AG to address or remedy. For example, Finland has requested the identification of possible restricted fishing areas and an assessment of the losses and adverse socio-economic impacts they and the pipeline in general will have on fisheries,[70] and points out that the forecasts of maritime transport used in the Espoo Report are inadequate.[71] The country also takes up the important issue of decommissioning the pipeline after the end of its life-cycle, urging the company to provide further information on comparing the environmental consequences of the alternatives of removing the pipeline or leaving it in place.[72] In addition, Finland suggests that the international coordination between all the Baltic Sea coastal states continue in the form of a monitoring program by exercising the option to commence post-project analysis provided for in Article 7 of the Espoo Convention.[73]

The Finnish national EIA procedure runs parallel to the international one. As stated above, in the next stage UREC gave its statement on 2 July 2009. As a liaison authority, UREC is obligated under national law to evaluate whether the EIS prepared by the company contains an adequate level of information from the viewpoint of Finnish national EIA legislation. UREC required additional studies, given the concerns it has had with the assessments prepared by the company thus far, and in particular the company's scoping document. The Finnish EIA will be completed when the liaison authority submits its statement on the quality of the EIS to the developer.[74] After the EIA procedure, the EIS and the opinion issued by UREC will be attached to the permit applications, and the process will enter the stage of making the final decisions. In the case of the BSGP, the final decisions will be first made by the Finnish Government[75] and then by the Western Finland Environmental Permit Authority.[76]

How should the BSGP transboundary EIA have been done?

From the legal perspective, there is no doubt that the parties to the Espoo Convention and the Russian Federation performed their duties better than the minimum requirements of the Espoo Convention stipulate. The Espoo Convention leaves it for each state to determine whether a proposed activity is likely to cause significant adverse transboundary impacts and, if so, on which countries. As noted above, this could have resulted in a 'fragmented' approach to the TEIA of the BSGP, with one or only some states notifying this or that state. This fragmented approach could have indeed been chosen, for if one looks only at the likely transboundary impacts, it is clear that in some sectors the pipeline would cause only minor harm to some of the littoral states.

Yet, the contracting parties abandoned this 'minimalist' reading of the Espoo Convention from the beginning in favour of a 'community' approach, making it possible to examine the environmental impacts of the BSGP in their totality, not only its impacts on each sector separately. In addition, the way they decided to apply the Espoo Convention allowed all individuals, associations and officials to comment via various fora on the stages of assessment of the entire BSGP project and each individual sector.

It should not be forgotten, however, that from the beginning of the TEIA procedure, criticism was heard from some of the affected states and, from, for example, the European Parliament.[77] The main focus of this criticism was the fact that the alternatives for the proposed activity (the BSGP) were not studied in their totality. Some of the states – especially the solely affected states – argued that the on-shore alternatives for routing the pipeline should have been studied as well,[78] and some, like Finland, argued that in general the alternatives should have been investigated in greater depth.[79]

Yet, the main criticism came from the Espoo Convention Secretariat. In the discussion paper it prepared on how complex activities should be better processed in the Espoo regime, there is veiled criticism of the way the BSGP TEIA was carried out.[80] Throughout the paper the Espoo Convention Secretariat uses the example of a gas pipeline crossing several jurisdictions ('[a]n example of the strategic dimension for a large-scale energy project concerning several countries might be the routing of a gas pipeline'),[81] while talking of the challenges faced by 'complex' activities.

The Espoo Convention Secretariat's discussion paper advances a rather different view on how the Espoo Convention should address proposed activities such as the BSGP. In its consideration, a strategic element is prevalent in cases which it terms 'complex activities', and many parts of the note clearly describe features of the BSGP. The paper advances a very broad idea as to what types of issues should be taken up in assessment where a strategic element is prevalent:

> *Geopolitical issues, economic interests and relations between Parties could influence the pipeline route, thus diminishing the alternatives considered in the EIA. In addition, the pipeline project could significantly influence the regional energy market, affecting countries' energy and climate policies and strategies. Finally, a large energy project could also make regional integration more difficult if there is no consensus among affected Parties. All these aspects form parts of the project's strategic dimension.*[82]

For the Espoo Convention Secretariat, ideally, the strategic dimension of such activities would be considered in the SEA of the plans, programmes or policies (PPP),[83] but 'in practice, a complex activity might not be planned on the basis of a PPP, or if it is, the PPP might not be subject to SEA. Complex projects might arise from an informal or non-statutory process, i.e., not following an established planning scheme of PPP'.[84] The solution to addressing the strategic dimension would be to broach it at the EIA stage: 'the [Espoo]Convention does not stop Parties from assessing strategic issues within EIA'.[85]

The discussion paper was an agenda item for the meeting of one of the sub-bodies of the Espoo Convention, the Working Group on EIA that met on 12 May 2009.[86] The EIA unit of the European Commission organized a special seminar related to 'complex activities' as part of the Working Group's meeting, although the seminar focused almost exclusively on how the BSGP TEIA was organized. The seminar started with a presentation by the origin states' contact points on why and how they organized the BSGP TEIA as they did. Overall, the civil servants felt that even though they encountered challenges throughout this TEIA – one unprecedented in scale, complexity and public attention – they had succeeded in organizing it in a sound manner.[87] After the seminar part of the Working Group's meeting, the Espoo Convention Secretariat proposed that a task-force be designated to examine guidance for these complex activities, and invited countries to join this group.[88] In the ensuing discussion, two starkly contrasting approaches to how the TEIA of projects like the BSGP should be arranged were aired: the one apparent in the Espoo Convention Secretariat's discussion paper and the one presented by the origin states of the BSGP TEIA.

The German delegate – who has been one of the key persons in organizing the TEIA of the BSGP – emphasized that if one raises issues related to the impacts of proposed activities on the geopolitics of the region and its energy and climate policies, this will be close to opening Pandora's Box.[89] In his view, a TEIA procedure is primarily an administrative process and therefore the only way to carry out a manageable TEIA is to focus on environmental issues, as was done in the BSGP TEIA. He also expressed the view that an EIA is an environmental tool to improve decision-making processes, and that the parties involved in the BSGP project took great care to avoid an in-depth discussion of political issues while defining the procedural steps of the TEIA. He asked how it would be possible to conduct an Espoo process appropriately if the impacts of proposed activities on a region's geopolitics and/or national energy and climate policies were also to be addressed during the procedure, given that it was already difficult enough to coordinate the BSGP TEIA with only environmental issues on the table.[90]

These two perspectives on the way the Espoo Convention should have been applied to the BSGP are fairly contradictory as to how ambitious an approach the Espoo regime should take in assessing such projects. The Espoo Convention Secretariat's discussion paper contains interesting, and to some extent realistic, suggestions. In the final part of that paper, where the Espoo Convention Secretariat suggests issues for future guidance, the following is recommended:

(b) Regarding methodology:
  (i) Define how to proceed when the strategic dimension of a complex activity was not assessed in a previous SEA, and give guidance on how to incorporate a formal analysis of strategic issues at the EIA stage. This would allow formal and transparent discussion of all important early decisions, such as the project's final aim and the means to achieve it, and for public participation at early stages of decision-making.[91]

From the legal viewpoint, this is easier said than done. Even if it can be argued that 'the Convention does not stop Parties from assessing strategic issues within EIA',[92] it should be made clear that in the case of the BSGP, the parties cannot legally obligate the company to assess strategic issues. A distinction must be made here between what the domestic officials are mandated to do in their national EIA procedures and what the contact points can require from the company in international coordination meetings on how to apply the Espoo Convention. In some national EIA systems the officials can in fact require a private company to examine more strategic issues in an EIA, although this is certainly the exception rather than the rule. Yet, the power of the contact points of the Espoo Convention does not extend to requiring Nord Stream AG to examine geopolitical, energy and climate policy issues as part of a TEIA.

To be sure, nothing would have prevented the four origin states and the Russian Federation at their meeting with the company in April 2006 from proposing to the company that it examine all these issues. But it is equally certain that the company had no legal obligation whatsoever to conduct such an extensive TEIA. The contact points of the Espoo Convention cannot dictate how the national EIA procedures, which are all carried out on the basis of national EIA legislation, should be conducted. In this very specific respect, the parties cannot assess strategic issues within EIA, and they have no bargaining power vis-à-vis the company by which they might induce it to conduct such studies. From the company's perspective, it would seem counterproductive to voluntarily commit itself to assessing the activities' potential impacts on the region's geopolitical situation or climate and energy policies. Why would it want to raise issues that would broaden the concerns associated with the BSGP?

But, perhaps even more crucially, why would these broad concerns, such as the activities' impacts on geopolitics, energy and climate policy, be addressed if they cannot be included in the working of the national permitting procedures linked to the EIA? The permitting procedures do not allow security or similar considerations to be included as part of the reasoning for whether a permit is granted for the construction of a gas pipeline and, if so, under what conditions.

CONCLUSION

It seems fair to argue that the way in which the responsible authorities organized the assessment of the BSGP was as ideal a way as the Espoo Convention could be applied in the specific circumstances of the case. Much depends in such cases on the willingness of the private company to conform to what the contact points of the Espoo Convention want it to do. If the company does not behave as required by international coordination meetings, this may have a negative influence on the decisions made by the national authorities when applying the national EIA and permitting procedures and may persuade the company to adhere to the soft guidance from these meetings. Such a consideration would leave at least some room for attempting to provide recommendations for 'complex activities' – a job now undertaken by the task force established by the Working-Group on EIA – and for using the lessons learned from the BSGP TEIA for this work.[93]

At least some lessons emerge when studying the BSGP TEIA. As argued above, the BSGP establishes an important and qualitative 'precedent' for future TEIAs concerning large-scale multi-jurisdictional proposed activities. After the BSGP TEIA, it is difficult for any contracting party to the Espoo Convention to avoid establishing a well-structured international coordination modelled on the procedure pioneered by the littoral states of the Baltic Sea, in particular the four origin states and the Russian Federation. The origin states and the Russian Federation chose to adopt a 'community' approach to TEIA, an ideal way to adapt the Espoo Convention to such a challenging project. Even though the BSGP TEIA was conducted in an innovative manner, this does not mean that it was executed perfectly, a realization that prompts suggestions on how to improve TEIAs for large-scale multi-jurisdictional activities.[94]

The present authors consider that the Espoo Convention Secretariat's discussion paper brings out interesting suggestions for future guidance, especially the possibility of including strategic elements in large-scale TEIAs. Yet, it does seem too much to try to include geopolitical scenarios and impacts on a region's climate and energy policy as part of TEIA. A more viable approach would be to emphasize those issues that are already part of TEIA, in particular the identification and rigorous comparison of alternatives (including zero action). Applied to the BSGP TEIA, the origin states and the Russian Federation could have asked the company at their meeting in April 2006 to make a broader international study and comparison of alternatives between land-based and sea-based routing alternatives.

Another lesson relates to harmonization of the data sets produced by national EIA procedures. As was discussed above, because of the differences in the national EIA procedures between the countries of origin and the Russian Federation, it was not possible to harmonize the way the procedures functioned, but only their timing. Yet, if large-scale multi-jurisdictional projects are planned that fall within the jurisdiction of countries with comparable national EIA procedures, it should also be possible to embark on more ambitious harmonization as regards the data sets produced by these procedures. This could take the form of an agreement between the origin states on the methods to be used in various assessments or on which issues need to be studied. If the national EIA procedures are very similar, it might even be possible to carry out a joint EIS, ideally using the strictest national standards to be found among the origin states as the basis for the quality requirements for the joint EIS. If an international steering committee could see to it that only one EIS is produced for the entire multi-jurisdictional proposed activity, it would certainly increase the credibility of the assessment results, especially if these are produced using the strictest of national standards available.

Overall, the BSGP transboundary EIA should be seen as a case where the Espoo regime functioned well, even though the Espoo Convention was not originally geared to these types of situations. The origin states and the Russian Federation opted from the beginning for a 'community' approach, enabling all the littoral states of the Baltic Sea and their public to influence the assessments made on the national and international levels of the environmental impacts of the BSGP. Even though critical comments have been voiced regarding how the BSGP transboundary EIA was organized, it will likely serve as an important precedent for future large-scale projects. The case has also prompted important discussions about the lessons to be learned from the BSGP which could be used to strengthen the regime over the long term. Some of these have been presented above.

The Espoo Convention clearly constitutes a very dynamic international environmental treaty regime, one capable of adjusting to new circumstances. This has been clearly demonstrated in the case of the BSGP transboundary EIA, where the various participants in the regime started to think about how to better regulate these 'complex activities' or how transboundary EIAs, such as that for the BSGP, require sub-regional normative development within the regime, a development that has been on-going for some time.[95]

The fact that there are clear conflicts between the various actors in the regime can be seen as a negative aspect. One such conflict was played out in the meeting of the Working Group on EIA, where the Espoo Convention Secretariat provided a discussion paper that advanced ideas that were clearly not shared by the parties who had organized the BSGP transboundary EIA. Yet, this can be seen as a positive sign as well. The fact that an international environmental treaty regime finds problematic elements in the BSGP transboundary EIA, which clearly was carried out in an innovative manner from the legal perspective, can only mean that the regime in question is a robust one, prepared to adapt itself to new circumstances and challenges.

Notes

This chapter was originally published as: Koivurova, T. and Pölönen, I. (2010) 'Transboundary Environmental Impact Assessment in the Case of the Baltic Sea Gas Pipeline', *International Journal of Marine and Coastal Law*, vol 25, pp151–181.

*. The chapter is based in part on the presentation on the topic by Timo Koivurova entitled 'Environmental Impact Assessment in a Transboundary Context', given at the conference 'The Nord Stream Pipeline: Legal, Economic and Environmental Issues', 20–21 February 2009, in Kiel, Germany (plenary room, Parliament of Schleswig-Holstein), available at www.internat-recht.uni-kiel.de/veranstaltungen/pipelineconference/Conference%20Programme% 20%28final%29.pdf (last accessed 11 June 2009).

1 Good examples of such treaty regimes are the near-universal conventions adopted at the 1992 Rio Conference on Environment and Development, namely the regime based on the United Nations Framework Convention on Climate Change, 9 May 1992, New York, in force 21 March 1994, 1771 *UNTS* 107 and the one created under the 1992 Convention on Biological Diversity, 5 June 1992, Rio de Janeiro, in force 29 December 1993, 31 *ILM* 818 (1992). This has also led to interesting suggestions from scholars, e.g., Churchill and Ulfstein perceive these types of regimes to be Autonomous Institutional Arrangements (AIA), entities comparable to intergovernmental organizations, with the concomitant possibility that the decisions of their plenary organs are binding on the states parties. See R. Churchill, G. Ulfstein (2000) 'Autonomous Institutional Arrangements in Multilateral Environmental Agreements: A Little-Noticed Phenomenon in International Law', 94 *American Journal of International Law* (AJIL) 623–659.

2 Two examples of such international environmental treaties are the 1974 Nordic Environment Protection Convention, 19 February 1974, Stockholm, in force 5 October 1974, 13 *ILM* 591 (1974) and the 1973 Agreement on the Conservation of Polar Bears, 15 November 1973, Oslo, in force 26 May 1976, 13 *ILM* 13 (1974). Both were negotiated, concluded and left at that, although the Polar Bear Agreement was to some extent developed under the IUCN Polar Bear Specialist Group. In addition, with the rising awareness of the problems polar bears will face with melting sea ice, Norway recently hosted the first formal treaty meeting between the five polar bear treaty partners. See N. Bankes, 'Climate Change and the Regime for the Conservation of Polar Bears' in: T. Koivurova, E. Carina, H. Keskitalo and N. Bankes (eds.) (2009) *Climate Governance in the Arctic* UK: Springer, 351–382.

3 The Convention on Environmental Impact Assessment in a Transboundary Context, 25 February 1991, 30 *ILM* 800 (1991) [hereinafter Espoo Convention], entered into force 10 September 1997. There is an increasing body of literature on TEIA; see, e.g., the special issue on transboundary EIA, 26 *Impact Assessment and Project Appraisal* (IAPR) (2008); *Theory and Practice of Transboundary Environmental Impact Assessment* (2008) K. K. Bast meijer, T. Koivurova (eds.) (2008) Leiden: Martinus Nijhoff Publishers; N. Craik, *The International Law of Environmental Impact Assessment, Process, Substance and Integration* Cambridge, UK. Cambridge University Press.

4 For an overview of the negotiations, see R. Connelly (1999) The UN Convention on EIA in a Transboundary Context: A Historical Perspective, 19 *Environmental Impact Assessment Review* 37–46.

5 Available at http://treaties.un.org/Pages/ViewDetails.aspx?src=TREATY&mtdsg_no=XXVII4-b&chapter=27&lang=en (last accessed 1.6.2009). The Protocol requires 16 parties for it to enter into force; currently 13 have accepted the Protocol as binding and it is thus not yet in force. (European Community approval cannot be counted for this purpose.)

6 Espoo Convention, Article 2(2, 3 and 7).

7 Ibid., Article 3(1, 2 and 3).

8 Ibid. The expression 'The public' was defined in Article 1(x) of the Espoo Convention simply as 'one or more natural or legal persons'. Due to the influence of the Aarhus Convention on Access to Information, Participation in Decision-making and Access to Justice in Environmental Matters (25 June 1998, Aarhus, Denmark, in force

30 October 2001, 38 *ILM* 517 (1999)), the second MOP, held in Sofia in 2001, adopted an amendment to the Espoo Convention, adding the following to Article 1(x): 'and, in accordance with national legislation or practise, their associations, organizations or groups'.

9 Ibid., Article 3(1, 2 and 3), Article 3(4–8) and Article 2(6) on non-discrimination.

10 Ibid., Articles 4 and 5.

11 Ibid., Article 6.

12 Ibid., Article 7 and Appendix V.

13 This emerges from the Second Review of Implementation of the Espoo Convention, (ECE/MP.EIA/11), available at www.unece.org/env/documents/2008/eia/ece.mp.eia.11.pdf (last accessed 2.6.2009).

14 'Exchange of Good Practices, Large-Scale Transboundary Projects, Application of the Espoo Convention to complex activities', Note by the Secretariat (ECE/MP.EIA/WG.1/2009/4, 2 March 2009), Economic Commission for Europe Meeting of the Parties to the Convention on Environmental Impact Assessment in a Transboundary Context, Working Group on Environmental Impact Assessment, Twelfth meeting, Geneva, 11–13 May 2009, Item 5(a) of the provisional agenda, available at www.unece.org/env/documents/2009/eia/wg.1/ece.mp.eia.wg.1.2009.4.e.pdf (last accessed 1.6.2009).

15 The project's name is Nord Stream and it is referred to here as the Baltic Sea Gas Pipeline (BSGP); it is being developed by the joint venture company Nord Stream AG.

16 Formally speaking, the note was prepared at the request of the Commission of the European Communities, further to the work plan adopted at the fourth MOP (ECE/MP.EIA/10, decision IV/7, annex).

17 'The Task Force on Complex Activities' is to begin its work in 2010 by examining the need for detailed recommendations on the application of the Espoo Convention to complex activities. If the Task Force concludes that detailed recommendations are needed, it will then draft them, taking into account: (a) lessons learned from past experiences in the application of the Espoo Convention; and (b) existing guidance prepared under the Espoo Convention. The Task Force will focus on joint EIA of large-scale transboundary activities, further to appendix VI, paragraph 2(g), of the Espoo Convention. The recommendations will be considered by the Working Group on EIA and, later, the MOP. The delegations of Romania, Ukraine, the European Commission and, subject to confirmation, Georgia and the Russian Federation, have indicated their willingness to join the Task Force. In addition, other EU Member States may agree to participate in the Task Force after the Baltic Sea sub-regional workshop in October 2009. The Working Group agreed that NGOs would be invited by the Task Force on a case-by-case basis to assist in its work. Information obtained from the Secretary of the Espoo Convention, Wiek Schrage, by email on 1 June 2009. The Seminar on Cooperation on the EIA Convention in the Baltic Sea sub-region (Vilnius, 22–23 October 2009) discussed the issue of whether there are more state parties willing to join the Task Force on Complex Activities, but no volunteers appeared. It was agreed that some of the Baltic Sea states could join the 2010 Moscow seminar on the topic. Information obtained from the official responsible in the Finnish Ministry of the Environment, via email (10.12.2009), on file with the author.

18 See note 14.

19 For instance, the European Parliament has taken a critical stance towards the project; see the European Parliament resolution of 8 July 2008 on the environmental impact of the planned gas pipeline in the Baltic Sea to link up the Russian Federation and Germany (Petitions 0614/2007 and 0952/2007) (2007/2118(INI)), INI reference available at www.europarl.europa.eu/oeil/FindByProcnum.do?lang=en&procnum=INI/2007/2118 (last accessed 24.01.2010).

20 A thorough overview of the various stages is provided in Chapter 2 of the Nord Stream so-called 'Espoo Report', the international EIS document (Espoo Report), available at www.nord-stream.com/fileadmin/Dokumente/eia_permitting/Chapter_02/Nord_Stream_Espoo_Report_English_Chapter_02.pdf (last accessed 2.6.2009).

21 These countries can be regarded as parties of origin because the BSGP passes through their territorial waters and/or their Exclusive Economic Zone (EEZ). See note 33.

22 See, however, E. Karm (2008) 'Environment and Energy: The Baltic Sea Gas Pipeline', 39 *Journal of Baltic Studies* (JBS) 99–121.

23 Espoo Convention, Article 3(1) and Appendix I, paragraph 8.

24 However, signature already requires some measures from the contracting states according to the Vienna Convention on the Law of Treaties, 23 May 1969, Vienna, Austria, in force 27 January 1980, 8 *ILM* 679 (1969) and, arguably, in the customary law of treaties. According to Article 18 of the Vienna Convention and the identical norm in the customary law of treaties: 'A state is obligated to refrain from acts which would defeat the object and purpose of a treaty when: (a) it has signed the treaty ... until it shall have made its intention clear not to become a party to the treaty'. In the case of the Espoo Convention, it can be argued that at least some kind of notification must be made to the potentially affected state even if no comprehensive TEIA procedure is organized; otherwise the object and purpose of the treaty, 'to prevent, reduce and control significant adverse transboundary environmental impact from proposed activities', would be frustrated. Also interesting is that when signing the Espoo Convention, the signatories adopted a resolution: 'to strive for the entry into force of the Convention as soon as possible and to seek to implement the Convention to the maximum extent possible pending its entry into force; 2. Consider that, pending the entry into force of the Convention, the necessary authority should be given to the Economic Commission for Europe and to its Executive Secretary to provide for a sufficient secretariat and, in the framework of the existing budgetary structure, for the appropriate financial means; 3. Continue to co-operate in bringing closer together their policies and strategies in relation to environmental impact assessment'. The resolution is on file with the authors.

25 Article 206, UNCLOS (10 December 1982, Montego Bay, Jamaica, in force 16 November 1996, 21 *ILM* 1261) [hereinafter the '1982 LOSC'], provides: 'When States have reasonable grounds for believing that planned activities under their jurisdiction or control may cause substantial pollution of or significant and harmful changes to the marine environment, they shall, as far as practicable, assess the potential effects of such activities on the marine environment and shall communicate reports of the results of such assessments in the manner provided in article 205'. Article 205, 1982 LOSC, then specifies how such results are to be published and communicated: 'States shall publish reports of the results obtained pursuant to

article 204 or provide such reports at appropriate intervals to the competent international organizations, which should make them available to all States'.

26 The 1992 Convention on the Protection of the Marine Environment of the Baltic Sea Area, 9 April 1992, Helsinki, in force 17 January 2000, 2099 *UNTS* 197. This Convention, with the amendments, can be found on the home page of the Helsinki Commission (Baltic Marine Environment Protection Commission), with the most recent amendment, available at www.helcom.fi/stc/files/Convention/Conv1108.pdf (last accessed 2.6.2009).

27 Article 7 stipulates: '1. Whenever an environmental impact assessment of a proposed activity that is likely to cause a significant adverse impact on the marine environment of the Baltic Sea Area is required by international law or supra-national regulations applicable to the Contracting Party of origin, that Contracting Party shall notify the Commission and any Contracting Party which may be affected by a transboundary impact on the Baltic Sea Area. 2. The Contracting Party of origin shall enter into consultations with any Contracting Party which is likely to be affected by such transboundary impact, whenever consultations are required by international law or supra-national regulations applicable to the Contracting Party of origin. 3. Where two or more Contracting Parties share transboundary waters within the catchment area of the Baltic Sea, these Parties shall cooperate to ensure that potential impacts on the marine environment of the Baltic Sea Area are fully investigated within the environmental impact assessment referred to in paragraph 1 of this Article. The Contracting Parties concerned shall jointly take appropriate measures in order to prevent and eliminate pollution including cumulative deleterious effects'. Even though HELCOM has not been directly in charge of the TEIA, Finland and other countries have sent all the pertinent documents to it.

28 Interestingly, however, there was an attempt within HELCOM to adopt practical guidance on how to apply Article 7 in the Baltic Sea. A working group was tasked to prepare this guidance and it produced a document entitled 'Guidance on the Practical Application of HELCOM Requirements on Conducting Environmental Impact Assessment', where specific recommendations were given. Interestingly, the guidance document based itself on the customary law principle of good neighbourliness and other relevant conventions, also mentioning the Espoo Convention. Oddly enough, it does not touch upon public participation at all, which is normally seen as an integral part of a TEIA. Another interesting aspect is that the recommendations envisage HELCOM points of contact and HELCOM itself playing a clear role in the international coordination of TEIA. There are even provisions requiring the HELCOM points of contact and HELCOM to be in touch with the Espoo points of contact and the Espoo Convention Secretariat where HELCOM coordinates the TEIA. All the Baltic Sea littoral states, with the exception of the Russian Federation, endorsed the guidelines document, but due to the Russian Federation's resistance they were not adopted. Guidance on the Practical Application of HELCOM Requirements on Conducting Environmental Impact Assessment, Helsinki Commission, HELCOM 26/2005, Baltic Marine Environment Protection Commission, 26th Meeting, Helsinki, Finland, 1–2 March 2005, Agenda Item 3, Matters arising from the subsidiary bodies, Document code: 3/1, Date: 14.1.2005 Submitted by: Executive Secretary. The document is on file with the authors.

29 The amendment to the EIA Directive (97/11/EC) incorporated the Espoo Convention requirements as part of the EC Law as the EC is a party to the Espoo Convention. There are some minor requirements of the EIA Directive on TEIA that go further than those of the Espoo Convention. For instance, the Espoo Convention requires in its Article 6(2) that the origin state deliver the final decision to the affected state, but it does not specifically require it to forward this information to the public of the affected state, a requirement that is contained in Article 9(3) of the EC EIA Directive. In addition, the EC EIA Directive states that the final decision on the proposed activity 'must' take into account the results of transboundary consultations and comments from the public of the affected state, whereas the Espoo Convention uses the weaker wording 'take due account'. Council Directive 97/11/EC of 3 March 1997 Amending Directive 85/337/EEC on the Assessment of the Effects of Certain Public and Private Projects on the Environment, OJ No. L 073, 14/03/1997.

30 See Chapter 3 of the Espoo Report, p.62, available at www.nord-stream.com/fileadmin/Dokumente/eia_permitting/Chapter_03/Nord_Stream_Espoo_Report_English_Chapter_03.pdf (last accessed 2.6.2009).

31 See note 24.

32 T. Koivurova, 'The Convention on Environmental Impact Assessment in a Transboundary Context', in: G. Ulfstein, T. Marauhn, A. Zimmermann (eds.) (2007) *Making Treaties Work*, Cambridge, UK: Cambridge University Press, 218–239.

33 Article 1 (ii and iii) provides: ' "Party of origin" means the Contracting Party or Parties to this Convention under whose jurisdiction a proposed activity is envisaged to take place; (iii) 'Affected Party' means the Contracting Party or Parties to this Convention likely to be affected by the transboundary impact of a proposed activity.

34 See p.32 of the Project Information Document (Espoo scoping document) by the company, available at www.nord-stream.com/fileadmin/Dokumente/1__PDF/2__PIDs/PID_ENGLISH.pdf (last accessed 2 June 2009).

35 The Russian Federation is not called an origin state because it is not a party to the Espoo Convention.

36 This meant that, in principle, with this decision a large number of origin state-affected state legal relations were created, which were then governed by the Espoo Convention (in the case of the Russian Federation, to the extent permitted by its national legislation).

37 See the dates and places of these meetings, available at www.nord-stream.com/fileadmin/Dokumente/Other_Publications/Nord_Stream_Espoo-meetings.pdf (last accessed 2 June 2009). There have also been two expert meetings; the experts were designated by the affected states. See Chapter 3 of the Espoo Report, available at www.nord-stream.com/fileadmin/Dokumente/eia_permitting/Chapter_03/Nord_Stream_Espoo_Report_English_Chapter_03.pdf (last accessed 2 June 2009).

38 For an overview of how the Espoo process and domestic procedures have proceeded, see Chapter 3 of the Espoo Report, available at www.nord-stream.com/fileadmin/Dokumente/eia_permitting/Chapter_03/Nord_Stream_Espoo_Report_English_Chapter_03.pdf (last accessed 2 June 2009).

39 Note by the Secretariat, 58(a.vi).

40 English version, November 2006, available at www.nord-stream.com/fileadmin/Dokumente/1__PDF/2__PIDs/PID_ENGLISH.pdf (last accessed 2 June 2009).

41 The document is on file with the authors.

42 Ibid.

43 Finland sent the notification letter, dated 14 November 2006, to all eight other affected states, available at www.ymparisto.fi/download.asp?contentid=76086&lan=fi (last accessed 2 June 2009).

44 Finland responded on 16 February 2007, indicating its willingness to participate in these countries' respective EIA processes, available at www.ymparisto.fi/download.asp?contentid=66885&lan=fi (last accessed 2 June 2009). All the affected countries informed Finland that they would participate in the Finnish EIA procedure.

45 The assessment program describes the strategy for the assessment of environmental impacts and public participation and it must also include, on a sufficient scale, alternatives for implementing the project which are studied during the EIA process. Specific content requirements for assessment programs are laid down in Section 9 of the Finnish EIA Decree (713/2006).

46 For an extensive case law analysis from the perspective of EIA quality, see I. Pölönen (2007) 'Ympäristövaikutusten arviointimenettely, Tutkimus YVA-menettelyn oikeudellisesta asemasta ja kehittämistarpeista ympäristöllisen vaikuttavuu den näkökulmasta' (Environmental Impact Assessment Procedure, The Legal Status of EIA and Improvement Needs from an Environmental Effectiveness Perspective), *Suomalaisen Lakimiesyhdistyksen julkaisuja*, A-sarja N:o 280. Jyväskylä, 181–186 and 192–204; I. Pölönen and T. Koivurova (2009) 'Rajat ylittävä ympäristövaikutusten arviointi – vaihtoehtotarkastelun riittävyys ja suhde lupapäätöksentekoon' (Environmental Impact Assessment in a Transboundary Context: Standards for Adequate Study of Alternatives and Their Relationship to Decision-making), 3 *Lakimies* 371–393.

47 Statement (27.2.2007, UUS-2006-R32–531) of the Uusimaa Regional Environmental Centre on the environmental impact assessment programme, Nord Stream, the Russian Federation-Germany offshore gas pipeline in the Finnish exclusive economic zone. Unofficial translation, available at www.ymparisto.fi/default.asp?contentid=320766&lan=EN (last accessed 2 June 2009).

48 Statement of the Uusimaa Regional Environmental Centre (27.2.2007, UUS-2006-R32531), p.16.

49 According to the Finnish EIA Act (468/1994), the MoE has a duty to coordinate Finland's reply to the notification.

50 The additional answer by the MoE is available at www.ymparisto.fi/download.asp?contentid=63992&lan=en (last accessed 2 June 2009).

51 Additional answer to the Russian Federation, which raises special concerns Finland has with the Russian Federation (Assessment of impacts in the Russian Federation that concern Finland), available at www.ymparisto.fi/download.asp?contentid=66886 &lan=fi (last accessed 2 June 2009).

52 For instance, seabed sediments, remains of weapons and munitions, bird life, impact on fish and fishing, extending the Natura 2000 network to the EEZ and shipwrecks. See note 43. Finland also submitted that the developer may want to consider setting up an international advisory board to act as an independent discussion forum for the project. Another proposal was for the developer to organize joint workshops with

NGOs that had shown interest in the TEIA procedure. Finland suggested various ways in which the parties of origin and the Russian Federation could harmonize their respective national EIA procedures, for example: in examining the alternatives, aiming for the same level of precision in the impact assessments, coming up with a monitoring programme and disseminating information and data to authorities, researchers and the general public.

53 Available at www.ymparisto.fi/download.asp?contentid=66889&lan=fi (last accessed 2 June 2009).

54 Under the EIA Act (Section 10), the developer is required to 'investigate the effects of the project and its various alternatives on the basis of the assessment programme and coordinating authority statement'.

55 The third document received in Finland was entitled 'The Status of the Nord Stream Pipeline Route in Denmark and in Germany' (October 2008). This document, which studied the alternative routings in Denmark and Germany, was also put on public display in Finland and a number of comments were received. These were only assembled by the MoE and notified to the other affected states. It is important that the MoE points out – in a letter dated 22 January 2009 – that because the data in the document are still very general, it can only properly review the adequacy of the EIA documentation when the EIS is completed. Available at www.ymparisto.fi/download.asp?contentid=97436&lan=fi (last accessed 2.6.2009), p.1.

56 Available at www.ymparisto.fi/download.asp?contentid=79195&lan=fi (last accessed 2 June 2009).

57 This alternative is located in the Russian Federation's territorial seas. However, the southern alternative also creates a small deviation for the pipeline route in the Finnish EEZ. Environmental Impact Assessment Report (EEZ of Finland), p.365.

58 Ibid.

59 On the requirement to consider project alternatives under the Espoo Convention, see P. Birnie, A. Boyle (2002) *International Law and the Environment*, 2nd ed, Oxford: Oxford University Press at 135; Craik, op. cit., see note 3, at 139–140, 158; Pölönen and Koivurova, op. cit., see note 46, at 382, 383.

60 Available at www.ymparisto.fi/download.asp?contentid=79195&lan=fi (last accessed 2.6.2009), p.2.

61 Tallinn, 21 September 2007, no. 60 (Refusal to give consent to conduct marine research), available at www.valitsus.ee/brf/failid/otsus_21_09_2007_eng.pdf (last accessed 2 June 2009).

62 Press release, only in Finnish, available at www.ymparisto.fi/default.asp?contentid=307254&lan=fi&clan=fi (last accessed 2.6.2009).

63 See pp.321–322 of the Espoo Report.

64 Press release, only in Finnish, available at www.ymparisto.fi/default.asp?contentid=313477&lan=fi (last accessed 2 June 2009).

65 See further note 74.

66 Comment released on 8 June 2009, 'Consultation in accordance with Articles 4 and 5 of the Convention on Environmental Impact Assessment in a Transboundary Context (Espoo Convention) for the Nord Stream Gas Pipeline' (dated 5 June 2009), available at www.ymparisto.fi/download.asp?contentid=104223&lan=fi (last accessed 10.6.2009).

67 Ibid. note 4. Finland is also concerned about the proposed detonations of munitions close to its border and requests further data – both from the Russian Federation and the company – on this, as well as on the possible mitigation measures available.

68 Ibid. note 3. 'With regard to the above impacts, Finland requests Sweden and the Russian Federation to provide information that is specific to the border locations, including information about the construction activities taking place in the vicinity of the border, the baseline data of the environment, an impact assessment of these areas, and individual mitigation measures'.

69 Ibid. note 4. Finland is also concerned about the planned routing of the pipeline in the Swedish and Danish sectors as regards the dispersal of wartime dumping sites of chemical weapons, in particular the worst-case scenario that the dispersal of such chemicals might accumulate in marine populations and thus have a very wide area of impact across the ecosystems. Therefore, Finland has asked Sweden and Denmark – in the case that these routes are chosen – to carry out 'proper risk assessments and management plans, including a description of transboundary impacts'. Ibid. note 6.

70 Ibid. note 5.

71 Ibid.

72 Ibid. note 7.

73 Ibid. note 7.

74 The UREC issued its statement on 2 July 2009. It stated that the EIS – as it concerns the part of the pipeline traversing the Finnish EEZ – is adequate, although it also requested clarifications and explanations from the company with respect to several issues to be submitted to the permit decision-maker, for instance, more specific risk assessment, in particular on how the pipeline relates to increasing vessel traffic in the Gulf of Finland; detailed assessment of the impacts on the fisheries over the whole life cycle of the proposed pipeline, including impacts on cross-border fisheries; and more specific monitoring programmes that need to be done together with the officials and expert organizations (altogether 15 clarification requests). Available in Finnish at www.ymparisto.fi/download.asp?contentid=105588&lan=fi (last accessed 10.12.2009).

75 This will be done on the basis of the Act on EEZ (1058/2004). The Finnish Government made this decision on 5 November 2009. It granted permission to the developer on the basis of the Act on EEZ, but subjected it to 11 conditions altogether, e.g., that the project must be implemented in accordance with the principle of precaution, paying particular attention to the fragility and vulnerability of the Baltic Sea. The Finnish Government expects that the developer can satisfy the requirements of the Finnish Water Act in these respects. The decision is to be taken in the near future by the Western Finland Environmental Permit Authority. See the decision in Finnish by the Finnish Government, available at www.tem.fi/files/25067/307_VN_NordStream_suostumus_051109.pdf (last accessed 10 December 2009).

76 This will be done on the basis of the Water Act (264/1961).

77 See note 19.

78 See, e.g., Estonia's response to the company's mid-term (October 2007) report (Status of the Nord Stream Pipeline Route in the Baltic Sea), No. 13-3-1/07/57953, 17 January 2008, available at www.ymparisto.fi/download.asp?contentid=80150&lan=fi (last accessed 2.6.2009) and Poland's response, DOOS-082/8412–3/08/IS, Warsaw, 18 January 2008, available at www.ymparisto.fi/download.asp?contentid=80161&lan=fi (last accessed 2 June 2009).

79 Ibid., available at www.ymparisto.fi/download.asp?contentid=79195&lan=fi (last accessed 2 June 2009).

80 Application of the Convention to complex activities, note by the Secretariat (ECE/MP. EIA/WG.1/2009/4, 2 March 2009, see note 14.

81 Ibid., paragraph 13.

82 Ibid.

83 The most important international convention here is the 2003 Protocol on Strategic Environmental Assessment (SEA Protocol; 21 May 2003, Kiev, Ukraine, available at http://untreaty.un.org/English/notpubl/27_4bE.pdf (last accessed 24.1.2010)) to the Convention on Environmental Impact Assessment in a Transboundary Context, which, even though not yet in force, will likely enter into force soon, given that out of the sixteen instruments of ratification (which will trigger the process for its entry into force according to Article 24), 13 have already been deposited (European Community approval cannot be counted for this purpose). Available at www.unece.org/env/eia/documents/legaltexts/protocolenglish.pdf (last accessed 10.12.2009). For a comprehensive study, see S. Marsden (2008) *Strategic Environmental Assessment In International & European Law: A Practitioner's Guide*, London: Earthscan Publications.

84 Ibid., paragraph 14.

85 Ibid., paragraph 16.

86 The provisional agenda is available at www.unece.org/env/documents/2009/eia/wg.1/ece.mp.eia.wg.1.2009.1.e.pdf (last accessed 2 June 2009).

87 One of the present authors (T.K.) was invited to present in this seminar; Observations by Timo Koivurova during the seminar in Geneva on 12 May 2009.

88 See note 15.

89 For a useful illustration as to how many ways there are to perceive the Baltic Sea Gas Pipeline project, see, e.g., R. L. Larsson, *Nord Stream, Sweden and Baltic Sea Security* FOI-R-2251 – SE, Base data report, March 2007, Defence Analysis. Available at http://lsa.umich.edu/UMICH/ceseuc/Home/ACADEMICS/Research%20Projects/Energy%20Security%20in%20Europe%20and%20Eurasia/Nord%20Stream,%20Sweden%20and%20Baltic%20Sea%20Security.pdf (last accessed 10 June 2009).

90 Observations by Timo Koivurova during the Working Group on EIA meeting on 12 June 2009. Confirmation received from Matthias Sauer from the Federal Ministry for the Environment, Nature Conservation and Nuclear Safety Division as to the correctness of the observation (on 3 June 2009).

91 Ibid., paragraph 58(b.i.).

92 Ibid., paragraph 16.

93 See note 17.

94 For interesting suggestions, see B. Flyvberg (2007) 'Policy and Planning for Large-Infrastructure Projects: Problems, Causes, Cures', 34 *Environment and Planning B: Planning and Design* (EPB) 578–597.

95 The sub-regions are the following: Baltic Sea (Denmark, Estonia, Finland, Germany, Latvia, Lithuania, Poland, Russian Federation, Sweden), Black Sea (Bulgaria, Georgia, Romania, Russian Federation, Turkey, Ukraine), Caspian Sea (Azerbaijan, Iran (Islamic Republic of), Kazakhstan, Russian Federation, Turkmenistan), Caucasus (Armenia, Azerbaijan, Georgia), Central Asia (Kazakhstan, Kyrgyzstan, Tajikistan, Turkmenistan,

Uzbekistan), Eastern Europe (Belarus, Republic of Moldova, Ukraine), Mediterranean Sea, South-Eastern Europe (Albania, Bosnia and Herzegovina, Bulgaria, Croatia, Greece, Montenegro, Romania, Serbia, The former Yugoslav Republic of Macedonia), available at www.unece.org/env/eia/subregions.html (last accessed 10.6.2009). For a recent discussion on this topic, see the paper prepared for the meeting of the Working Group on EIA (ECE/MP.EIA/WG.1/2009/32 March 2009), available at www.unece.org/env/documents/2009/eia/wg.1/ece.mp.eia.wg.1.2009.3.e.pdf (last accessed 10 June 2009).

Part IV

Protocol Law and Potential

10

Strategic Environmental Assessment and Sustainability in Europe: How Bright Is the Future?

Simon Marsden and Jan De Mulder

INTRODUCTION

Strategic environmental assessment (SEA) is the evaluation of the impacts of *policies*, *plans* and *programmes* upon the environment. SEA is closely related to environmental impact assessment (EIA), which is the evaluation of the impacts of *projects* upon the environment. SEA should strengthen EIA (which is often applied late in the decision-making process), because it addresses impacts at higher or earlier levels of the process, and thereby aims to avoid them at the lower levels or at a later stage. It also specifically advocates integrating environmental factors into decision-making, in order to advance sustainability.

EIA and SEA were first required under the 1969 National Environmental Policy Act (NEPA) in the US,[1] and requirements for EIA are now common worldwide. Requirements for SEA have also been adopted by a number of countries, sometimes together with those for EIA, sometimes linked with those for EIA and sometimes separately. The purpose of this chapter is to look specifically at two of these requirements: Council Directive 2001/42/EC of 27 June 2001 on the Assessment of the Effects of Certain Plans and Programmes on the Environment (SEA Directive);[2] and the Protocol on Strategic Environmental Assessment (SEA Protocol)[3] to the United Nations Economic Commission for Europe's (UNECE) Espoo Convention on Environmental Impact Assessment in a Trans-boundary Context (Espoo Convention).[4]

Both of these legal instruments have been a long time coming. The SEA Directive has its origins in interest in the European Commission going back over a decade; there has been a similar interest among parties to the UNECE (which includes the Member States of the European Community (EC), as well as non-EC countries and a few others), with the publication of an influential SEA guide over one decade ago.[5]

Whilst the institutionalization of SEA in these legal instruments means a considerable step forwards to scrutinize governmental plans and programmes, administrations and 'experts' continue to struggle to develop appropriate methodological approaches to implement these instruments. It seems to be difficult for SEA practitioners to get out of the shadow of the classical EIA tradition. One of the reasons might be that the provisions of both legal SEA instruments reflect a great similarity with the stages of the EIA process: screening, scoping, drafting the report, reviewing the report and its use by the public during the inquiry about the permit application, and the final decision about the proposed activity, after which monitoring may follow. It should also be mentioned that the five-year review of the EIA Directive by the European Commission[6] has revealed that implementation remains problematic with respect to a number of issues, such as scoping, the study of alternatives and the quality of reports. So it is no surprise that the EIA Directive relates to the main areas of EC legislation in which complaints cases and the resultant infringements are found. As of May 2004, there were 225 open complaints cases, of which there were 68 infringements. This implementation deficit should not be neglected and may be considered as a warning signal regarding the legal future of SEA.

One should also not overlook the fact that in the US, NEPA was not very successful, which may hold lessons for the success of SEA in Europe. Underwood and Alton state:

> *... Like the SEA-Directive, NEPA created a broadly scoped and dynamic environmental accounting tool within the US. However, the US governmental agencies have struggled with the application of NEPA. There has been an extreme reluctance to utilize NEPA's full capacity for over three decades ...*[7]

SEA, ENVIRONMENTAL POLICY INTEGRATION AND SUSTAINABLE DEVELOPMENT

Noble has rightly identified the key component of SEA as strategy, which he defines as 'the art of the general; the prelude to the beginning; the determination of objectives and means, and the adoption of courses of action to achieve specific ends'.[8] Based upon an application of strategic characteristics to specific SEAs, Noble defines SEA as follows, emphasizing the importance of alternatives and outcomes:

> *SEA is the proactive assessment of alternatives to proposed or existing policies, plans and programs, in the context of a broader vision, set of goals, or objectives to assess the likely outcomes of various means to select the best alternatives(s) to reach desired ends.*[9]

SEA can play a significant role in enhancing the integration of environmental considerations in policy and planning processes because it is directed at strategic decision-making. The general benefits of SEA are that it can help decision-makers by first of all achieving environmentally sound and sustainable development, and furthermore by:

- strengthening policy, plan and programme-making processes;
- saving time and money by avoiding costly mistakes; and
- improving good governance and building public trust and confidence in decision-making.[10]

The European Commission, in the lead up to the SEA Directive coming into force, took a keen interest in 'these advantages', commissioning a three-volume report into 'the way in which environmental considerations are included in policy, plan and programme decision-making in all sectors ... rather than simply raising the profile of environmental policies within government and institutional agendas'.[11] The report is concerned with a wide range of matters, including existing experience with SEA worldwide, but the main objective of the report is arguably to inform implementation strategies for the Directive.

Among many others, Dalal-Clayton and Sadler have argued that SEA is needed to advance the sustainability agenda.[12] Quite what this means is open to debate. Dovers has argued that what is required is for SEA 'to play a significant part in the policy and institutional challenge of sustainability',[13] what he calls 'deep SEA'. Deep SEA can be contrasted with 'shallow SEA' in that the latter is concerned with scoping environmental impacts of classes of developments or regional resource and environmental management issues, often from the private sector. It does this by simply applying EIA to plans and programmes, rather than being concerned to challenge the process itself, particularly to address the policy process and the institutional restraints to which it is subject. Dovers distinguishes between first-order (direct) and second-order (indirect) impacts: an example of first order impacts is biodiversity loss; an example of second-order impacts is land clearance. There is little point in preventing biodiversity loss if land is still being cleared; without a habitat many species will simply end up in zoos or on isolated reserves, if not extinct. Deep SEA addresses the latter as well as the former:

> *As well as dealing with second-order impacts, a key characteristic of deep SEA is that the main focus is on government policies and actions, not on development proposals from the private sector. This is because ...*

> *public policy, institutions and law are enormously influential in shaping society and hence are crucial targets for assessment ... it is doubly important to establish as a target for SEA, because broad government policy directions are more likely to escape proper assessment by government than any private sector proposals.*[14]

While the SEA Directive and Protocol are primarily concerned with direct effects from government plans and programmes (the Protocol also can focus on policies and legislation), indirect effects must also be addressed; illustrating elements of both shallow and deep SEA.[15] In this context, Partidário and Clark suggest that SEA can play a role towards sustainability if a number of conditions (outlined in Table 10.1) are met.

A national sustainable development strategy is commonly used to set the policy framework and to ensure that environmental protection requirements are integrated into policies, plans and programmes for specific sectors. Sustainable development indicators are sometimes used to monitor the effects of decisions upon the environment. Although a lot of work has already been done on indicators,[17] their further development is essential to assess and report on progress in achieving environmental goals. In relation to specific environmental issues and/or media, indicators are required that focus on the state of the environment, driving forces and pressures. Assessing progress in relation to environmental integration across economic sectors requires the development of tailored sets of indicators (such as along the lines of Transport and Environment Reporting Mechanism (TERM) in the EU).[18] Similar work is being undertaken by the European Environment Agency in relation to other sectors, such as agriculture, tourism and energy.[19]

Table 10.1 *Conditions needed for SEA to promote sustainability (Partidário and Clark)*[16]

- A policy framework is in place, establishing the articulation across sectoral policies and institutional contexts.
- Credible and feasible alternatives [are considered] that allow evaluation of a decision based on comparable rather than absolute values.
- [There is] recognition that uncertainty characterizes any policy and planning decision.
- Simple though pragmatic indicators [are used] that can assist monitoring of the decisions to determine actual effects.
- Good communications mechanisms [exist] to ensure that all partners in the SEA process are adequately involved and their perspectives contemplated.

Both these and other conditions will be considered below in relation to the SEA Directive and SEA Protocol. A recent publication from the Organization for Economic Cooperation and Development (OECD) on the results of sustainable development has clearly indicated that EIA is playing and SEA is expected to play a role in such a policy:

> *All OECD countries have procedures for environmental impact assessments of large projects, usually combining scientific evaluation and extensive public consultation, before a final decision can be made. While countries are increasingly using strategic environmental assessment, much remains to be done to ensure that sectoral policies (such as agricultural and energy policies) take into account their environmental consequences.*[20]

SEA Directive

The SEA Directive contains 20 recitals, 15 articles and two annexes. It is binding upon the 25 Member States of the EC. For the purposes of this chapter, the important provisions are Recitals 1 and 2 (which provide the context of sustainability), and Article 1, which set out the objectives.

Recital 1 of the Directive refers to Article 174 of the Treaty on European Union (TEU),[21] which provides that Community policy on the environment is to contribute to the preservation, protection and improvement of the quality of the environment, the protection of human health and the prudent and rational utilization of natural resources and that it is based on the precautionary principle. Recital 1 also refers to Article 6 of the TEU, which provides that environmental protection requirements are to be integrated into the definition of EC policies and activities, with a view to promoting sustainable development. Recital 1 therefore provides the policy framework for SEA to play a role towards sustainability. In basing itself firmly on the precautionary principle, the EC policy on the environment also recognizes that uncertainty characterizes any policy and planning decision.

Recital 2 of the Directive refers to the Fifth Environment Action Programme (EAP): *Towards Sustainability – a European Community Programme of Policy and Action in Relation to the Environment and Sustainable Development*,[22] as supplemented by the Council decision on its review, which affirmed the importance of SEA.[23] Also, the Sixth EAP pays attention to SEA (and EIA),[24] requiring the full and effective use and implementation of EIA and SEA as one of the further efforts for the integration of environmental protection requirements into the preparation, definition and implementation of EC policies and activities (Article 3, paragraph 3).

Article 1 of the SEA Directive, in setting out objectives, recognizes the importance of the sustainable development context provided by Articles 174 and 6 of the Treaty, as cited in the recitals. Article 1 states:

> *The objective of this Directive is to provide for a high level of protection of the environment and to contribute to the integration of environmental considerations into the preparation and adoption of plans and programmes with a view to promoting sustainable development, by ensuring that, in accordance with this Directive, an environmental assessment is carried out of certain plans and programmes which are likely to have significant effects on the environment.*

Annex I(e) of the Directive states that the environmental report that is required under Article 5 must outline 'the environmental protection objectives established at international, EC or Member State level, which are relevant to the plan or programme and the way those objectives and any environmental considerations have been taken into account during its preparation'. This is significant because it highlights that EC environmental policy cannot be viewed in a vacuum; rather it is subject to international law and policy, where relevant. This is seen also in Recitals 3 and 7 of the Directive, which recognize the role of SEA in international law.[25] Recital 3 refers to the Convention on Biological Diversity (CBD),[26] which requires parties to integrate as far as possible and appropriate the conservation and sustainable use of biological diversity into relevant sectoral and cross-sectoral plans and programmes. Recital 7 refers to the Espoo Convention (discussed briefly below).

Annex II of the Directive outlines criteria for determining significant effects, where SEA under the Directive is not mandatory (under Article 3(2)), but discretionary (under Article 3(3)–(4)). These criteria include 'the relevance of the plan or programme for the integration of environmental considerations in particular with a view to promoting sustainable development' and 'the relevance of the plan or programme for the implementation of Community legislation on the environment' (e.g. plans and programmes linked to waste management or water protection).

The application of the Directive is limited in scope, but is closely related to Directive 85/337/EEC (as amended) on the Assessment of the Effects of Certain Public and Private Projects on the Environment (the EIA Directive).[27] This is explained in the SEA Directive Guidance.[28] Not only are policies excluded, but there are strict criteria established in Articles 2 and 3, which emphasize that plans and programmes must be formally required and be prepared for listed sectors (which must, in turn, set the framework for future development consent of projects under the EIA Directive), or must be plans that must be assessed under Directive 92/43/EEC on the Conservation of Natural Habitats and of Wild Flora and Fauna (the Habitats Directive). The sectors listed in SEA Directive Article 3 usually requiring assessment are: agriculture, forestry, fisheries, energy, industry, transport, waste management, water management, telecommunications, tourism, town and country planning or land use. These are subject to review, together with the types of plans and programmes, as Article 12(3) of the Directive contains a provision that is designed to further improve integration. This states:

> *With a view further to integrating environmental protection requirements, in accordance with Article 6 of the Treaty, and taking into account the experience acquired in the application of this Directive in the Member States ... a report [on the application and effectiveness of this Directive will be provided before 21 July 2006, and] will be accompanied by proposals for amendment of this Directive, if appropriate. In particular, the Commission will consider the possibility of extending the scope of this Directive to other areas/sectors and other types of plans and programmes.*

De Mulder states that the SEA Directive is 'an important step towards the realization of a concrete integration approach'.[29] He draws attention to the integration of environmental concerns in planning processes that the SEA Directive will contribute towards and states that 'a more integrated system of planning means that environmental criteria are incorporated throughout the planning process, which could help to implement the concept of sustainable development'.[30]

The question remains, of course, whether this opportunity will be used by the Member States when transposing and applying the SEA Directive. Article 4(2) of the SEA Directive stipulates that:

> *The requirements of this Directive shall either be integrated into existing procedures in Member States for the adoption of plans and programmes or incorporated in procedures established to comply with this Directive.*

From a formal perspective and based on efficiency considerations, it is obvious that the 'integration track' in Article 4(2) offers advantages, but this approach does not guarantee an effective outcome with respect to the integration of environmental considerations. Communications between experts at EU-level meetings indicate that the ongoing discussions on preparing draft SEA legislation in the different Member States are very much about the tension between the environment administration and the other policy fields regarding the assurance of an effective SEA process within the existing planning procedures.

The Commission's SEA Directive Guidance remains extremely formal when elaborating the meaning of Article 4(2) and refers only to 'planning procedures' and the opportunity to avoid overlap.[31] Also, the fact that the SEA Directive Guidance describes 'the environmental report' as required by Article 5 of the SEA

Directive as the central part of the environmental assessment has to be considered as a legacy of the EIA approach.[32] The Guidance continues as follows:

> ... *The environmental report is an important tool for integrating environmental considerations into the preparation and adoption of plans and programmes since it ensures that their likely significant effects on the environment are identified, described and assessed and taken into account in that process.*[33]

However, as the EIA practice of the past decades offers numerous examples of the fact that even a good report or environmental impact statement (EIS) does not ensure that environmental concerns are taken into account in the final decision, so there is no reason to believe that the SEA process would be more successful. On the contrary, SEA experts keep on repeating that an effective SEA requires approaches that are different. Partidário favours a so-called top-down approach.[34] This means that the scope of a SEA is designed or developed around national strategies or environmental and sustainability objectives. In the next stage, the particular objectives, options and alternatives are the subject of an assessment. The assessment has to result in the formulation of a context for the development of solutions in order to address the initial policy problems. This approach differs from a SEA (bottom-up) approach (similar to the EIA approach) that is done after the policy proposals – including alternatives – have been formulated. Such an approach leads primarily to mitigation of the effects of proposed measures.

The top-down approach also requires other skills and methods of work and focuses the assessment work differently. Nitz and Brown have described this as follows:

> *EA practitioners need to become familiar with the policy process, and there is a body of literature and practice within the policy sciences to assist with this. Environmental assessment practitioners must:*
>
> - *understand the stages of policy making and identify the activities and issues that will be addressed within each stage;*
> - *identify when and by which actors, decisions are made within different stages of the policy-making process (decision scoping), and the appropriate form and content of environmental information that should be available at these decision points; and*
> - *mould the content and form of SEA to contribute to these decisions. This moulding will have to be specific to each particular policy making context.*[35]

Returning to Partidário and Clark's conditions, which must be met if SEA is to play a role towards sustainability, aside from the sustainable development framework and the recognition of uncertainty (discussed above), each of these conditions is provided for by the Directive itself. Alternatives must be analysed in accordance with Article 5, which concerns the environmental report. 'Reasonable alternatives' to the likely significant effects on the environment of implementing the plan or programme must be considered, 'taking into account the objectives and the geographical scope of the plan or programme'. Annex I of the Directive provides more detail on what the environmental report must contain. Annex I(h) includes a requirement to provide an 'outline of the reasons for selecting the alternatives dealt with, and a description of how the assessment was undertaken including any difficulties (such as technical deficiencies or lack of know-how) encountered in compiling the required information'.

Article 6 of the Directive is concerned with consultation, which is the last of Partidário and Clark's conditions. Article 6 implements the UNECE Aarhus Convention on Access to Information, Public Participation in Decision-Making and Access to Justice in Environmental Matters, which has been a significant influence upon the SEA Directive.[36] The Directive requires that authorities designated by Member States on the basis of their specific environmental responsibilities, must be consulted in determining whether certain plans or programmes are to be assessed in determining the scope and level of information to be contained within the environmental report, and on the draft plan or programme and environmental report. The public must also be consulted on the draft plan or programme and environmental report if the public is 'affected by, or having an interest in, the decision-making'.[37] With regard to the public participation requirements of the SEA Directive, there has been criticism for a failure to comply with Article 7 of the Aarhus Convention.[38]

Finally, in relation to Partidário and Clark's conditions set out in Table 10.1, Article 10 of the Directive provides for significant environmental effects of the implementation of plans or programmes to be monitored, in order that unforeseen adverse effects can be identified at an early stage, and in order to carry out remedial action where needed. Annex I(i) requires that monitoring measures must be described as part of the environmental report, and Annex I(g) requires the report to outline 'the measures envisaged to prevent, reduce and as fully as possible offset any significant adverse effects ...'. With regard to the use of indicators to monitor effects, these can be based on the criteria that the Directive sets out to determine the likely significance of effects. The criteria are set out in Annex II, and are of two types: characteristics of the plans and programmes, and characteristics of the effects and the area likely to be affected.

With regard to the decision-making process itself, Articles 8 and 9 are of relevance. Article 8 provides that the environmental report, opinions expressed on it and results of any transboundary consultations must be taken into account during the preparation of the plan or programme. Article 9 requires that Member States must inform the designated authorities and public of the plan or programme subsequently adopted, including information on how environmental considerations have been integrated into the plan or programme, and how the environmental report, opinions and consultations have been taken into account. Information must also be disclosed on the reasons for choosing the plan or programme adopted, and of the monitoring measures decided upon. These are significant provisions as they are directed to transparent and accountable decision-making.

The first review by the European Commission of the SEA Directive is scheduled for 2006. The question remains whether there will by then be sufficient practical experience with the application of the requirements. As Risse et al observed: 'the general requirements prescribed by the Directive are not restrictive and leave ample room for creativity, flexibility and adaptability to suit each Member State's context'.[39] At the same time, the Commission has initiated a study on the relationship between the EIA and SEA Directives and, when complete, an experts' working group will recommend possible amendments to the EIA Directive. This could be a first move towards integrating the two directives. As Sheate has suggested: '... This coincidence of the review processes provides an early opportunity to consider the consolidation of both directives as a coherent whole'.[40]

An additional consideration regarding the possible merging of the directives is the way the Commission deals with these instruments. This can be illustrated by the approach of DG Transport and Energy and in particular the new TEN-T Guidelines adopted by the European Parliament and the Council on 21 April 2004.[41] According to Article 8.2 of the Guidelines, the Commission shall:

> *... By 21 July 2004 ... in agreement with Member States, develop suitable methods for implementing the strategic environmental assessment with the objective of ensuring, inter alia, appropriate coordination, avoiding duplication of effort, and achieving simplification and acceleration of planning processes for cross-border projects and corridors.*[42]

SEA PROTOCOL

The SEA Protocol is binding upon the parties to the Espoo Convention that have signed it. These include 37 states and the EC. In common with the link between the SEA and EIA Directives, the SEA Protocol is linked with another international instrument, the Espoo Convention, which shares the same Secretariat, and the Aarhus Convention, the requirements of which are largely incorporated into the SEA Protocol. While Espoo is concerned specifically with transboundary impacts from projects, the SEA Protocol is not, as some may suspect, limited to transboundary impacts from plans and programmes. Rather, it operates as a stand-alone instrument, and is concerned with impacts from plans and programmes within a state, with consideration of transboundary effects being a secondary concern.

The Protocol contains eight recitals, 26 articles and five annexes. The first recital recognizes the importance of integrating environmental, including health, considerations into the preparation and adoption of plans and programmes and, to the extent appropriate, policies and legislation. The second recital commits the parties to promoting sustainable development, based in particular upon the conclusions of the United Nations Conference on Environment and Development (UNCED), especially Principles 4 and 10 of the Rio Declaration on Environment and Development,[43] together with the outcome of the third Ministerial Conference on Environment and Health[44] and the World Summit on Sustainable Development.[45] These developments arguably provide the policy framework needed because they place the objective of the Protocol (described below) within a sustainable development context agreed by the international community, including the emphasis placed by the Protocol upon environmental health. However, as seen below, the framework is significantly weaker than that provided for the SEA Directive.

The third and fifth recitals recognize the importance of the Espoo and Aarhus Conventions respectively, with the sixth emphasizing the importance of public participation, which is the final criterion in Partidário and Clark's criteria in Table 10.1. The fourth recital stresses the role of SEA:

> Recognizing *that strategic environmental assessment should have an important role in the preparation and adoption of plans, programmes, and, to the extent appropriate, policies and legislation, and that the wider application of the principles of environmental impact assessment to plans, programmes, policies and legislation will further strengthen the systematic analysis of their significant environmental effects ...*

The objectives of the SEA Protocol are found in Article 1, which states:

> *The objective of this Protocol is to provide for a high level of protection of the environment, including health, by:*
>
> *(a) ensuring that environmental, including health, considerations are thoroughly taken into account in the development of plans and programmes;*
>
> *(b) contributing to the consideration of environmental, including health, concerns in the preparation of policies and legislation;*
>
> *(c) establishing clear, transparent and effective procedures for strategic environmental assessment;*
>
> *(d) providing for public participation in strategic environmental assessment;*
>
> *(e) integrating by these means environmental, including health, concerns into measures and instruments designed to further sustainable development.*

With regard to the role of the SEA Protocol in 'further[ing] sustainable development', outlined in Article 1(e), De Mulder is highly critical. He states that:

> *Although the considerations in the preamble refer to the WSSD [World Summit on Sustainable Development] outcome (Johannesburg, 2002), the objectives of the SEA Protocol are not clearly situated within a policy framework aimed at sustainable development. Given the absence of SEA in the meagre WSSD Plan of Implementation, perhaps one should not be surprised.*[46]

This is significant because while some international instruments recognize the importance of SEA (such as the CBD), and while the SEA Directive is clearly situated in a sustainability framework of the Treaty and Action Programmes of the EU, which recognizes the role of SEA in taking the agenda forward,[47] this may not the case with the SEA Protocol, especially given the much-criticized WSSD outcomes.[48]

Of other provisions, and as also seen in the recitals, of immediate note is that while the Protocol is directed at plans and programmes (see Article 4), Article 13(1) provides that:

> *Each Party shall endeavour to ensure that environmental, including health, concerns are considered and integrated to the extent appropriate in the preparation of its proposals for policies and legislation that are likely to have significant effects on the environment, including health.*

While this is a discretionary provision, it is given weight because Article 13(4) requires that:

> *Each Party shall report to the Meeting of the Parties to this Convention serving the Meeting of the Parties to this Protocol on its application of this article.*

De Mulder comments on Article 13(4) that it 'is intended to push parties towards the implementation of this article'.[49]

The title of the Protocol gives additional substance to this conclusion. The 'SEA Directive' is so-called by those seeking to shorten the title and with a view to emphasize the original broad aspirations of its early drafters. While plans and programmes are strategic decisions, they are not as strategic as policies and legislation, as they are not the highest levels of decision-making. In Dovers' terminology, they are 'shallow' rather than 'deep' targets for SEA. The SEA Protocol therefore takes assessment of strategic decision-making a stage further, albeit on a discretionary basis, as it can be applied to policies and legislation.[50] However, there have been criticisms of the Protocol (especially among non-governmental organizations (NGOs)), for failing to go as far as it could;[51] indeed the final version of the Protocol is far more limited than earlier drafts.[52]

Procedural provisions of the SEA Protocol are very similar, if not identical, to those of the SEA Directive. For example, the definitions of plans and programmes are identical, although in the Directive 'authority' is defined as an authority 'at national, regional and local level'.[53] The definition of SEA is also very similar to the definition of EA under the SEA Directive, although the SEA Protocol goes further and includes the evaluation of likely environmental and health effects.[54] The Protocol must be applied to the same plans and programmes as the Directive, although regional development is an additional sector, and mining is part of the industry sector. Sectoral plans and programmes must set the framework for future development consent of projects for which an EIA is required under national or international law; Annex I (based on Appendix I of the Espoo Convention)[55] and Annex II (based on Annex 2 of the EIA Directive), set out relevant projects. Similar provisions to those discussed above regarding the SEA Directive also allow discretionary SEA under the Protocol.

Jendroska and Stec state that even minor modifications to supplement the obligations envisaged in the SEA Directive were not accepted by the EU Member State negotiators during the work on the draft SEA Protocol.[56] An example concerns the fact that a Norwegian-inspired effort to introduce the requirement of a notification document was deleted at the end of the negotiation process, even though such a document is part of EIA practice in a number of countries, including EC Member States.

With regard to consultation and public participation, there are some differences. The SEA Protocol does not contain a provision for providing information on the decision taken regarding the plan or programme. Unlike the Directive, the Protocol includes an (albeit discretionary) provision in Article 6(3) to involve the public to decide on the content of the environmental report (scoping). However, as with the SEA Directive, the SEA Protocol has arguably also failed to adequately implement the Aarhus Convention, because it has not included the definition of 'the public concerned', found in Article 2(5).

Of the other conditions that must be satisfied, reasonable alternatives must be considered in the environmental report in accordance with Article 7(2) – a very similar provision to Article 5 of the SEA Directive; and Annex IV(8) emphasizes the importance of Article 7(2) – Annex IV(8) being almost identical to Annex I(h) of the SEA Directive. Under Article 11(2), during the decision-making process, reasons must be given for adopting a plan or programme 'in the light of the reasonable alternatives considered' – again this is very similar to Annex I(h) of the Directive. With regard to the acknowledgement of uncertainty, Annex III, paragraph 5, provides that where there is discretion as to whether plans and programmes are to be assessed, one of the criteria for determining likely significant effects is 'the nature of the environmental, including health, effects such as probability, duration, frequency, reversibility, magnitude and extent ...'. Further, paragraph 6 cites 'the risks to the environment, including health'. Similar provisions are included in Annex II, paragraph 2 of the Directive.

Finally, with regard to monitoring, Article 12(1) of the Protocol is almost identical to the requirement in Article 10 of the Directive, as it requires each party to monitor effects 'to identify, at an early stage, unforeseen adverse effects and to be able to undertake appropriate remedial action'. Monitoring measures must be stipulated in the environmental report in accordance with Annex IV(7) – again, this is very similar to Annex I(g) of the Directive. Provisions concerning the decision-making process are similar to the SEA Directive in that Article 11 of the Protocol requires the same as Article 9(1) of the Directive, with monitoring results to be made public, in accordance with Article 12(2).

The SEA Protocol is subject to review by the parties under its Article 14, which deals with the MOP to the Convention. There are a number of elements to this, outlined in paragraph 4, which include the need to improve procedures, exchange information, and consider and adopt proposals for amendment.

A likely future amendment might concern the improvement of the quality control provision. Article 12 of the SEA Protocol is considered weaker than the requirement in the SEA Directive.[57] This is rather surprising, as even the Commission's representative admitted during the negotiations that this requirement in the SEA Directive could have been stronger, and the five-year review of the EIA Directive has proven the need for such a provision.[58] In an European Environmental Bureau seminar report it was rightly stated that quality control is a major problem with project level EIA, and it was concluded that:

> *We learn from this experience in implementing the SEA Directive. Key issues will be who is responsible for quality control, what power do they have over the decision-making process, do they have competence in environmental matters, and do they have sufficient resources for the job?*[59]

These questions are also relevant regarding the application of the SEA Protocol. But, given the fact that the Implementation Committee of the Espoo Convention has only started the real work quite recently, one might conclude that compliance issues regarding the SEA Protocol, such as the quality of future environmental reports, are not an immediate concern.

A background paper for a work plan on the Protocol that was circulated by the UNECE at the end of 2003 indicated that the UNECE was keen to encourage broad application of the Protocol.[60] The first Meeting of the Signatories, however, which was part of the third meeting of the parties of the Espoo Convention, reflected a more cautious approach. Due to financial constraints, the part of the approved (Espoo) work plan that deals with the SEA

Protocol has only three items on the schedule, with guidance subsequently released on public participation, subregional cooperation and on good practice. Priority will be given for developing the necessary institutional arrangements for the first meeting of the parties of the SEA Protocol. Some already initiated activities concern capacity building (in collaboration with United Nations Development Program (UNDP) and the elaboration of a manual.[61]

Towards Sustainability Impact Assessment (SIA)?

The need to implement the SEA Directive has drawn more attention to the linkage between 'environmental' assessment and other types of appraisal that include aspects of the two other 'sustainable development' pillars: economic and social considerations.

The UK's 2003 guidance document on applying the SEA Directive to the planning system has built on experience from ten years of sustainability appraisals:

> *The integrated treatment of economic, environmental and social issues is a key principle of the Government's Sustainable Development Strategy, set out in 'A Better Quality of Life: A strategy for sustainable development for the UK'. Planning authorities are already required to have regard to economic, environmental and social considerations when preparing development plans, and Policy Planning Guidance notes and other ODPM [Office of the Deputy Prime Minister] publications reflect this approach ... Subject to approval by Parliament, Sustainability Appraisal will be mandatory for Local Development Documents (LDDs) and Regional Spatial Strategies (RSSs) under the Planning and Compulsory Purchase Bill. It is intended to define the future requirements for Sustainability Appraisals so as to make clear to authorities that in carrying them out, they must fully meet the requirements of the SEA Directive. ODPM will produce guidance for authorities on sustainability appraisal when the new planning system is brought into effect. This will give further details on social and economic issues to be addressed and on the relationships between SEA and Sustainability Appraisal ... Meanwhile, the present guidance gives some indications of how an SEA can form part of a Sustainability Appraisal which also examines the social and economic effects of a plan. These parts of the guidance do not aim to be comprehensive, but only to suggest what types of effect could be considered beyond the scope of SEA and how they might be measured and presented.*[62]

This excerpt reveals how complex the institutional setting for a policy towards sustainable development – including the application of SEA – may become.

These complexities could already be foreseen, given some of the theoretical background.

Hardly two years after Agenda 21 was adopted in Rio, Dalal-Clayton suggested that a framework for Sustainability Analysis needed, inter alia, to comprise a suite or 'tool kit' of methodologies and approaches, which:

> *explicitly focus on the trade-offs between the biophysical, social and economic aspects of projects, programmes and policies, recognizing that these take place within a framework of political decision making;*
>
> *are undertaken in a systematic, integrative and transparent way;*
>
> *are participative (not just consultative), to the extent possible and practicable in the context of prevailing socio-political circumstances;*
>
> *need to operate within a set of defined criteria and guidelines for sustainable development, recognizing that these may often be best practice approximations; and*
>
> *recognize that environmental assessment is a major point of departure because it is a process which is well institutionalized in policy and law.*[63]

So it is no wonder that both practitioners and scholars try to find more manageable approaches, like George who stated, after examining the UK's objectives-based sustainability appraisal practice, that a criterion-based approach might offer advantages regarding the practicality of the appraisal or assessment.[64] Others, like Fischer, however, stick to the objectives-led approach and fear too much flexibility and adaptability:

> *... Some of the main reasons for conducting SEA ... include a better consideration of the environment in decision making for sustainable development which is not likely to be achieved in the absence of objectives-led pre-structured approaches.*[65]

At the latest annual conference of the International Association for Impact Assessment (Vancouver, April 2004), Dalal-Clayton and Sadler gave an overview of the current status of the 'next generation' of more integrated assessment tools. They indicated that there is a lot of diversity between tools that can be considered as 'sustainability appraisal–approaches'. Concepts, theoretical work and workshops are flourishing all over the world, but they pose the question of whether there is anything happening in reality. The major challenge, according to both experts, will be to sharpen the focus and to apply these tools effectively so there is a real impact on decision-making.

The latest developments in the EU suggest the need to be more realistic about how far SEA will be implemented. Further to the June 2001 Göteborg

European Council that called for 'mechanisms to ensure that all major policy proposals include a sustainability impact assessment (SIA) covering their potential economic, social and environmental consequences',[66] a system of SIA was to be introduced, which also includes cumulative impact assessment (CIA), cost–benefit analysis (CBA) and health impact assessment (HIA).[67] This was not that new, as already in May 1998 the Commission in a Communication had proposed that all its major policy proposals would be subject to an EIA. This was endorsed by the European Council in June 1998 in Cardiff.[68]

The EU's Action Plan 'simplifying and improving the regulatory environment' that was made public on 5 June 2002, however, made clear that the intention was to come to a 'consolidated and proportionate instrument for assessing the impact of its legislative and policy initiatives, covering regulatory impact assessment and sustainable development (in the economic, social and environmental fields) and incorporating the existing instruments and methods'.[69] The subsequent communication from the Commission on Impact Assessment[70] did not elaborate on SIA. The Commission's impact assessment process consists of two stages: a preliminary assessment; and an extended impact assessment. In the latter stage, the scope should include an in-depth analysis of the potential impacts of the proposal on the economy, the society and on the environment.[71] Annex 2 of this communication offers some additional information on the main components of the extended impact assessment, but does not go beyond some vague generalities.

It appears that the obvious need to integrate the several existing separate impact assessment procedures has survived as the only objective of the Commission. While such a streamlined approach is probably essential to offset 'assessment fatigue', the primary focus on sustainable development – and specifically on environmental sustainability – seems to be lost. As SIA differs significantly from regulatory or business impact assessments, clear guidelines should have been developed spelling out its minimum essential characteristics. But this has not been done, and even the EIA approach raises questions. Concerning the latter, the EU's Sixth EAP places a legal obligation on the Commission to give 'consideration, prior to their adoption, of whether actions in the economic and social fields contribute to and are coherent with the objectives, targets and time frame of the Programme'.[72] This suggests the need to set minimum environmental requirements within an integrated impact assessment system, and an 'objectives-led' approach reflecting the priorities set by the Sixth EAP Thematic Strategies.

When reading an extended impact assessment on a proposal for a directive such as on services in the internal market, it is amazing how brief the analysis of the environmental impact can be.[73] After a previous report[74] indicated the potential risks of introducing a poorly prepared and framed SIA-approach, Wilkinson et al reviewed in a recent Commission report[75] whether and how sustainable development considerations have been addressed in the extended impact assessments done in 2003. Their conclusions are quite negative and include the poor quality and limited number of the assessments undertaken. Factors that have contributed to this situation include the absence of a formal quality control mechanism and the lack of an institutional framework within which 'learning by doing' can take place in practice. Furthermore, a lack of transparency in the impact assessment system is mentioned, as well as a non-systematic way of selecting Commission proposals for an impact assessment. The report states also that none of the impact assessments fully followed the Commission's Impact Assessment Guidelines,[76] whilst the sustainable development issues in the Guidelines are described as: 'brief, not easy to understand, and insufficiently detailed to be of practical use in an extended impact assessment'.[77]

The impact assessment system, as being undertaken at present by the Commission, is, however, not the only one, as DG Trade started a process in 1999 to apply sustainability impact assessments to its trade policies.[78] A major focus of this process was and still is to develop a methodological framework next to the 'experimental' application of SIA. At present, it is being acknowledged that the methodology still needs improvement. Again, the similarity to the basic EIA approach is striking.[79] The EC is not alone in this approach, as the US and Canada started similar exercises, built on their experiences with the Commission for Environmental Cooperation as part of the North American Free Trade Agreement (NAFTA).[80] Having learnt some lessons, key figures in these North American assessments remain, however, prudent about the potential of the instrument.[81] Given the complexity of the trade issue, it could be questioned if SIA in this context will ever go beyond the academic environment.[82]

In order to address some of the major problems concerning the application of impact assessment by the Commission, one might wonder if a defined legal environment could be helpful. On the other hand, one cannot dismiss the fact that the legal foundations of EIA have not been able to guarantee the proper implementation of the instrument, so it is questionable if a legal framework would be convincing enough for a more adequate impact assessment approach.

CONCLUSIONS

The SEA Directive and SEA Protocol are both significant requirements for SEA. Both meet the conditions argued by Partidário and Clark to be necessary for SEA to contribute to sustainability, albeit to varying degrees. The SEA Directive has a far better policy framework in place for sustainable development, including explicit recognition of the precautionary principle. However, this is not surprising, given the emphasis put on this by the EU, particularly under the Treaty and in its series of Action Programmes. While both the Directive and Protocol are fairly limited in application to plans and programmes, the Protocol goes further in encouraging application to some policies and legislative proposals as well. The Directive leaves it up to Member States to go further than its requirements if desired, and some European countries have already applied SEA to policies and legislation.[83]

It is reasonable to assume that as experience is gained under both instruments there will be changes that may well result in greater exposure of higher order decisions to public input and evaluation. It is hard to deny the formal changes that the Aarhus Convention has already brought in this direction, with the European Directive on Public Participation,[84] and moves by the European Commission to subject all major policy and regulatory initiatives to assessment of economic, social and environmental impacts. The outcome of the future implementation of the Public Participation Directive remains to be seen, but the above-mentioned review of the Commission's impact assessment system gives not much reason for great optimism. If environmental policies are currently not as integrated as they could be, further, possibly 'deeper' application of SEA could be on the agenda of the future. But the question remains whether there is enough political willingness to do so.

The current application of impact assessment relies heavily on discretion and seems unable to gather much institutional support. Even the newly established SEA Directive seems to be already out of fashion. It might be an indication that in the review of the five-year work programme by the EU Commissioner for the Environment, there is no mention of the adoption of the SEA Directive.[85] Also, a recent OECD publication on 'implementing sustainable development' contains rather worrying language:

> *The vast majority of OECD countries do not subject environmental protection laws to a systematic review of their economic consequences. Moreover, there have been instances when cost–benefit analysis has been poorly integrated into policy decisions. In contrast to the lack of systematic economic assessments of environmental programmes, the integration of environmental concerns in public decision-making processes at the project level is better established.*[86]

This observation is contrary to one of the conclusions of a recent Institute for European Environmental Policy (IEEP) study stating that the Commission's impact assessments gave most attention to economic impacts and only limited attention to environmental and social impacts.[87] The Commission Working Document of 1 June 2004 acknowledges that the Cardiff Process has produced mixed results and enumerates several shortcomings, such as a general lack of consistency and insufficient political commitment.[88] This working document concludes with a number of suggestions to increase the effectiveness of the process. Impact assessments at Community level are mentioned as potential powerful instruments, whilst stringent implementation of the EIA and SEA Directives is considered as a key to advancing environmental integration.[89] However, it remains doubtful if these assessment instruments can reverse the trend. As Sheate observes:

> *... the Cardiff Process – which philosophically supports EIA and SEA and an environmental perspective of sustainable development – appears to be struggling against the overwhelming dominance of the Lisbon Process. The Cardiff Process' own integration with the Lisbon Process risks it becoming the Cinderella of sustainable development, and is perhaps a metaphor for the risks to environmental protection of complete absorbance into (rather than integration with) the socio-economic aspects of policy making.*[90]

Beyond the EU, within the UNECE, the Cavtat Declaration adopted on the occasion of the third meeting of the parties of the Espoo Convention is also not ambitious, and a draft paragraph mentioning this Convention and the SEA Protocol as cornerstones on which to build more holistic, integrated forms of impact assessment and environmental management, as identified in the Johannesburg Plan of Implementation, did not survive, in part due to resistance from many EU Member States and the Commission.[91] Without being unduly pessimistic, it may well be that sustainability will remain a word with different meanings, while SEA, rather than integrating environmental considerations into policies, plans and programmes, may end up trading them off with economic and social ones, as in SIA.

NOTES

This chapter was originally published as: Marsden, S. and De Mulder, J. (2005) 'SEA and Sustainability in Europe: How Bright Is the Future?', *Review of European Community and International Environmental Law*, vol 14, no 1, pp50–62.

1 National Environmental Policy Act (1969), 42 USC 4321–4347, found at www.nrca.org/eias/index.htm.
2 Council Directive 2001/42/EC of 27 June 2001 on the Assessment of the Effects of Certain Plans and Programmes on the Environment, [2001] OJ L197 (SEA Directive), found at http://europa.eu.int/comm/environment/eia/sea-legalcontext.htm#legal.
3 Protocol on Strategic Environmental Assessment to the Convention on Environmental Impact Assessment in a Transboundary Context (Kiev, 23 May 2003).
4 Convention on Environmental Impact Assessment in a Transboundary Context (Espoo Convention) (Espoo, 25 February 1991).
5 United Nations Economic Commission for Europe, *Application of Environmental Assessment Principles to Policies, Plans and Programmes* (United Nations, 1992).
6 Report from the Commission to the European Parliament and the Council on the Application and Effectiveness of the EIA Directive (Directive 85/337/EEC, as amended by Directive 97/11/EC): How Successful are the Member States in Implementing the EIA Directive? (European Commission, undated).
7 P. B. Underwood and C. A. Alton (2003) 'Could the SEA Directive Succeed Within the United States?', 23:3 *Environmental Impact Assessment Review*, 260.
8 B. F. Noble (2000) 'Strategic Environmental Assessment: What is it? And What Makes it Strategic?', 2:2 *Journal of Environmental Assessment Policy and Management*, 206.

9 Ibid., at 215.

10 J. Dusik, T. Fischer, and B. Sadler, *Benefits of a Strategic Environmental Assessment* (UNDP-REC, 2003), at 5.

11 ICON, *SEA and Integration of the Environment into Strategic Decision-Making* (Imperial College Consultants Ltd, 2001), vol 1, at i.

12 B. Dalal-Clayton and B. Sadler, 'Strategic Environmental Assessment: A Rapidly Evolving Approach', working paper (International Institute for Environment and Development, undated). B. Dalal-Clayton and B. Sadler are currently working on an overview of SEA, with a special focus on developing countries and countries in transition, and its potential role as a key tool for strategies for sustainable development and poverty reduction. See B. Dalal-Clayton and B. Sadler, 'The Status and Potential of Strategic Environmental Assessment (SEA)' (draft of 17 September 2003), found at www.iied.org/spa/sea.html.

13 S. Dovers (2002), 'Too Deep a SEA: Strategic Environmental Assessment in the Era of Sustainability', in S. Marsden and S. Dovers, *Strategic Environmental Assessment in Australasia* (The Federation Press), at 43.

14 Ibid., at 26.

15 'Effects' under both the SEA Directive (Annex I) and Protocol (Annex IV) include 'secondary, cumulative, synergistic, short-, medium- and long-term, permanent and temporary, positive and negative effects'.

16 M. R. Partidário, and R. Clark, 'Introduction', in M. R. Partidário and R. Clark (eds), *Perspectives on Strategic Environmental Assessment* (CRC-Lewis, 2000), at 5.

17 See, for instance, the UN-CSD Theme Indicator Framework, found on the CSD website at www.un.org/esa/sustdev/natlinfo/indicators/isdms2001/table_4.htm.

18 European Environment Agency (1999), *Towards a Transport and Environment Mechanism for the EU – Part 2: Some Preliminary Indicator Sheets* (EEA), found at http://reports.eea.eu.int/TEC18-2/en.

19 European Environment Agency, *Indicators* (EEA, undated), found at http://themes.eea.eu.int/indicators/.

20 Organization for Economic Cooperation and Development, *Implementing Sustainable Development: Key Results 2001–2004* (OECD, May 2004), at 10, found at www.oecd.org/dataoecd/25/35/31683750.pdf.

21 Treaty on European Union (Maastricht, 7 February 1992), [2002] OJ C325.

22 Commission Communication of 27 March 1992, *Towards Sustainability: A European Community Programme of Policy and Action in relation to the Environment and Sustainable Development*, COM (92) 23 final.

23 Decision No 2179/98/EC of the European Parliament and of the Council of 24 September 1998 on the Review of the European Community Programme of Policy and Action in relation to the Environment and Sustainable Development, *Towards Sustainability*, [1998] OJ L275.

24 Commission Communication of 24 January 2001 *on the Sixth Environment Action Programme of the EC*, COM (2001) 31 final.

25 S. Marsden (2002), 'SEA and International Law: An Analysis of the Effectiveness of the SEA Protocol to the Espoo Convention, and of the Influence of the SEA Directive and Aarhus Convention on its Development', 1 *Environmental Law Network International Review* at 1.

26 Convention on Biological Diversity (Rio de Janeiro, 5 June 1992).

27 See the European Commission, *Environmental Impact Assessment* (European Commission, undated), found at http://europa.eu.int/comm/environment/eia/eia-legalcontext.htm#legalcontext. See also generally, L. Feldmann, M. Vanderhaegen, and C. Pirotte, 'The EU's SEA Directive: Status and Links to Integration and Sustainable Development', 21:3 *Environmental Impact Assessment Review* (2001), 203.
28 Representatives of Member States and the Environment Directorate-General of the European Commission, *Implementation of Directive 2001/ 42 on the Assessment of the Effects of Certain Plans and Programmes on the Environment* (23 September 2003).
29 J. De Mulder, 'The New Directive on Strategic Environmental Assessment', 1 *Environmental Law Network International Review* (2001), at 19.
30 Ibid., at 14–15.
31 See Representatives of Member States, n. 28 above, at 21–22.
32 Ibid., at 23.
33 Ibid., at 23.
34 M. Partidário, 'SEA Methodological approaches', BEACON Workshop, Brussels, (24 June 2004).
35 T Nitz and A. L. Brown (2001) 'SEA must learn how policy making works', 3:3 *Environmental Assessment Policy and Management*, 340. See also L. Brown and T. Nitz, 'Applying SEA to policy making: The policy cycle model and the Queensland Policy Handbook' in S. Marsden and S. Dovers, note 13 above, at 84–98.
36 See J. De Mulder, note 29 above, at 17–18. See the Convention on Access to Information, Public Participation in Decision-Making and Access to Justice in Environmental Matters (Aarhus, 25 June 1998).
37 See SEA Directive, note 2 above, Article 6.1 and 6.4.
38 See A. Mathiesen, 'Public Participation in Decision Making and Access to Justice in EC Environmental Law: The Case of Certain Plans and Programmes', 12:2 *European Environmental Law Review* (2003), 46. Under Article 7 of the Aarhus Convention, each party must introduce appropriate practical and/or other provisions for the public to participate during the preparation of plans and programmes relating to the environment, within a transparent and fair framework, having provided the necessary information to the public. The relevant public authority must designate the public, taking into account the objectives of the Convention, before it is able to participate. To the extent appropriate, each party must also endeavour to provide opportunities for public participation in the preparation of policies relating to the environment.
39 N. Risse et al (2003) 'Implementing the European SEA Directive: The Member States' margin of discretion', 23:4 *Environmental Impact Assessment Review*, 467.
40 W. Sheate, 'The EC Directive on Strategic Environmental Assessment: A much-needed boost for environmental integration', 12:12 *European Environmental Law Review* (December 2003), at 347.
41 Regulation No 807/2004 of the European Parliament and of the Council of 21 April 2004 amending Council Regulation (EC) No 2236/ 95 laying down general rules for the granting of Community financial aid in the field of trans-European networks, [2004] OJ L143.
42 When completed, the Guidelines will be available at www.europa.eu.int/comm/environment/eia/sea-support.htm.

43 See the Rio Declaration on Environment and Development, UN Doc. A/CONF. 151/26/Rev. 1 (Vol I) (Rio de Janeiro, 1992).

44 European Commission and World Health Organization Regional Office for Europe, Declaration of Third Ministerial Conference on Environment and Health (London, 16–18 June 1999).

45 See Johannesburg Declaration, *Report of the World Summit on Sustainable Development* (Johannesburg, South Africa, 26 August–4 September 2002), UN Doc. A/CONF. 199/20, Resolution 1, Annex; and Johannesburg Plan of Implementation, *Report of the World Summit on Sustainable Development* (Johannesburg, South Africa, 26 August–4 September 2002), UN Doc. A/CONF.199/20, Resolution 2, Annex.

46 J. De Mulder (2003) 'The New UNECE Protocol on Strategic Environmental Assessment', 2 *Environmental Law Network International Review*, 13.

47 See Commission Communications, notes 22 and 24 above.

48 These include a watering down of environment protection in favour of economic development, including further corporate-led globalization, such as the privatization of energy, water and health; a failure to deal with issues of global debt and poverty, without which sustainable development cannot be realized; and a lack of concrete proposals for action, such as new environmental treaties.

49 See J. De Mulder, note 46 above, at 9.

50 SEA provisions in some jurisdictions do this currently. See S. Marsden, 'Why is Legislative EA in Canada Ineffective and How Can it be Enhanced?', 18:3 *Environmental Impact Assessment Review* (1998), 241. In Europe see the Dutch Environmental Test: S. Marsden, 'Legislative EA in the Netherlands: The E-Test as a Strategic and Integrative Instrument', 9:3 *European Environment* (1999), 91.

51 See J. De Mulder, note 46 above, at 2–3. Mandatory application to policies and legislation could have been required, for example.

52 See S. Marsden, note 25 above.

53 See SEA Directive, note 2 above, Article 2(a).

54 See SEA Protocol, note 3 above, Article 2(6).

55 This Annex has been extended (and became almost identical to Annex I of the EIA Directive) by the amendment of the Espoo Convention during the Third Meeting of the Parties (Cavtat, 1–4 June 2004).

56 J. Jendroska and S. Stec, 'The Kyiv Protocol on Strategic Environmental Assessment', 33:3–4 *Environmental Policy and Law* (2003), 105–110.

57 Ibid., at 108.

58 Commission of European Communities, *Five-Year Report to the European Parliament and Council on Application and Effectiveness of EIA Directive* (European Commission, 2003).

59 P. Hamblin, 'Summary of the Strategic Environmental Assessment seminar', *SEA Making a Difference*, Seminar report (EEB, March 2004), at 39.

60 UNECE, Preparatory Meeting for the First Meeting of the Signatories to the Protocol on Strategic Environmental Assessment, Background Paper Presenting Possible Elements for a Work Plan for the SEA Protocol (Geneva, 27–28 November 2003).

61 Ibid.

62 Strategic Environmental Assessment Directive: Guidance for Planning Authorities. Practical guidance on applying European Directive 2001/ 42/EC 'on the assessment

of the effects of certain plans and programmes on the environment' to land use and spatial plans in England (Office of the Deputy Prime Minister, October 2003), at 9, found at www.odpm.gov.uk/stellent/groups/odpm_planning/documents/page/odpm_plan_026670.pdf. See also R. Therivel and P. Minas (2002) 'Ensuring effective sustainability appraisal', 29:2 *Impact Assessment and Project Appraisal*, 81.

63 D. B. Dalal-Clayton, S. Bass et al, *National Sustainable Development Strategies: Experience and Dilemmas*, Environmental Planning Issues No. 6 (IIED, 1994).

64 C. George (2001) 'Sustainability appraisal for sustainable development: integrating everything from jobs to climate change', 19:2 *Impact Assessment and Project Appraisal*, 103. See also R. Therivel, and P. Minas, note 62 above.

65 T Fischer (2003) 'Strategic Environmental Assessment in Post-Modern Times', 23:2 *Environmental Impact Assessment Review*, 167.

66 Presidency Conclusions, Goteborg European Council (15–16 June 2001). SIA has been defined by the Commission in the context of the EU sustainable development strategy as follows: 'The elaboration and implementation of all EU policies must be informed by the principles of the EU sustainable development strategy. This will be achieved, inter alia, by introducing sustainability impact assessment for all major policy proposals, analysing their economic, social and environmental consequences including those for developing countries'. See European Commission, The EU and the World Summit on Sustainable Development, *Frequently Asked Questions About the EU and its Position on the World Summit on Sustainable Development in Johannesburg*, (European Commission, undated), found at http://europa.eu.int/comm/environment/wssd/qa_general_en.html. The disadvantages of SIA are similar to sustainable development, as the environment is (in theory at least) given an equal weighting in the process, which may result in trade-offs in favour of economic issues especially. Many prefer to retain EIA (or SEA for policies, plans and programmes) tools, which were originally intended to advance environmental protection.

67 These are well-established tools of impact assessment, some of which are required under legislation. Cumulative impact assessment (CIA) commonly refers to the assessment of environmental impacts from a collection of associated developments whereby the impact in combination is often much greater than the impact of the individual developments. Cost–benefit analysis (CBA) is a long-established tool used to determine the economic feasibility of a proposal. Health impact assessment (HIA) is a process to determine the effects of development upon human health.

68 D. Grimeaud, 'The integration of environmental concerns into EC policies: A genuine policy development', 9:7 *European Environmental Law Review* (July 2000), 210. The 'Cardiff Process' is the name given to the process launched by the European Council at their meeting in Cardiff, in June 1998, requiring different Council formations to integrate environmental considerations into their respective activities, putting Article 6 of the EC Treaty into practice. The related 'Lisbon Strategy' is a commitment to bring about economic, social and environmental renewal in the EU. In March 2000, the European Council in Lisbon set out a 10-year strategy to make the EU the world's most dynamic and competitive economy. Under the strategy, a stronger economy will drive job creation, alongside environmental and social policies that ensure sustainable development and social inclusion.

69 Commission Communication of 5 June 2002, *Action Plan Simplifying and Improving the Regulatory Environment*, COM (2002) 278 final, at 7.
70 Commission Communication of 5 June 2002 on impact assessment, COM (2002) 276 final.
71 See Presidency Conclusions, note 66 above.
72 See Commission Communication, note 24 above.
73 In this case it is only two pages. See proposal for a Directive of the European Parliament and of the Council on services in the internal market, SEC (2004) 21, and Commission Staff Working Paper of 13 January 2004, *Extended Impact Assessment of Proposal for a Directive on Services in the Internal Market*, COM (2004) 2 final, at 52, found at http://europa.eu.int/comm/internal_market/en/services/services/docs/2004-impact-assessment_en.pdf.
74 A. Kraemer and D. Wilkinson, *EU Environmental Governance: A Benchmark of Policy Instruments with a Focus on Agriculture, Energy and Transport* (ECOLOGIC–IEEP, June 2002). This report refers briefly to SEA when dealing with the issue on how to involve Member States in impact assessments.
75 D. Wilkinson et al, *Sustainable Development in the European Commission's Integrated Impact Assessments for 2003* (Institute for European Environmental Policy (IEEP), April 2004).
76 See Commission Communication, note 70 above.
77 See D. Wilkinson et al, note 75 above.
78 See European Commission, *Sustainability Impact Assessment* (January 2004), found at http://europa.eu.int/comm/trade/issues/global/sia/index_en.htm.
79 C. Kirkpatrick and N. Lee, *Further Development of the Methodology for a Sustainability Impact Assessment of Proposed WTO Negotiations* (IDPM–Manchester, 5 April 2002), at 10.
80 The North American Commission for Environmental Cooperation was created under the North American Agreement on Environmental Cooperation.
81 *Impact Assessment of International Trade* (IAIA conference, Vancouver, 27 April 2004). See also P. B. Underwood, and C. A. Alton, note 7 above.
82 N. Lee and C. Kirkpatrick (2001) 'Methodologies for sustainability impact assessments of proposals for new trade agreements', 3:3 *Journal of Environmental Assessment Policy and Management*, 409–410.
83 These countries include the Netherlands and Finland.
84 Council Directive 2003/35/EC of the European Parliament and Council of 26 May 2003 providing for public participation in respect of the drawing up of certain plans and programmes relating to the environment and amending with regard to public participation and access to justice Council Directives 85/337/EEC and 96/61/EC, [2003] OJ L156/17.
85 European Commission, *Five Years as European Commissioner for the Environment* (European Commission, June 2004), found at http://europa.eu.int/comm/commission_barroso/dimas/policies/policies_en.htm.
86 Organization for Economic Cooperation and Development, *Implementing Sustainable Development: Key Results 2001–2004* (OECD, 2004), at 4.
87 Institute of European Environmental Policy, *The Future of EU Environment Policy: Challenges and Opportunities* (IEEP, December 2004), at 25.

88 European Commission, *Integrating environmental considerations into other policy areas – A stocktaking of the Cardiff process*, working document, COM (2004) 94 final, at 31.

89 Ibid., at 35–36.

90 W. Sheate (2003) 'Changing conceptions and potential for conflict in environmental assessment – Environmental integration and sustainable development', 3:5 *Environmental Policy and Law*, 229.

91 See UNECE, note 60 above.

11

The SEA Directive and the SEA Protocol Adopted to Spatial Planning: Similarities and Differences

Gernot Stoeglehner and Gerald Wegerer

Introduction

Strategic environmental assessment (SEA) was introduced in the Member States of the European Union (EU) when the 'Directive, 2001/42/EC of the European Parliament and of the Council on the assessment of the effects of certain plans and programmes on the environment' (Strategic Environmental Assessment (SEA)) went into force on 27 June 2001 and had to be implemented by the Member States until 21 July 2004.

In addition, the Protocol on Strategic Environmental Assessment as part of the Convention on Environmental Impact Assessment in a Transboundary Context (SEA Protocol) was elaborated by an ad hoc working group of the United Nations Economic Commission for Europe (UNECE) and adopted on 21 May 2003, at the Fifth Ministry Conference 'Environment for Europe' in Kiev. The SEA Protocol will enter into force 'on the ninetieth day after the date of deposit of the sixteenth instrument of ratification, acceptance, approval or accession' (Article 24, Paragraph 1, SEA Protocol). Finland, the Czech Republic and Albania were the first States to ratify the SEA Protocol in 2005 (UNECE, 2006). In Estonia and Germany, ratification is expected in 2006 (UNECE, 2005, p.2).

Previous work of a former official of the EU Directorate-General of the Environment claims that the SEA Protocol basically adopts the SEA Directive, extending the validity area to the UNECE Member States, enlarging the scope to policies on a voluntary basis and equating health issues with environmental issues (Feldmann, 2003, p. 92). This statement suggests that the SEA Protocol will not change the SEA practice essentially within the EU Member States. The aim of this chapter is first to clarify whether this result can be confirmed in a detailed investigation and second to estimate if further expenses and benefits in spatial planning can be expected from the Protocol implementation. The expenses and benefits will be analysed on behalf of spatial planning in Austria.

The topic of this chapter is developed in four steps: first, the SEA Directive of the EU will be introduced in detail as base for the following comparison. Second, differences from the SEA Protocol in relation to the SEA Directive will be outlined. According to this analysis, we will indicate the necessary amendments for implementing the SEA Protocol in those laws of the EU Member States, which are already fulfilling the requirements of the SEA Directive. Third, we estimate the additional expenses and benefits of integrating the SEA Protocol using the example of spatial planning in Austria, where SEA was introduced by the EU Directive. Fourth, we draw some conclusions about the ratio of expenses and benefits of the SEA Protocol in comparison to the SEA Directive.

THE SEA DIRECTIVE

In the following section, the elements and contents of an SEA are introduced. An SEA contains the preparation of an environmental report, consultations and the taking into account of the environmental report and the results of the consultations in decision-making. The public, environmental authorities and neighbouring EU Member States that are likely to be affected by the environmental impacts of the plan or programme have to be consulted. As the SEA Directive contains aspects of the SEA process and SEA contents, this section is structured in the same way.

SEA process

The SEA process includes eightelements (Articles 4–10, SEA Directive) that do not have to follow each other stringently but can be carried out in parallel to some extent (Stoeglehner, 2003a, p.11; Wegerer and Stoeglehner, 2004, p.196):

- *Screening*: The screening determines whether an SEA has to be executed. Environmental authorities shall be consulted and the public has to be informed especially about the reasons why an SEA is not carried out. An SEA according to the SEA Directive has to be applied to plans and programmes as well as their modifications (Article 2 SEA Directive):
  - which are subject to preparation and/or adoption by an authority at national, regional or local level, or which are prepared by an authority for adoption through a legislative procedure by Parliament or Government, and
  - which are required by legislative, regulatory or administrative provisions.

For plans and programmes that comply with this definition, the scope of the SEA is determined in Article 3 of the Directive. All plans and programmes that likely have significant environmental effects shall be subject to an environmental assessment (EA). Two specifications of the scope can be distinguished (Stoeglehner, 2003b, p. 46):

- *The obligatory scope* has to be applied for plans and programmes:
  - which are prepared for agriculture, forestry, fisheries, energy, industry, transport, waste management, water management, telecommunications, tourism, town and country planning or land use, and which set the framework for future development consents of projects listed in Annexes I and II to EU Directive 85/337/EEC (Environmental Impact Assessment (EIA) Directive); or
  - which, in view of the likely effect on sites, have been determined to require an assessment pursuant to Article 6 or 7 of EU Directive 92/43/EEC (Fauna-Flora-Habitat Directive).
- *The non-obligatory scope* can affect:
  - plans and programmes that set the framework for projects other than listed in the EIA Directive;
  - minor modifications of plans and programmes pursuant to the obligatory scope;
  - plans and programmes that determine the use of small areas on local level.

The Member States can determine the non-obligatory scope either through case by case examination, by specifying types of plans and programmes or by combining these approaches. In any case, the criteria of Annex II of the SEA Directive have to be taken into account. Therefore, the Member States cannot decide arbitrarily about the non-obligatory scope: if the criteria of Annex II are fulfilled by the plan or programme, an SEA has to be carried out. Furthermore, exceptions from the scope are granted in the SEA Directive for plans and programmes with the purpose to serve national defense or civil emergency only, for financial or budget plans and programmes.

- *Scoping*: The scope of the environmental analysis and assessment including aspects of contents, methods, time and space of the appraisals is determined. Environmental authorities have to be consulted.
- *Appraisal of alternatives*: The development, appraisal and assessment of reasonable alternatives are mandatory for plans and programmes that have to undergo an SEA. 'Reasonable' alternatives take the objectives and the geographical scope of the plan or programme into account.
- *Environmental report*: The environmental report comprises the results of the environmental surveys and assessments according to Annex I of the SEA Directive (see SEA contents, p.239).

- *Consultations*: Certain consultation partners shall have the opportunity to express their opinion on the environmental report and the draft plan or programme during the SEA process. Partners for the consultations are:
  - the public, including the public affected or likely to be affected by, or having an interest in the decision-making of the plan or programme, comprising relevant non-governmental organizations (NGOs);
  - the environmental authorities, which by reason of their specific environmental responsibilities, are likely to be affected by the environmental consequences of the plan or programme implementation; and
  - EU Member States that are likely to be significantly affected by the environmental impacts of the plan or programme, as far as they request transboundary consultations.
- *Decision-making*: The environmental report and the results of the consultations shall be taken into account during the preparation of the plan or programme before the adoption or submission to the legislative procedure. The decision-making bodies are not bound by the results of the SEA.
- *Information on the decision*: All participants involved in the SEA process shall be informed about the adopted plan or programme and shall be given access to a statement, summarizing how environmental considerations have been integrated into the plan or programme, and how the results of the SEA process have been taken into account in the decision as well as the associated monitoring measures.
- *Monitoring*: The significant environmental effects of the plan or programme implementation shall be monitored in order to, inter alia, identify unforeseen adverse effects and to be able to undertake appropriate remedial action.

The Directive gives leeway to the Member States for implementing SEA in new legal norms, e.g. in a separate SEA law, or integrating it in existing laws, e.g. in the planning laws and procedures. Furthermore, it is not finally determined, whether SEA is an examination tool that follows the planning process, or if SEA is a planning tool that can be applied within the planning processes to permanently consider environmental issues. The Member States can choose between the two SEA strategies. The majority of aspects of the SEA Directive favour the second interpretation as an integrated planning tool. First, the main goal of the Directive seems not to be a strict control of planners and decision-makers, but a strategic consideration of environmental concerns, a self-reflection of planners towards the promotion of sustainable development and a public discussion (Scholles, 2001, p.127). Second, as the system of spatial planning in Austria is already taking environmental considerations into account, SEA shall be implemented in existing processes of spatial planning (OEROK, 2004, p.13). Until now, eight out of nine spatial planning laws of the Austrian provinces were amended according to the SEA Directive. The legislative bodies followed this recommendation, although not all aspects of the SEA Directive were fully covered in the national implementation (Proebstl et al, 2006; Stoeglehner, 2004a, p.184ff.).

SEA contents

Annex I of the SEA Directive defines the contents of the SEA by the information that has to be included in the environmental report. These contents can be categorized in the following way (Stoeglehner, 2004b, p.62):

- *Information about the plan or programme*: The environmental report shall describe the contents of the plan or programme, main objectives and the relationship with other relevant plans and programmes. The SEA also introduces a mandatory comparison of alternatives, so the reasons for selecting the alternatives have to be outlined (lit. a, h Annex I SEA Directive).
- *Description of the applied methods and procedures*: This chapter not only includes the applied methods and procedures but also the difficulties encountered in compiling the required information (lit. h Annex I SEA Directive).
- *Description of environmental objectives*: Relevant environmental objectives established at the international, community or member state level, and the way how these objectives are considered have to be described (lit. e Annex I SEA Directive).
- *Description of the environmental situation*: Here the relevant aspects of the current environmental situation, the environmental characteristics of the areas likely to be significantly affected and relevant existing environmental problems – especially related to Natura-2000 areas – have to be outlined. Also a prognosis of the environmental development without implementation of the planning measures has to be included (lit. b, c, d Annex I SEA Directive). This can be addressed as the 'zero alternative' for the environmental assessment.
- *Environmental appraisal*: The likely significant effects on different environmental issues have to be appraised (lit. f Annex I SEA Directive). These issues are listed and can be categorized as human ecological, biotic, abiotic and anthropogenic subjects of protection (Stoeglehner, 2003a, p.57) as well as the interrelationships among them (see Table 11.1). The environmental issues have to be elaborated with the methods and level of detail determined during the scoping. Another challenge is given by the fact that secondary, cumulative, synergistic, short, medium and long term, permanent and temporary, positive and negative effects shall be regarded.
- *Monitoring and compensation measures*: Monitoring measures have to be described as well as measures envisaged to prevent, reduce or compensate significant adverse environmental effects of the plan or programme implementation as fully as possible (lit. g, i Annex I SEA Directive).
- *Non-technical summary*: The environmental report shall include a non-technical summary of its content (lit. j Annex I SEA Directive).

Table 11.1 *Differences of the environmental aspects to be considered between SEA Directive and SEA Protocol*

Environmental aspects	*SEA Directive*	*SEA Protocol*
Human–ecological	Human health	Human health
	Population	–
Biotic	Flora	Flora
	Fauna	Fauna
	Biodiversity	Biodiversity
	Landscape	Landscape
	–	Natural sites
Abiotic	Soil	Soil
	Water	Water
	Air	Air
	Climate factors	Climate
Anthropogenic	Material assets	Material assets
	Cultural heritage	Cultural heritage
	Architectural heritage	–
	Archaeological heritage	–
Interaction	Among these aspects	Among these aspects

A great variety of methods is suitable to assess the environmental effects of plans and programmes as is extensively explained in literature (e.g. Jacoby, 2000). The fact is emphasized that, according to the EU Directive, the SEA shall not cause a significant additional effort, e.g. for spatial research or analysis, but that the planning methods shall be applied according to the state of the art, which are necessary anyway to soundly balance the fundamentals for decision-making. The processing of the environmental considerations shall follow the procedure and framework of the contents given by the SEA Directive.

AMENDMENTS BY THE SEA PROTOCOL

The SEA Directive served as foundation for the SEA Protocol. Yet, by looking carefully at the details, some major differences might be revealed. The applied method is a comparative analysis of the two legal instruments based on the SEA Directive. Only differences of the SEA Protocol will be discussed, overlapping regulation will not be further mentioned. First, we analyse the SEA process and, second, the SEA contents. The aim of this section is just to describe, not to evaluate the differences. The latter will be done later in the chapter.

SEA process

The analysis of the SEA process is carried out in the eight SEA steps as discussed earlier in the chapter. According to the objectives of the SEA Protocol two relevant changes were made which affect the whole SEA process: first, the SEA Protocol accents health aspects, which leads to an involvement of health authorities in all aspects where environmental authorities had to be consulted so far. The emphasis of health issues was highly promoted by the participating WHO – World Health Organization (UNECE, 2003a and b, p.4). Second, public participation shall be exaggerated during the whole SEA process. In detail, the following results can be gained:

- *Screening*: The most important difference is that the SEA Protocol addresses not only plans and programmes, but also policies and legislation. Concerning policies and legislation, the Protocol is not as stringent as for plans and programmes. While for policies and legislation, the signing states 'shall endeavour to ensure that' (Article 13, Paragraph 1 SEA Protocol) an SEA is carried out, the regulations for plans and programmes are very closely related to the SEA Directive, including the obligatory and the non-obligatory scope. Two differences can be revealed within the obligatory scope of the Protocol:
    - Within the specific fields of planning, an SEA shall be carried out for two further issues in the SEA Protocol: mining and regional development.
    - There is no link to the Habitat Directive or comparable areas of nature protection formulated in the SEA Protocol.

  Furthermore, a relevant difference appears in the aspect of public participation: according to the EU Directive the public has to be informed about the results of the screening. According to the Protocol, public participation is extended: 'each party shall endeavour to provide opportunities for the participation of the public concerned in the screening of plans and programmes' (Article 5, Paragraph 3 SEA Protocol). Concerning the consultation of authorities, not only environmental authorities, but also health authorities have to be involved.

  Moreover, the criteria for the non-obligatory scope to be considered are slightly different: according to Annex II of the SEA Directive, inter alia, the 'cumulative nature of effects' has to be considered. This criterion is not determined in the SEA Protocol.
- *Scoping*: Analogous to the screening public participation and the involvement of health authorities are introduced to the scoping. This major change means that each state shall endeavour that the public concerned shall take part when determining the relevant information to be included in the environmental report (Article 6, Paragraph 3 SEA Protocol).

- *Environmental report*: In the environmental report, the state of the art of assessment methods as well as the level of detail of the plan or programme within a hierarchy shall be addressed. The SEA Protocol adds two more aspects that shall be taken into account:
  - the interests of the public; and
  - the information needs of the decision-making body.

Also, some aspects of the SEA Directive cannot be found in the SEA Protocol: the regulations concerning the duplication of assessments within, inter alia, a hierarchy of plans and programmes or with other assessment instruments do not appear in the SEA Protocol.

- *Consultations*: The SEA Protocol does not refer to 'consultations' any longer, but calls a very similar mechanism public participation. Also 'public participation' includes the rights to be informed and to give statements during the SEA process. More sophisticated methods of public participation like workshops, etc., are not addressed by the SEA Protocol.

Still some details are of relevant difference: first, the scope of public participation according to the SEA Protocol is extended to screening and scoping on a more or less voluntary basis. Second, the invited public is restricted compared to the SEA Directive. The SEA Directive addresses the public to be affected or likely to be affected by the environmental effects of the plan or programme as well as the public having an interest in the decision-making process (Article 6, Paragraph 4 SEA Directive). On the contrary, the SEA Protocol only addresses the public concerned, which can be interpreted that only the affected public, but not the merely interested public, is allowed to be involved in 'public participation'.

Concerning the transboundary consultations, one detail is more precise in the SEA Protocol: the state that is adopting the plan has to inform the neighbouring states about the draft plan and the environmental report. The affected state can examine the documents and declare if it wishes consultations. The protocol allows for the state adopting the plan to indicate a reasonable time schedule for the consulted state to declare if consultations are wanted. Measures to prevent, reduce or mitigate adverse effects shall be properly addressed during consultation.

- *Decision-making*: The decision-making is formulated more strictly: the results of the SEA shall be taken into 'due' account. Special emphasis shall be laid on the measures to prevent, reduce or mitigate the adverse effects identified in the environmental report.
- *Information on the decision*: Compared to the SEA Directive, a similar statement on the decision has to be made but without addressing the monitoring measures.

- *Monitoring*: The procedures and aims of the monitoring are the same between SEA Directive and SEA Protocol, but one major difference appears concerning the publicity of the monitoring results. Whereas in the SEA Directive no publicity is provided, according to the SEA Protocol, it shall be guaranteed that the environmental and health authorities as well as the public shall be informed.

SEA contents

The analysis of the SEA contents is executed by comparing the information to be included in the environmental report. Basically, two major differences can be found: the list of environmental aspects to be considered and the connection to areas of nature protection.

First, in the list of environmental aspects that have to be regarded, the SEA Protocol emphasizes health effects. Likely significant environmental and health effects of the reasonable plan or programme alternatives have to be addressed in the environmental report (Article 7, Paragraph 2 SEA Protocol). In detail, there are a small number of differences in the definition of the 'environmental aspects' between the SEA Directive and the SEA Protocol as exposed in Table 11.1.

First difference is the definition of the human–ecological environmental aspects, where the 'population' is not considered. This means that the effects of a certain plan or programme on a group of persons, their extent and spatial allocation as well as their social interaction are not explicitly addressed in the SEA Protocol. In the SEA Protocol, additional 'natural sites' among the biotic aspects are integrated. The third difference is the definition of cultural heritage, which is not further determined in the SEA Protocol. In the SEA Directive, architectural and archaeological heritage are explicitly named as aspects of cultural heritage.

Second, concerning the content of an SEA, the specific information to be included in the environmental report differs. In terms of the SEA Directive, existing environmental problems relevant to the plan or programme, including in particular those, 'relating to any areas of a particular environmental importance', such as designated Natura-2000 areas (Annex I SEA Directive), have to be surveyed and described. According to the SEA Protocol, no direct links to protected areas are established. Only a simple indication – the mention of 'natural sites' – exists in the SEA Protocol that problems related to any areas of a particular environmental importance have to be considered (Article 2, Paragraph 7 SEA Protocol).

Benefits and expenses of the SEA Protocol in spatial planning

As benefits and expenses of SEA differ from between countries and planning issues, we focus our study on spatial planning in Austria, which covers the fields of regional development, town and country planning or land use according to Article 4, Paragraph 2 SEA Protocol. The spatial planning system in Austria is quite complex because the responsible legal body is not the federal state but the provinces. Therefore, we have nine spatial planning regulations. Typically, the planning hierarchy consists of six different levels of plans that cover the provincial, regional, sub-regional and municipal level, and all aspects of spatial planning, including environmental, social, economic and cultural issues. Planning processes already consist of SEA elements such as the consultation according to the SEA Directive, the elaboration of planning reports, the rules on decision-making and the announcement of the decision.

As before, we will introduce the benefits and expenses that arise from the implementation of the SEA Directive according to recent surveys carried out by the authors, then we discuss whether additional benefits and expenses can be expected due to the implementation of the SEA Protocol.

SEA benefits

In general, independent of the fields of planning addressed by an SEA, several benefits can be gained in aspects of planning procedures, planning contents and awareness raising, transparency and acceptance (Weber and Stoeglehner, 2001, p.232f):

- aspects of planning procedures such as:
    - coordination with environmental (and health) authorities early in the process;
    - the right to give statements in the case of transboundary environmental effects caused by a neighbour state;
    - potential relief of the examination load on environmental impact assessment (EIA) by SEA.
- aspects of planning contents such as:
    - search for environmentally friendly alternatives;
    - better integration of environmental precaution in planning processes;
    - mitigation of planning mistakes and follow up environmental costs;
    - better decision-making base due to setting of minimum standards;
    - widening the knowledge base via involvement of environmental authorities early in the planning processes;
- aspects of awareness raising, transparency and acceptance such as:
    - support of decision-making bodies to adopt environmentally friendly and sustainable solutions;
    - promotion of public participation, transparency, awareness raising and responsibility in decision-making processes;
    - increasing of the acceptance of planning decisions by public participation.

In spatial planning, we especially want to stress the potential to improve the quality of planning by the mechanisms of an SEA according to the Directive (Stoeglehner and Wegerer, 2004, p.57f):

- As we found out in our case studies in two Austrian provinces (Stoeglehner, 2003a; Wegerer and Stoeglehner, 2004), all planning processes contained the elaboration of variants. Until now, most of the variants were discussed among planners and decision-makers, but hardly in public. Therefore, the mandatory appraisal of variants and documentation in the environmental report will improve the transparency and traceability of planning processes. Above all, this obligatory appraisal of alternatives might support the process of searching for environmentally friendly variants and strengthen environmental precaution.
- A second issue of quality assurance in planning processes is the outlining of a minimum content of the environmental report. A structured analysis and appraisal of significant environmental effects will be necessary considering environmental issues according to Annex I of the Directive. The level of facts and values have to be separated. A framework for the structure of environmental appraisals is indicated that secures or increases the transparency and traceability of planning processes, and improves the decision-making base.
- The knowledge base of decisions can be broadened by the consultation of environmental authorities and the public, e.g. when NGOs with special environmental knowledge are involved.
- In our opinion, the statement explaining the decision is a very effective tool to raise the quality of planning: the planning authorities have to show clearly why the decision was made and how the environmental effects according to the environmental report and the statements collected during the consultations with environmental authorities, the public and likely affected neighbouring EU Member States have been considered. The public will be empowered to control decision-making and to claim that environmental issues are taken into account.
- Monitoring is an instrument of self-control if all negative environmental effects have been predicted correctly and if the objectives of the planning processes have been reached.

Some of the positive effects can be intensified by the SEA Protocol:

- The knowledge base can be further widened by including health authorities.
- By extending public participation to screening and scoping, the public will be involved earlier than according to the SEA Directive. The public will have more rights to deliver statements starting from the first step of an SEA. This will further increase the transparency and traceability of the planning processes.

- The importance of monitoring will be greatly improved, because it will be developed from an instrument of control by decision-making authorities to an instrument of control by the public. If planning authorities do not take appropriate remedial action, the public is empowered to enforce measures via democratic processes.

One benefit of the SEA Protocol will be the possibility to extend SEA to policies. In Austria, many decisions that influence the spatial development and the consumption of the environment are made at policy level without substantial appraisal of effects on the environment or spatial development. SEA at this level might reveal undesired negative effects and raise the awareness within the political bodies that decide upon such matters.

SEA expenses

According to Stoeglehner (2004b, p.71), the SEA Directive can be implemented without major difficulties into the existing procedures of spatial planning in Austria. Some steps of the planning processes must be specified in more detail, which sometimes only means formalizing existing informal planning procedures. Four main expenses can be identified:

- The appraisal of alternatives might cause the biggest operating expense of an SEA, because without SEA only one planning solution has to be elaborated in detail and sent for public consultation and to the decision-making process. With an SEA, additional alternatives have to be prepared thoroughly and at a quality high enough to be discussed in public.
- The environmental report causes greater expense, because the contents according to the SEA Directive are not fully covered in recent planning reports: methodologies were not described enough; although comprehensive environmental considerations have been a substantial part of all evaluated planning processes, the concluding and summarizing description of planning contents and their specific environmental effects were not included in former planning reports (Stoeglehner and Wegerer, 2004, p.57).
- Transboundary consultations will cause significant expenses, because on the one hand they lead to a greater delay in the planning procedure, and on the other hand they create costs because of necessary translation work of the draft plan and the environmental report.
- Monitoring demands a continuous observation of the environmental effects of a spatial plan. Although some kind of monitoring was carried out until now, the structure and procedures of the monitoring have to be formalized and the use of new indicators will be necessary.

In addition to these four expenses, which are quite reasonable in comparison to the obtainable benefits from the implementation of the SEA Directive, we expect some more expenses because of the SEA Protocol. These expenses are due to the enlarged public participation and the increased integration of health aspects into the planning process.

Public participation will or should be extended to consultations during screening, scoping and information about monitoring. A special problem for the organization of planning processes will be the consultation phases during screening and scoping. With the well-known consultation phase for the draft plan and environmental report, this adds up to three consultation phases. Planning authorities have to provide for reasonable time frames for the invitation and the handing in of statements and for dealing with the statements in a traceable way. Due to these new procedures, we estimate a minimum delay in the planning processes of at least three to five months. This might cause serious problems especially for amendments of plans or programmes. Additionally, resources will be required to explain the abstract terms 'screening' and 'scoping', the legal framework and methodologies to the public in a way that enables the public to understand the processes and give valuable statements. Experiences in Austria show that only a small part of the public is interested in consultations about spatial plans. Therefore, we think that these additional consultations bear the risk that the public may lose interest in taking part in the planning process before the crucial consultations concerning the draft plan and the environmental report, which might even undermine the objectives of SEA.

The public information about monitoring means that the monitoring results have to be edited and explained in a way that the public understands their importance for their daily life. This might cause more efforts for the authorities than just preparing data for internal use and presentation to decision-makers.

The emphasis on health aspects might cause a major difficulty taking the consultation of health authorities into account. Human health was always a concern in spatial planning and is also an environmental aspect of SEA according to the Directive. Until now, health aspects have had to be considered by the planning and environmental authorities already involved in all planning processes, no matter if an SEA has to be carried out or not. These persons use methods of spatial planning and environmental appraisals that include health impacts in multi criteria decision-making. With the SEA Protocol, it is intended that health professionals shall undertake the appraisal of health effects rather than environmental or social scientists (UNECE, 2003a, b, p.4). While the SEA Directive does not introduce authorities into planning processes that never had to deal with complex decision-making problems specific for spatial planning new contacts with health authorities have to be established according to the SEA Protocol. The SEA Directive implementation did not require new capacity building for the spatial planning or environmental authorities, but within health authorities new specifically trained and educated personnel resources have to be provided to fulfil this new task.

This problem is also addressed by WHO in a paper that links together Health Impact Assessment (HIA) and the SEA Protocol (WHO, 2001, p.22).

Stressing the importance of human health within an SEA might bear the risk that environment will mainly be seen from an anthropocentric point of view that thinks little of other issues like fauna and flora. The environmental considerations in decision-making might be focused on human health only.

CONCLUSION

The evidence presented demonstrates that the SEA Directive and the SEA Protocol are quite similar on the one hand but show a few significant differences on the other hand: the different scope of the Directive and the Protocol, more public participation in the Protocol but with a narrower definition of 'the public', the accentuation of health issues including the consultation of health authorities, no direct links to protected areas established in the SEA Protocol, and the requirement to provide information to the public about the monitoring results.

These differences lead to a significantly different balance of expenses and benefits between the SEA Directive and SEA Protocol applied in spatial planning. Additional benefits of the SEA Protocol might be the extended scope to policies and legislation, the widening of the knowledge base by consultation of health authorities, more consultation rights for the public especially in screening and scoping, and the provision of information to the public about the monitoring results. All measures serve to enlarge the transparency and traceability of planning processes.

On the other hand, we expect greater expenses from the implementation of the SEA Protocol because of the potentially increased public participation and the inclusion of health impacts in planning processes. From these two amendments, the following threats to the planning processes might arise:

- The increased public participation will lead to considerable delays in the planning processes which might be a special problem for amendments of plans or programmes. As the matters of screening and scoping are more abstract than the environmental report, commenting on these issues in the early stage of the planning process might make the public tired of the participation process and even deter the public from commenting on the crucial environmental report. The definition of 'the public' is narrowed in the SEA Protocol. Only the public concerned is eligible to consultations; in the SEA Directive, it is the public concerned and the public interested.
- The inclusion of health authorities might lead to a considerable expense in the capacity building of health authorities to participate in spatial planning processes. The stressing of health might cause a shift of interest from environmental to human health issues. This could mean that environmental issues like fauna and flora might lose value in the decision-making processes.

In our opinion, the balance of expenses and benefits of an SEA will be changed by the implementation of the SEA Protocol in comparison to the SEA Directive. Previous surveys done by the authors stated that, given the appropriate implementation of the SEA Directive, SEA will lead to a better preparation of the fundamentals of decision-making in spatial planning in Austria. This will further contribute to raising the quality of spatial plans as well as the legal certainty, without causing major delays or expenses in the spatial planning processes. SEA also should yield a major effort to increasingly integrate environmental considerations into the planning processes, to promote preventive, integrative environmental protection and also to facilitate sustainable development. This overall positive summary of SEA could be put into another perspective by the implementation of the SEA Protocol, because, according to our results, comparatively modest additional benefits might be accompanied by quite high additional expenses.

This chapter was originally published as: Stoeglehner, G. and Wegerer, G. (2006) 'The SEA Directive and SEA Protocol Adapted to Spatial Planning: Similarities and Differences', *Environmental Impact Assessment Review*, vol 26, no 6, pp586–599.

References

Feldmann L. (2003) Das Kiew-SUP-Protokoll zur Espoo-Konvention – Seine Querverbindungen zu bestehenden Instrumenten und seine Auswirkungen. The Kiev-SEA Protocol of the Espoo-Convention – its cross-connections to existing instruments and its impacts. In: UVP-report 2/2003, p.90–2.

Jacoby C. (2000) Die Strategische Umweltprüfung (SUP) in der Raumordnung: Instrumente, Methoden und Rechtsgrundlagen für die Bewertung von Standortalternativen in der Stadt-und Regionalplanung. Strategic Environmental Assessment in spatial planning: instruments, methodology and legal framework for assessing alternatives of sites in urban and regional planning. Berlin.

OEROK – Österreichische Raumordnungskonferenz. Methodenpapier zur Umsetzung der Richtlinie 2001/42/EG des Europäischen Parlaments und des Rates vom 27 Juni 2001 (SUP-Richtlinie) in die Raumordnungspraxis Österreichs. A methodological guide for implementing the SEA Directive into the spatial planning practice in Austria. Vienna.

Proebstl U., Jiricka A.and Stoeglehner G. (2006) Die SUP-Umsetzung in der örtlichen Raumordnung in Österreich. The SEA-implementation in municipal spatial planning in Austria. In: UVP-report 1/2006.

SEA Directive: Directive 2001/42/EC of the European Parliament and of the Council of 27 June 2001 on the assessment of the effects of certain plans and programmes on the environment. Official Journal L 197/30 of the European Communities of 21.7.2001.

Scholles F. (2001) Szenarien zur Umsetzung der SUP-Richtlinie in Deutschland Scenarios for implementing the SEA Directive in Germany. In: UVP-report 2/2001; p.127–130.

Stoeglehner G. (2003a) Die Strategische Umweltprüfung in der nominellen Raumordnung Oberösterreichs. Strategic Environmental Assessment and spatial planning in Upper

Austria. Dissertation carried out at BOKU – University of Natural Resources and Applied Life Sciences Vienna.

Stoeglehner G. (2003b) Integrating Strategic Environmental Assessment into structural funds programming. In: Environmental Center for Central and Eastern Europe, editor. Linking development with the environment: perspectives from the EU and accession countries. Bratislava ISBN: 80-968850-4-9, p.43–54.

Stoeglehner G. (2004a) Die Strategische Umweltprüfung im Salzburger Raumordnungsgesetz-Haare in der SUP? (Strategic Environmental Assessment in the spatial planning regulation of the province of Salzburg, UVP-report 4/2004/), p.183–6.

Stoeglehner G. (2004b) Integrating Strategic Environmental Assessment into community development plans – a case study from Austria. European Environment, 14. The Journal of European Environmental Policy, p.58–72.

Stoeglehner G. and Wegerer G. (2004) Die Strategische Umweltprüfung – Ein Planungsinstrument zur Qualitätssicherung in der Raumordnung. SEA – an instrument for providing quality in spatial planning. DISP, p.52–9.

UNECE – United Nations Economic Commission for Europe. (2003a) Meeting of the parties to the Convention on Environmental Impact Assessment in a Transboundary Context, Preparatory Meeting for the First Meeting of the Signatories to the Protocol on Strategic Environmental Assessment: Background paper presenting possible elements for a work plan for the SEA Protocol. First Meeting, Geneva 27 and 28 November 2003.

UNECE – United Nations Economic Commission for Europe. (2003b) Protocol on Strategic Environmental Assessment to the Convention on Environmental Impact Assessment in a Transboundary Context.

UNECE – United Nations Economic Commission for Europe (2005) Meeting of the parties to the Convention on Environmental Impact Assessment in a Transboundary Context, Meeting of the Signatories to the Protocol on Strategic Environmental Assessment. Report of the Second Meeting, Geneva 25 and 26 April 2005.

UNECE – United Nations Economic Commission for Europe (2006) Status of the signatories of the protocol on Strategic Environmental Assessment. www.unece.org/env/eia/protocol_status.html Last accessed 4 March 2006.

Weber G. and Stoeglehner G. (2001) Integrationsmöglichkeiten der strategischen Umweltprüfung in die nominelle und funktionelle Raumordnung – dargestellt an ausgewählten Beispielen. Possibilities of integrating strategic environmental assessments into the planning system – shown on selected examples. Wien, published in www.lebensministerium.at/umwelt/

Wegerer G. and Stoeglehner G. (2004) Implementing Strategic Environmental Assessment (SEA) into spatial planning – an ex-post survey of spatial development concepts of Lower Austria. SUSTAIN Life – Secure Survival II. Conference Proceedings Prague, 22–25 September 2004 ISBN: 80-213-1197-5.

WHO – World Health Organization, Regional Office for Europe. (2001) Health impact assessment as a part of Strategic Environmental Assessment. A review of health impact assessment concepts, methods and practice to support the development of a protocol on Strategic Environmental Assessment to the Espoo Convention, which adequately covers health impacts (November).

12

The Institutional Context for Transboundary Environmental Impact Assessment in Belgium: Multi-level Setting – A Matter of Smooth Governance?

Jan De Mulder

Environmental impact assessment (EIA) and strategic environmental assessment (SEA) procedures are prescribed in laws and regulations. These formal instruments are supplemented by more informal administrative practices. An EIA or SEA process always involves a number of categories of participants or stakeholders (including institutions). Given the objective of this process, describing the relationships between these stakeholders in the impact assessment process as an 'administrative negotiation process' is quite acceptable (De Hemptinne, 1994). So the EIA/SEA serves multiple purposes (Cashmore et al, 2004). As the participants or stakeholders have different expectations regarding the process and its outcome given their interests, perceptions and societal values, their opinions about a 'good quality EIA' might also differ. From this participatory perspective, one may agree with observations in the literature that EIA professionals – but also the other stakeholders in the process – should come to grips with the facts that EIAs (and, even more, SEAs, given their 'strategic nature') are not science and will always contain unexamined and unexplained value assumptions (Beattie, 1995).

In a transboundary context, these value assumptions might even be culturally influenced (Van Dijk and Nijsten, 2004), as even 'administrative cultures' in neighbouring states may differ (Berkenbosch, 2002; Van Schie and Raessen, 2001).

Within some European Union (EU) Member States, the devolution of policies (by giving subnational entities growing and sometimes even legislative competencies) has created different institutional and administrative approaches. So certain EIA or SEA procedures within one state may involve transboundary negotiations or even the need for an institutionalized arrangement. Otherwise more centralized states may have institutional distrust towards emerging subnational – but fully competent – impact assessment approaches in a neighbouring state. All these developments and factors indicate that the growing practice of transboundary impact assessment (TIA) undoubtedly has a number of specific features and encounters peculiar problems. On the other hand, TIA is also very much a product of the application of national and subnational impact assessment systems based on regulations and administrative practices. As such TIA will reflect the strengths and weaknesses of these systems.

The Belgian impact assessment systems are quite diverse also regarding the transboundary and quality assurance issues. This chapter aims to introduce these systems and illustrate current practices.

Belgium is a federal state, made up of three communities (language-related: Flemish-, French- and German-speaking) and three regions (territory-related: Flanders, Wallonia and Brussels, the capital). In Flanders, the community-related and territorial aspects have been merged into one regional entity.

Constitutional reforms in 1970, 1980, 1988, 1993 and 2001 have resulted in devolution of competences. Substantial powers in the cultural, social, economic and environmental spheres have been transferred from the federal level to the regional level (Deketelaere and Schutyser, 2000). However, as the heart of the federal state is an economic and monetary union, it is no surprise that quite a number of socio-economic issues remain under federal competence, contrary to the major part of environmental policy-making which came in the hands of the regional policy-makers (Lavrysen, 1999). The legislative power of the communities and the regions is exercised by the enacting of decrees (ordinances in Brussels). Decrees or ordinances have force of law throughout the territory for which they are intended. In order to improve cooperation between the different policy levels and to avoid expected deadlock situations, constitutional reform has also introduced 'co-operation agreements'. Some agreements may have direct effect, others require additional approval by the different parliaments (federal and regional). Already a number of these agreements (e.g. on EIA in a 'trans-regional' context) exist but their enforcement remains questionable due to a lack of enforcement mechanism.

Particular features of the Belgian federal system are the absence of a hierarchy of norms as well as the possibility for regions or communities to conclude international agreements in the allocated areas of responsibility (e.g. most environmental issues).

The societal evolution and the historical devolution of competencies have resulted in a multi-actor policy approach at different policy levels: municipal, provincial, regional and national (federal). Competencies regarding particular policy fields, such as environment, are often not attributed to one policy level. The application of policy instruments in such a framework leads to complex processes and regulatory frameworks for decision-making within Belgium.

The transposition of the consecutive EU directives (EIA Directive 85/337/EC as amended) has resulted in a growing EIA practice. EIA approaches and requirements are found in horizontal as well as in specific sectoral legislation. Most EIA legislation, however, is to be found at the regional level, except for the projects in the Belgian marine environment and also for nuclear installations, as the decision-making for both categories was kept under federal competency.

Later on, the transposition of the SEA EU directive of 2001 (SEA Directive 2001/42/EC) revealed a more profound 'impact' on decision-making processes. The final adoption of federal and regional SEA legislation happened in the course of 2006–2008.

The application of the impact assessment frameworks has raised questions about the coherence and both proponents and authorities have to deal with these institutional features. Institutions provide not only for frameworks, they are also stakeholders in decision-making and have an interest in impact assessment (Nooteboom, 2007).

This contribution gives a brief overview of the major relevant legal frameworks at the federal and the Flemish-regional level. Given that the Flemish region borders the Netherlands, some examples of TIA approaches are presented and situated within the Belgian institutional setting.

LEGISLATION AT THE BELGIAN FEDERAL LEVEL

As a limited number of environmental policy items – such as the protection against ion radiation, transit of waste, protection of the marine environment, and the policy concerning product norms – remained within the federal competence, instruments that have to be applied in order to fulfil the implementation of legal requirements are addressed at this level.

So at the federal level, EIA and SEA regulations and provisions have been adopted concerning:

1. Protection against ionizing radiation from nuclear installations;
2. Protection of the marine environment; and
3. A limited number of plans and programmes.

Given the scope of activities, all three regulatory sets are important from a potential transboundary impact perspective.

Nuclear installations

The Belgian sites for nuclear plants are situated in border areas so, given the amplitude of radioactive impacts, the establishment and (mostly the further) exploitation of such installations is subject to transboundary impact assessment requirements.

The Act of 15 April 1994 on the protection of the population and the environment against ionizing radiation from nuclear installations and the Federal Agency for Nuclear Control (FANC) is the basic act and replaced a similar act of 1958 which became outdated.

This act classifies nuclear installations in different categories. The permit application for installations of category I must include an EIA which contains:

- Information as stipulated by the recommendations of the European Commission of 6 December 1999 (1999/829/Euratom) concerning the application of Article 37 of the Euratom Treaty;
- The necessary information to provide and assess the effects on the environment related to ionizing radiation;
- A draft of the most important alternatives, including a justification of the final choice with respect to the effects on the environment.

The permitting regulation concerning nuclear issues stipulates that the EIA is the responsibility of the developer who has to appoint natural persons or a legal person to draft the EIS. These people can do this work only after the developer receives approval from the FANC that has based its opinion on a document submitted by the developer, containing information about the consultant(s), such as their technical competencies and other relevant references.[1]

International consultation

The Royal Order on the permit procedure contains a specific provision (Article 6.3.2) on international consultations. In cases as provided for by Article 37 of the Euratom Treaty, the FANC has to consult the European Commission for advice. Also the Scientific Board of FANC may consult the European Commission on, for example, the effects on the environment. If the Scientific Board is of the opinion that the planned installation may have significant effects on the environment of one or more states that are party to the European Economic Area (EEA), or on the request of one or more of these parties which have the opinion that the planned installation may have significant effects on their environment, the FANC will send a summary of the EIA to these states at the same moment that the permit application file is submitted to the concerned municipalities.

Belgium has two major nuclear sites, one in Wallonia, the other in Flanders. The nuclear park in Flanders is at Doel (near the Dutch border) where four reactors are located. In 2009, the steam generators of one of the reactors will be replaced, and the permit procedure started in 2007. Public consultation is part of this procedure as well as advice from the municipalities which are located within 5km of the site. This permit application was open for consultation in two municipalities in The Netherlands, so foreign citizens and local governments were able to submit comments and objections.

Marine developments

Offshore activities for energy production are an example of marine developments that lead to environmental effects and other impacts, including transboundary ones.

Activities in the small Belgian part of the North Sea became rather controversial given Belgium's rather limited capacity for land-based wind farms and the need for renewable energy resources. Currently, the proportion of renewable energy in Belgium is 2.2 per cent, and the projection is 7 per cent by 2013. The EU requires Belgium to use 13 per cent of its energy from renewable sources by 2020 and according to studies off-shore wind farms could produce up to 27 per cent of current energy consumption. The European Wind Energy Association has calculated that this means a need for 1100 new windmills. Over the past eight years, the Flemish Government has been able to issue 167 construction permits for (land-based) windmills. Like a number of other off-shore activities, the construction of wind farms is subject to federal legislation.

The act of 20 January 1999 on the protection of the marine environment in sea areas under Belgian jurisdiction, the 'MMM Act' (changed by the acts of 17 September 2005 and 21 April 2007 and implemented through a number of Royal Orders) establishes the legal basis for the protection of the Belgian part of the North Sea against sea-related pollution and for the conservation, restoration and development of nature. This act summarizes some general principles of environmental law and transposes international legal obligations:[2]

- The prevention principle: prevention is better than cure.
- The precautionary principle: preventive measures must be taken if there are grounds for concern regarding pollution.
- The principle of sustainable management: human activities must be managed in such a way that the marine ecosystem remains in a condition that ensures the continued use of the sea.
- The polluter pays principle: the costs of measures to prevent and fight pollution are to be borne by the polluter.
- The principle of restoration: if the environment is damaged or disrupted, the marine environment must be restored to its original condition as far as is possible.

The principle of objective liability is also established: in the event of any damage to or disruption of the environment in sea areas as a result of an accident or an infringement of the law, the party having caused the damage to or disruption of the environment is obliged to remedy this, even if they are not at fault.

A general obligation is established, as regards activities for which a permit is required in advance, to prepare a report on the environmental effects (the 'classical' EIA at the initiative of the proponent) and to undertake EA before and during these activities (carried out by the Government). The legal requirements of the MMM Act are elaborated in a number of Royal Orders, including: the Royal Order of 7 September 2003 on the procedure for permits required for certain activities in sea areas and the Royal Order of 9 September 2003 on the assessment of environmental effects.[3] Article 6 of the Royal Order of 9 September 2003 requires that a co-ordinator is in charge of the supervision of the EIA drafting. This co-ordinator may be employed by the proponent and, if such is the case, the coordinator gets certain safeguards in order to assure independence. This Royal Order transposes the procedural requirements of the EIA Directive but refers also to the Espoo Convention (the UNECE Convention on EIA in a Transboundary Context) in the definition of 'activity with a transboundary dimension' that is to be understood as any activity that falls under the scope of application of the Espoo Convention or the EIA Directive.

Next to the legal requirements regarding the contents of an EIA, the proponent may address the competent authority (the Management Unit of the North Sea Mathematical Models and the Scheldt Estuary, MUMM) for scoping advice. The review or quality control is done by the MUMM. The MUMM investigates if the proponent's EIS is complete and of a sufficient quality. It may require additional information or research to be done by the proponent or the MUMM may commission the additional research or do it on its own. The MUMM presents a report (assessment of the proponent's EIS) about its investigation and issues an advice to the minister who is responsible for granting or refusing the permit.

Article 23 of the above-mentioned Royal Order concerns also transboundary impacts, but from an ex-post perspective when the activity has been permitted. Such activities are being monitored and a kind of permanent EIA is being executed if the MUMM has stated that the activity may have significant effects on the marine environment of another EU member state, or a party to the Espoo Convention or on request from such a state or party. In case the permanent EIA indicates that significant negative effects may occur, the concerned state or party is informed immediately in order to start consultations about necessary measures.

This provision implements Article 7 of the Espoo Convention on post-project analysis. Though this article has a non-mandatory character, it offers some opportunities. The guidance on the practical application of the Espoo Convention clarifies this provision and indicates that a post-project analysis has to analyse, as a minimum, both the activity as well as its potential adverse transboundary impacts. A post-project analysis is typically based on the monitoring of the activity and its impacts.[4]

The MMM Act does not deal with plans and programmes, so SEA is not an issue under this Act. However, given the size of at least the wind farm projects and their role in an overall energy strategy, one may question the appropriateness of this project-level approach. This is also recognized in the literature; however, the need or usefulness of SEA is not always obvious (Douvere et al, 2007). SEA for marine developments is being addressed by the federal SEA Act (see below).

There are three projects for the construction of windmills in the Belgian part of the North Sea: C-Power on the Thorntonbank, Belwind on the Bligh Bank and Eldepasco on the Bank zonder naam (Bank without name) (Figure 12.1). The C-Power wind farm has 60 windmills and is located 27km off Zeebrugge. The park will be built in water 6.25m deep and the turbines will be installed in an area of 18 km^2. Work started on land in 2007 with the construction of the

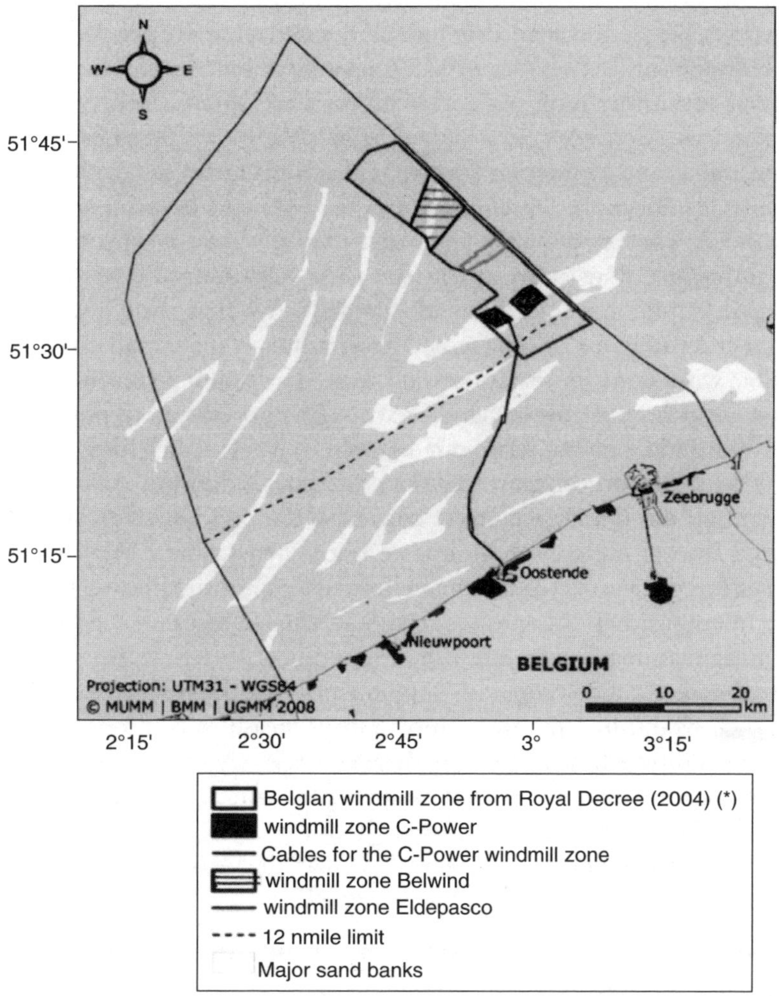

Figure 12.1 *Windfarms in the Belgian part of the North Sea*

Note: * Royal decree, 17 May 2004; Belgian official journal, 29 June 2004

first six foundations. In spring 2008 work started at sea. With a capacity of 330 megawatts, the Belwind farm will deliver energy to 330,000 households. The wind farm, which is located 42km off the coast of Zeebrugge, is the world's farthest offshore wind farm. The park will be built in water 20-35m deep and the turbines will be installed in an area of 35 km^2.[5]

Consultation with the Dutch authorities about the Belwind farm[6]

The concession for this wind farm is located 530m from the Dutch maritime border. So the application for this project as well as the relevant information about the procedure was sent to the Dutch authorities on 1 August 2007. The MUMM received a request for consultation on 17 September 2007 but received no remarks or objections before the final deadline of the consultation period (24 October 2007). However, on 18 October 2007 a consultation meeting between the Belgian and Dutch authorities took place in Brussels. The Dutch Minister of Transport and Water sent a letter on 23 October 2007 to indicate possible problems for maritime traffic and requesting further research on fauna issues. In its opinion to the federal minister in December 2007, the MUMM gave an overview of the transboundary consultation and indicated that all elements mentioned by the Dutch authorities were addressed in the EIA. Regarding the Dutch concerns, the MUMM opinion is quite remarkable as it says that the Dutch objections apply not specifically to the concerned project but to the cumulative impacts of all possible wind farms in the concession zone. The opinion continues by stating that these concerns can only be handled at a higher governmental level. This point of view illustrates that an SEA can be seen as the missing link between the planning of the wind-farm zone and the project-level decision-making.

Regarding the Dutch ecological concerns, the opinion states that all issues mentioned during the consultation were researched in the EIA that concluded that the effects were expected to remain limited and not significant. The MUMM has the opinion that the proposed monitoring measures – particularly on avifauna, sea mammals, noise and fish – should be sufficient to register and assess unforeseen ecological damage with consequences for the nature in the Voordelta area. If such would be the case, compensatory measures could be taken, but at present, no additional measures were deemed necessary or proposed.

Regarding the economic loss for the shipping sector, the opinion sees no differences between the Dutch and Belgian interests. It also puts the need for environmental friendly energy developments on a higher level than the maritime transport interests.

Plans and programmes and strategic environmental assessment

The field of application of the federal SEA Act (Act of 13 February 2006 on the EA of certain plans and programmes and the participation of the public in the development of certain environmental plans and programmes and its implementing

Royal Orders of 22 October 2006 and 5 June 2007) is rather limited. A SEA is mandatory for only six types of plans. For some of these types, earlier sectoral legislation (e.g. on the organization of the electricity market and on the transport of gas products) already provided for a SEA requirement. One category concerns plans and programmes for the exploration and exploitation of non-living resources in the Belgian territorial sea and the continental shelf. To introduce a SEA requirement for other types of plans or programmes, a Royal Order has to be approved. The SEA Act stipulates also that, for certain plans or programmes, the proponent has to provide for a SEA when the Council of Ministers has decided so after consulting an Administrative Advisory Committee (established by this SEA Act).

The Administrative Advisory Committee is also involved in SEA scoping. The proponent has to present a draft scoping document to the committee. The committee gives an advice and the proponent decides on the scope but has to inform the committee about the decision. Members of this committee belong to several federal departments, such as energy, mobility, economy and also sustainable development. When the proponent relies on consultants for drafting the SEA, they need to ensure there is no conflict of interest. Contrary to the regional approaches there is no obligation to contract certified consultants. The proponent (and consultants) needs to follow or apply the advice of the committee.

The proponent has to submit the draft plan or programme and SEA to the Advisory Committee as well to the Federal Council on Sustainable Development and the regional governments for advice.

Regarding the consultation of foreign authorities when the implementation of the plan or programme might have a transboundary environmental impact, Article 13 of the SEA Act refers to the UNECE Espoo Convention.[7] This is quite intriguing given the existence of the Kiev Protocol on SEA to this Convention (not yet into force; signed but not yet ratified by Belgium), but might be explained by the provision in the convention that encourages the application of its principles to policies, plans and programmes.

Transboundary SEA

The Royal Order of 5 June 2007 makes a distinction between different parties of origin of the plan or programme.[8]

In case of a *Belgian plan or programme*, this federal approach contains two phases:

1. The notification phase: The proponent of the plan or programme needs to provide the following to the possibly affected party:
   - The draft plan or programme and the SEA (environmental report).
   - A description of the planning and decision-making procedure for the plan/programme.
   - The arrangements for the consultation of the public, including the starting date as well as the duration.

2. The consultation and information phase: The possibly affected party needs to inform the proponent about the organization of national consultation.
   - Within 45 days after the notification has been sent; if this is not done, the internal procedure continues.
   - The proponent and the possibly affected party agree on a reasonable timing regarding the organization of the national consultation.
   - The possibly affected party gives its advice to the proponent within the agreed deadline.
   - The possibly affected party receives a copy of the plan or programme and the environmental report within 10 days after publication in the official journal.

In case of a *foreign plan or programme*, this federal approach contains the following steps:

- In case a federal service (administration, institution, minister) receives a foreign draft plan or programme and SEA on the presumption that it may have significant effects on the Belgian environment (which means in practical terms only the Belgian part of the North Sea as the other 'territorial environments are governed by the regions'), the information is forwarded to the federal Minister of the Environment, the federal Minister of the North Sea, the Federal Council on Sustainable Development, the regional governments and the Administrative Advisory Committee.
- The minister competent for the protection of the marine environment decides if the proposed plan/programme might have a significant environmental effect
- If the minister decides there might be a significant effect, he organizes consultation of the public.
- After the public consultation, the minister competent for the protection of the marine environment sends the comments and advice to the competent authorities of the party of origin and informs the federal Minister for the Environment and the Administrative Advisory Committee.

If no significant effects are expected, all involved authorities are informed immediately.

Legislation at the Flemish regional level

All three Belgian regions have EIA and SEA systems. The Walloon Region has the oldest EIA system. It had a quite elaborate system that has since been amended.[9] The general provisions on EIA and SEA are found in the fifth part of book I of the Walloon Environmental Code. The Brussels EIA procedure is also elaborated.[10] As both regional EIA and SEA regimes have to transpose and implement the EU directives they also contain provisions on transboundary impact assessment which provide for information exchange and consultation.

The EIA/SEA decree of 18 December 2002 introduced the first comprehensive set of provisions on EA at the Flemish level. Through this decree, EIA, SEA (see Table 12.1) and safety reporting (as required by the Seveso Directive) became part of the framework decree on general provisions regarding environmental policy.

Table 12.1 *The Flemish SEA procedure: Steps and involvement of actors*

Step	*Action*	*Actors*	*Delay – target date*
0	Informal pre-consultation	Proponent, competent SEA authority, consultants	
	PM screening opportunity	Proponent, competent SEA authority, consultants	30 days (+ 60 days in case of transboundary procedure)
1	Notification (including documentation)	Competent SEA authority	Date = N
2	Declaration of completeness of notification is sent to proponent	Competent SEA authority	Receipt N + 20 days Date = Y
	Scoping phase = > requirements about SEA contents		
3	Start of consultation Gathering of advices, organization of publicity of notification and documentation, discussions, meeting (optional)	Competent SEA authority, governmental services, local authorities, foreign authorities, the public, proponent, consultants	Y + 30 days
	Comments are analysed and processed, Scoping decision	Competent SEA authority	Y + 50 days (+ 60 days in case of transboundary procedure)
	Decision is sent to proponent*		
	Middle phase		
4	SEA is being drafted Intermediate consultation (including meetings) is possible	Proponent, consultants	(months)
	Final phase – *quality control*		
5	Submission of the final SEA	Proponent	
6	Review of the SEA Approval or rejection of SEA	Competent SEA authority	50 days
	Decision is sent to proponent*	Competent SEA authority	

Note: * After receiving this decision the proponent may decide to introduce an 'appeal' to an administrative advisory commission.

The decree of 27 April 2007 replaced the SEA chapter completely.[11] The current EIA and SEA provisions were supplemented by the implementing orders of the Flemish Government of 10 December 2004 (lists of projects for which an EIA is mandatory, directly or after screening), 12 October 2007 (on SEA, mainly consultation requirements)12 and 18 April 2008 (on the SEA integration in the physical planning procedures). Based on the relevant provisions of the decree of 27 April 2007, Articles 6 and 11 of the implementing order of 12 October provide for (but mainly repeat) the requirement for transboundary consultation in the screening and scoping phases and also in the public inquiry phase of the decision-making on the plan or programme.

The interregional EIA cooperation agreement

During the first decade of regional EIA practices in Belgium, the lack of information exchange between the regions became obvious, especially for projects located at the regional 'borders' or for major (even cross-border) projects. For that reason the three regions agreed on a cooperation agreement on 4 July 1994.[13] This agreement has been in force since 4 September 1994. The federal level was (and is) not involved in this agreement. The contents of this agreement reveal many similarities with transboundary 'international' EIA arrangements, as based on the Espoo Convention.

Article 2 of this agreement concerns the scope, which includes every project that requires an EIA. The government of the affected region is informed about a project for which a permit application is submitted. The agreement does not include any criteria that might help a government to assess whether a project may lead to transboundary effects. A common interpretative declaration to this agreement does not really clarify this issue. The government of the region of origin has to send a copy of the EIS to the government of the affected region. The latter has to receive this information before the public inquiry in the region of origin is organized and has to be informed about the practical modalities of this inquiry. The inhabitants of the affected region have the same rights to participate in the public inquiry as the inhabitants of the region of origin. The affected region may hold its own public inquiry on its own territory and send the outcome of this inquiry to the government of the region of origin.

However, this public inquiry in the affected region may not lead to a prolongation of the duration of the public inquiry as provided for in the legislation of the region of origin. This means that, for example, in Flanders an EIS will contain a chapter on transboundary impacts in case the competent environment administration (EIA-SEA Unit) has concluded on the basis of the notification documentation that the project could lead to transboundary effects on the environment of the neighbouring region. However, this cooperation agreement is not implemented strictly, and its existence has revealed a number of problems and enhanced political discussions.

The approach in this agreement remains rather vague and complex as the principle actors are the regional governments and the agreement does not assign a particular role to the competent environmental administrations. Also the legal nature of this kind of internal agreement (it was not endorsed by the regional parliaments) is not helpful to ensure its application.

The common declaration to this agreement included an evaluation before the end of 1995. This intention was repeated in an interregional agreement of 6 April 2000. A couple of years later a revision of this cooperation agreement came on the political agenda due to a serious case of (transboundary) air pollution (caused by a fire in Brussels) that raised questions even from Germany. Given the expanding EIA and SEA regulations within Belgium at the different policy levels, one might suppose that a coherent proposal – involving all covered policy fields and levels – would be an option, but to date this cannot be confirmed. Some years ago, the FANC started a separate initiative for a cooperation agreement on nuclear installations that has as yet not been finalized.

FLEMISH–DUTCH ARRANGEMENTS AND AGREEMENTS

A bilateral EIA arrangement

On the basis of the Espoo Convention and responding to growing pressure from NGOs some 15 years ago, the Flemish and Dutch environmental administrations agreed on an information and consultation approach on the basis of the existing EIA legislation in both jurisdictions. This approach was endorsed at the political level – through an exchange of letters – in both countries (but it is not to be considered as an agreement or treaty) and has been applied since 1 September 1995. Initially planned for only one year, the approach has not been amended up to now (Spring 2008). However, in the course of the past years, several bilateral meetings have taken place and elements for amendments were raised, including particular text proposals, inter alia due to the changes to the legislation in both countries, such as the introduction of SEA.

As this arrangement had to incorporate the requirements of two sets of legislation, the different steps to be followed in case of its application contain two stages:

1. The EIA-process stage; and
2. The permit application stage.

In case a proponent plans an activity (project) in Flanders for which an EIA is mandatory, and the project may have potential significant adverse transboundary effects in The Netherlands, the bilateral arrangement requires the following actions (steps):

1. The Flemish EIA-SEA Unit has approved the team of EIA-experts and determines that significant adverse transboundary impacts may occur. Decision to initiate and apply the transboundary EIA-procedure.
2. Simultaneously, the following actions take place:
   a. The Flemish EIA-SEA Unit sends the notification file to the relevant Flemish provincial and municipal government.
   b. The Flemish EIA-SEA Unit sends the notification file to the Dutch focal point: the relevant provincial government, with an accompanying letter containing the following information:
      - An explanation of the EIA procedure, including an indication of the possibilities for the Dutch governmental bodies for involvement;
      - A list of possible involved Dutch governmental bodies;
      - A list of involved Dutch governmental bodies receiving the notification-file (including the EIA start document that is a kind of scoping report).

The notification file contains:

- A brief description of the proposed activity;
- A brief description of the location of the project;
- The name of the proponent;
- The composition of the team of experts.

3. The Flemish Minister of the Environment notifies the Dutch Minister of the Environment that the TEIA procedure has been initiated.
4. The Dutch focal point reacts without delay in case not all involved Dutch governmental bodies received a copy of the notification-file (including the EIA start document). Furthermore, the focal point outlines the means for informing the public, including the possibility for inspection in The Netherlands:

   - The means of publicity in The Netherlands of an EIS which has a certificate of conformity;
   - The means of publicity of the public enquiry in Flanders of the permit application for which an EIS is mandatory.

5. The proponent is informed about the application of the TEIA procedure. The Flemish EIA-SEA Unit sends the EIA start document to the involved Dutch governmental bodies and invites them, together with the Flemish bodies, to a meeting about the EIA start document.
6. The Flemish EIA-SEA Unit invites the relevant Dutch governmental bodies for a discussion about the draft EIS, including the opportunity for comments.
7. Immediately after the Flemish EIA-SEA Unit has sent the EIS (including its approval) to the proponent, it sends a certified copy to:

a. The involved Flemish provincial and municipal governments; an accompanying letter explains that it concerns a project for which the transboundary EIA procedure is being applied.
b. The focal point and the involved Dutch governmental bodies; an accompanying letter indicates how information about the EIS may be obtained on written request, as well as the period within which consultation as provided for by Article 7 of EIA directive may be initiated.

8. The Flemish Minister of the Environment informs the Dutch Minister of the Environment about the sending of the EIS to the provincial government.

The next set of steps are beyond the formal EIA procedure as they are part of the environmental permit application procedure:

9. The Flemish Environment administration takes care of the publicity about the EIS by way of announcement in Dutch newspapers and the EIS is open to the public through the services of the Dutch focal point.
10. One of the involved Dutch governmental bodies may ask for consultation within the specified period on the basis of the EIS and as provided for by Article 7 of the EIA directive.
11. The consultation takes place between the requesting parties and the Flemish EIA-SEA Unit.
12. Within the framework of the public inquiry about the environmental permit application for a project for which an EIS is mandatory, the competent mayor takes care of the publicity about the EIS by:
    a. Bill posting.
    b. The announcement of the public inquiry in at least two Flemish and two Dutch newspapers (including one local publication in each country).
    c. The publicity (open for inspection) of the permit application including the EIS and annexes.
    d. The organization of at least one information meeting in Flanders, which has to be announced in at least two Flemish and two Dutch newspapers with, for each, one with a local character.

The competent mayor informs the Dutch focal point about these initiatives.

13. The competent authority sends a certified copy of its decision about the environmental permit application to the Dutch focal point.
14. On written request, the competent authorities are obliged to provide all information about the inspection possibilities concerning the permit.

The Scheldt River developments and agreements

Flanders and the Netherlands share the Scheldt Estuary, which is situated on both Dutch and Flemish territory. It stretches over some 160km between the city of

Ghent in Flanders and the river mouth in the Netherlands. The management of this river has always been a sensitive issue given the historical relationship between the countries, as Belgium was once part of the Netherlands (e.g. Secession Treaty of 1839). Given the major economic importance of the harbour of Antwerp for the Belgian economy, accessibility is an ongoing concern. The estuary is regularly subject to storm floods from the North Sea; it provides maritime access to important ports, such as Antwerp; and its ecological value is considerable with both fresh and salt-water tides.

In 1998, the Netherlands and Flanders decided to develop a joint long-term vision for the Scheldt Estuary and its functions of flood safety, ports accessibility and important natural ecosystem. In this, the basic idea was: 'Developing a healthy and multifunctional water system, supporting human needs in a sustainable way'. The Technical Scheldt Commission (TSC, established in 1948 by the governments of Belgium, Luxemburg and the Netherlands) took the lead in drawing up this holistic vision and was able to present the result to the competent government representatives in January 2001. The long-term vision was summarized in an overall target for the year 2030. This target has five characteristics:

- The preservation of the estuary's physical system characteristics is the point of departure for management and policy.
- Maximum safety against flooding is an important condition of existence for both countries.
- As driving forces for welfare, the Scheldt ports require optimal accessibility.
- The estuarine ecosystem is healthy and dynamic.
- The Netherlands and Flanders cooperate in the administrative–political and operational fields.

In 2001, the governments of both countries adopted this overall target and already by 2002, the drawing up of the 2010 Development Outline (DO) for the Scheldt Estuary had started. The aim of the 2010 DO was to define those projects and measures which, in a first stage, must be started up no later than 2010 to ensure the realization of the long-term vision in 2030. A special project organization, ProSes (Project management Scheldt Estuary), was created in order to draw up this DO in close consultation with all stakeholders and under the supervision of the TSC.[14] In December 2004, the official version was presented to government representatives, after intensive communication with the stakeholders and a consultation with the general public on their views on the outline. On 11 March 2005, both governments agreed to start the implementation of the full 2010 DO. The main projects in the 2010 DO are:

- Safety against flooding: implementation of the updated Sigmaplan (a plan dating from 1977 that aims to reduce the risk of floods in Flanders).

- Accessibility: deepening and widening of the shipping lane to the port of Antwerp.
- Ecosystem: development of 600 ha of estuarine nature along the Western Scheldt in The Netherlands and 1100 ha of estuarine nature and wetlands along the Sea Scheldt in Flanders.
- Both countries will jointly monitor the evolution of the estuary and the effects of the implemented projects in order to extend knowledge of the estuary and to facilitate possible corrections.

Besides these large-scale projects, a considerable number of smaller administrative and legal measures were defined. Regarding the bilateral institutional framework, particular Scheldt treaties were concluded on 21 December 2005. Only at the end of 2007 were these treaties approved by the Dutch legislature. The treaty on the policy and management of the Scheldt Estuary includes provisions that very much stress the need and willingness to align the existing decision-making procedures for plans, programmes and projects and even possibility of specific bilateral regulation (Article 5). Another important provision in this treaty concerns the monitoring of the morphological evolution of the Scheldt Estuary but also the ex post evaluation of effects of, for example, projects (Article 6).

This treaty establishes also a Flemish–Dutch Scheldt Commission. The role of this bilateral commission is to foster the structural engagement of local authorities and relevant societal stakeholders in the implementation of treaty's objectives. This commission is also the liaison with the International Scheldt Commission established by the Scheldt Treaty (Ghent, 2 December 2002).[15] The treaty on the implementation of the 2010 DO contains a number of more detailed provisions, inter alia concerning technical, financial and administrative arrangements, including administrative and physical monitoring issues.

Strategic environmental assessment

Given the nature of the overall goal of the 2010 DO, it was obvious that a SEA was necessary. It was decided to create a joint SEA but that meant also developing a procedure acceptable for both countries. The existing regulations in both countries revealed the need to address the following challenges:

- Who is the joint authority in a joint procedure?
- How to agree on and formulate identical guidelines for the environmental report.
- Procedural timing differences.

The SEA was the first basic step and was followed by several TEIAs and (national) EIAs and SEAs (see Box 12.1).

The consultants who steered the SEA of the 2010 DO afterwards published some reflections about their experiences, which included the following:

- The quality of research and decision-making can be improved by connecting them. Stakeholders should have their say in formulating research objectives. They should be involved in discussing intermediate results, and consultants should not be afraid of using unfinished results in the debate (Zanting et al, 2004).
- The scoping phase should focus on the information necessary for the decision-making. Choices should be made, justified and communicated. Be aware of the risk of researching all interesting issues but in a superficial way.
- An integrated approach is not always better. Thematic approaches or separate studies on specific measures can be sufficient. This was done in the SEA and only in the last phase was 'expert judgement' used to evaluate the combined effects of sets of measures.
- The research should be focused as much as possible on the 'principles of alternatives'. If alternatives are changed afterwards, the research conclusions may remain valid. In order to get a grip on these principles, the use of representative 'example projects' is recommended.

Box 12.1 EIA for the deepening of the Scheldt River

For the decision-making on the project for the deepening of the Scheldt River (as agreed and preliminarily researched in the SEA on the 2010 DO) a transboundary EIA was made and approved by the competent EIA authorities in both countries. Contrary to the Flemish EIA approach, which is purely internal-administrative, the Dutch EIA approach includes an external, independent review institution, the EIA Commission, which uses a pool of experts.

In The Netherlands, a specific commission is established for each EIA and the Scheldt EIA Commission published its advice on 1 February 2008 (Commissie-MER, 2008).

Elements of an EIA evaluation

In its evaluation, the commission seeks to identify shortcomings with respect to the requirements as formulated at the start. Next the commission will assess the shortcomings. The central question concerns the availability of all necessary information to give the decision-makers the opportunity to take fully into account the environmental concerns. If this is not possible, the commission shall indicate an essential shortcoming and advice that the required information needs to be made available before the decision-making. Other shortcomings of the EIA will also be addressed in the commission's advice as far as they can be formulated as clear recommendations for the competent authority. This means that the commission's advice is limited to important issues.

This particular advice was positive about the EIA (e.g. on the research of alternatives) and was also based on additional information that was provided by the proponent. The commission also stressed the uncertainty about the predictions regarding the morphological developments which also have an impact on the effect-predictions for some environmental factors. Furthermore, the advice confirms the need for monitoring, evaluation and ongoing research.

Under the title 'particularities', the commission remarked that, at the time of its advice, the Flemish authorities had already approved the EIA, so the EIA could be used in further decision-making in Flanders.

Differences in the EIA-legislation are the underlying reason and reveal 'another' transparency approach:

- In the Netherlands, the commission gives its advice after the EIA has been made public (for comments);
- In Flanders, the EIA becomes public after the approval. In this case, the commission had already given comments on the draft EIA, to assure a substantial congruency and allow the finalization of the Flemish EIA procedure.

CONCLUSION

In general, the Belgian federal and regional assessment systems reveal a certain degree of diversity with respect to the specific procedural elaboration of the main EIA and SEA phases as provided for in the EU EIA and SEA Directives. This intra-Belgian diversity is the result of the creation of regional jurisdictions, the inevitable outcome of the deconstruction of the unitary Belgian state. But also at the federal level, the different sectoral EIA systems have their own particularities, though they are mainly influenced by the international legal requirements. The public and advising institutions may also play a (rather limited) role at this stage. All systems implement the supranational requirements on TEIA and even SEA. The final review of the EIS or environmental report remains in Flanders an 'in house' operation for the environment administration that approves or rejects the EIS or environmental report, while in the Walloon region, some advising institutions can comment on the quality of the report.

In general terms, there is no coherent national system into which the different federal and regional EIA and SEA systems can fit. The poor implementation of the 1994 cooperation agreement on EIA illustrates a strong internal institutional border. As even the highest administrative court, the Council of State, is mostly functioning with single language chambers, there is no central Belgian public law 'guardian' of the impact assessment policy and practices, except for the Constitutional Court.[16]

It seems that the range of Belgian EIA and SEA systems have weakened the practice and definitely not supported the perception among outsiders of a 'strong EIA system'. The example of the Netherlands illustrates, nevertheless, that another approach, namely a consistent national EIA system that is branded as an effective tool in decision-making even without using accredited consultants, can enhance the export of internal knowledge, expertise and capacity.[17]

As a result of the experiences with the Flemish–Dutch arrangements (including the role of national institutions like the Dutch EIA Commission), it is obvious that transboundary impact assessment can provide 'institutional' inspiration even to remediate internal failures. In the current debate in Belgium about greater efficiency in public administrations, one may question the added value of the different EIA/SEA systems. In the search for acceptable solutions towards smooth governance, the creation of a 'joint body' as suggested in the Guidance on the Practical Application of the Espoo Convention and confirmed in practice (e.g. Albergaria and Fidelis, 2006), might be an option. This new institution could be the subject of a new cooperation agreement on impact assessment. To avoid the risk that the role of this joint body would be reduced to the management of procedures, a more substantial function should be the aim. We should not forget the fundamental role of impact assessment in increasing early transparency and accountability in decision-making (Nooteboom, 2007), so a Belgian Quality Board for Impact Assessment is desperately needed.

Notes

This chapter was originally published as: De Mulder, J. (2008) 'The Institutional Context for Transboundary Environmental Impact Assessment in Belgium: Multi-level Setting – A Matter of Smooth Governance?', *Impact Assessment and Project Appraisal*, vol 26, no 4, pp277–288.

1. Royal Order of 20 July 2001, see www.fanc.fgov.be/download/reglementation_20_07_2001_fr.pdf, last accessed 19 August 2008. The act was amended on 15 May 2007.
2. The key instruments in this context are the OSPAR Convention for the Protection of the Marine Environment of the North-East Atlantic (Paris, 1992), the Bonn Agreement for cooperation in dealing with pollution of the North Sea by oil and other harmful substances (1983) and the system of International Conferences on the Protection of the North Sea.
3. Royal Order of 7 September 2003, see www.mumm.ac.be/Downloads/MBBS170903 pp46101-46111.pdf, last accessed 19 August 2008.
4. ECE/MP.EIA/8, Guidance on the practical application of the Espoo Convention, UN, Geneva, 2006, 20.

5. www.mumm.ac.be/EN/Tools/viewimage.php?Pic=Atlas2007/WindmillsCurrent_large_EN.gif, last accessed 15 April 2008.
6. Advies van het bestuur aan de minister bevoegd voor de bescherming van het mariene milieu betreffende de machtigings-en vergunningsaanvraag van de n.v. Belwind voor de bouw en exploitatie van een windmolenpark op de Bligh Bank in de Noordzee (Advice/opinion from the federal marine protection administration regarding the EIA for the Belwind project to the competent federal minister), December 2007, available at www.mumm.ac.be, last accessed 20 May 2008.
7. Belgium approved its ratification on 9 June 1999, published in the Official Journal of 31 December 1999 (the Espoo Convention on EIA in a transboundary context came into force on 10 September 1997).
8. Published in Official Journal of 21 June 2007.
9. See http://wallex.wallonie.be/indexMain.html, last accessed 19 August 2008; look at: 'evaluation des incidences sur I'environnement – droit interne'.
10. Ordinance of 5 June 1997 (art. 70–78), see www.ibgebim.be/, last accessed 15 April 2008.
11. Official Journal of 20 June 2007; this decree amended the Decree on General Provisions regarding Environmental Policy, as amended by the Decree of 18 December 2002; see www.mervlaanderen.be, last accessed 20 May 2008.
12. Official Journal of 7 November 2007.
13. Official Journal of 11 August 1994.
14. ProSes was abolished as a separate organization by the end of 2007 and split into a Working Group 2010 DO and as also incorporated in the TSC.
15. www.isc-cie.com/index_nl.asp, last accessed 19 August 2008.
16. e.g. Judgement of 14 September 2006.
17. Private sector consultancies as well as experts working for or with the Dutch 'M.e.r. or EIA Commission' are worldwide active; an overview of their contributions in international reports and journals and their contributions to international meetings such as the IAIA illustrates this clearly.

REFERENCES

Albergaria, R. and T. Fidelis (2006) Transboundary EIA: Iberian experiences. Environmental Impact Assessment Review, 26, 614–632

Beattie, R. (1995) Everything you already know about EIA (but don't often admit). Environmental Impact Assessment Review, 15, 109.

Berkenbosch, R. (2002) Langetermijnvisie Schelde-estuarium. KenMERken, 9.

Cashmore, M., R. Gwilliam, R. Morgan, D. Cobb and A. Bond (2004) The interminable issue of effectiveness: substantive purposes, outcomes and research challenges in the advancement of environmental impact assessment theory. Impact Assessment and Project Appraisal, 22, 297.

Commissie-MER (2008) Verruiming vaargeul Beneden-Zeeschelde en Westerschelde Toetsingsadvies over het mi-lieueffectrapport, Utrecht, 1 februari / rapportnummer 1702–204.

De Hemptinne, J. (1994) La negociation: outil d' aide a la prise de decision et de reglement des conflits environnementaux. Revue Interdisciplinaire d' Etudes juridiques, 129–161.
Deketelaere, K. and F. Schutyser (2000) International Encyclopaedia of Laws: Environmental Law: Belgium. The Hague: Kluwer.
Douvere, F., F. Maes, A. Vanhulle and J. Schrijvers (2007) The role of marine spatial planning in sea use management: the Belgian case. Manne Policy, 31,182–191
Lavrysen, L. (1999) De bevoegdheidsverdeling in Belgie: Leefmi-lieu en waterbeleid. Brugge: Die Keure.
Nooteboom, S. (2007) Impact assessment procedures for sustainable development: a complexity theory perspective. Environmental Impact Assessment Review, 27, 645–665.
Van Dijk, J. and R. Nijsten (2004) Inspraak over grenzen. KenMERken, 21.
Van Schie, R. and M. Raessen (2001) Grensoverschrijdend adviseren. KenMERken, 11.
Zanting, H., P. Weijers and E. van Essen (2004) Strategisch MER Schelde-estuarium. KenMERken, October, 5.

Appendix 1:

Convention on Environmental Impact Assessment in a Transboundary Context, Espoo (Finland), 25 February 1991

The Parties to this Convention,

Aware of the interrelationship between economic activities and their environmental consequences,

Affirming the need to ensure environmentally sound and sustainable development,

Determined to enhance international co-operation in assessing environmental impact in particular in a transboundary context,

Mindful of the need and importance to develop anticipatory policies and of preventing, mitigating and monitoring significant adverse environmental impact in general and more specifically in a transboundary context,

Recalling the relevant provisions of the Charter of the United Nations, the Declaration of the Stockholm Conference on the Human Environment, the Final Act of the Conference on Security and Co-operation in Europe (CSCE) and the Concluding Documents of the Madrid and Vienna Meetings of Representatives of the Participating States of the CSCE,

Commending the ongoing activities of States to ensure that, through their national legal and administrative provisions and their national policies, environmental impact assessment is carried out,

Conscious of the need to give explicit consideration to environmental factors at an early stage in the decision-making process by applying environmental impact assessment, at all appropriate administrative levels, as a necessary tool to improve the quality of information presented to decision makers so that environmentally sound decisions can be made paying careful attention to minimizing significant adverse impact, particularly in a transboundary context,

Mindful of the efforts of international organizations to promote the use of environmental impact assessment both at the national and international levels, and taking into account work on environmental impact assessment carried out under the auspices of the United Nations Economic Commission for Europe, in particular results achieved by the Seminar on Environmental Impact Assessment (September 1987, Warsaw, Poland) as well as noting the Goals and Principles on environmental impact assessment adopted by the Governing Council of the United Nations Environment Programme, and the Ministerial Declaration on Sustainable Development (May 1990, Bergen, Norway),

Have agreed as follows:

Article 1

Definitions
For the purposes of this Convention,

- (i) 'Parties' means, unless the text otherwise indicates, the Contracting Parties to this Convention;
- (ii) 'Party of origin' means the Contracting Party or Parties to this Convention under whose jurisdiction a proposed activity is envisaged to take place;
- (iii) 'Affected Party' means the Contracting Party or Parties to this Convention likely to be affected by the transboundary impact of a proposed activity;
- (iv) 'Concerned Parties' means the Party of origin and the affected Party of an environmental impact assessment pursuant to this Convention;
- (v) 'Proposed activity' means any activity or any major change to an activity subject to a decision of a competent authority in accordance with an applicable national procedure;
- (vi) 'Environmental impact assessment' means a national procedure for evaluating the likely impact of a proposed activity on the environment;
- (vii) 'Impact' means any effect caused by a proposed activity on the environment including human health and safety, flora, fauna, soil, air, water, climate, landscape and historical monuments or other physical structures or the interaction among these factors; it also includes effects on cultural heritage or socio-economic conditions resulting from alterations to those factors;
- (viii) 'Transboundary impact' means any impact, not exclusively of a global nature, within an area under the jurisdiction of a Party caused by a proposed activity the physical origin of which is situated wholly or in part within the area under the jurisdiction of another Party;
- (ix) 'Competent authority' means the national authority or authorities designated by a Party as responsible for performing the tasks covered by this Convention and/or the authority or authorities entrusted by a Party with decision-making powers regarding a proposed activity;
- (x) 'The Public' means one or more natural or legal persons.

Article 2

General Provisions

1. The Parties shall, either individually or jointly, take all appropriate and effective measures to prevent, reduce and control significant adverse transboundary environmental impact from proposed activities.
2. Each Party shall take the necessary legal, administrative or other measures to implement the provisions of this Convention, including, with respect to proposed activities listed in Appendix I that are likely to cause significant adverse transboundary impact, the establishment of an environmental impact assessment procedure that permits public participation and preparation of the environmental impact assessment documentation described in Appendix II.
3. The Party of origin shall ensure that in accordance with the provisions of this Convention an environmental impact assessment is undertaken prior to a decision to authorize or undertake a proposed activity listed in Appendix I that is likely to cause a significant adverse transboundary impact.
4. The Party of origin shall, consistent with the provisions of this Convention, ensure that affected Parties are notified of a proposed activity listed in Appendix I that is likely to cause a significant adverse transboundary impact.
5. Concerned Parties shall, at the initiative of any such Party, enter into discussions on whether one or more proposed activities not listed in Appendix I is or are likely to cause a significant adverse transboundary impact and thus should be treated as if it or they were so listed. Where those Parties so agree, the activity or activities shall be thus treated. General guidance for identifying criteria to determine significant adverse impact is set forth in Appendix III.
6. The Party of origin shall provide, in accordance with the provisions of this Convention, an opportunity to the public in the areas likely to be affected to participate in relevant environmental impact assessment procedures regarding proposed activities and shall ensure that the opportunity provided to the public of the affected Party is equivalent to that provided to the public of the Party of origin.
7. Environmental impact assessments as required by this Convention shall, as a minimum requirement, be undertaken at the project level of the proposed activity. To the extent appropriate, the Parties shall endeavour to apply the principles of environmental impact assessment to policies, plans and programmes.
8. The provisions of this Convention shall not affect the right of Parties to implement national laws, regulations, administrative provisions or accepted

legal practices protecting information the supply of which would be prejudicial to industrial and commercial secrecy or national security.

9. The provisions of this Convention shall not affect the right of particular Parties to implement, by bilateral or multilateral agreement where appropriate, more stringent measures than those of this Convention.
10. The provisions of this Convention shall not prejudice any obligations of the Parties under international law with regard to activities having or likely to have a transboundary impact.

Article 3

Notification

1. For a proposed activity listed in Appendix I that is likely to cause a significant adverse transboundary impact, the Party of origin shall, for the purposes of ensuring adequate and effective consultations under Article 5, notify any Party which it considers may be an affected Party as early as possible and no later than when informing its own public about that proposed activity.
2. This notification shall contain, inter alia:

   (a) Information on the proposed activity, including any available information on its possible transboundary impact;
   (b) The nature of the possible decision; and
   (c) An indication of a reasonable time within which a response under paragraph 3 of this Article is required, taking into account the nature of the proposed activity;

and may include the information set out in paragraph 5 of this Article.

3. The affected Party shall respond to the Party of origin within the time specified in the notification, acknowledging receipt of the notification, and shall indicate whether it intends to participate in the environmental impact assessment procedure.
4. If the affected Party indicates that it does not intend to participate in the environmental impact assessment procedure, or if it does not respond within the time specified in the notification, the provisions in paragraphs 5, 6, 7 and 8 of this Article and in Articles 4 to 7 will not apply. In such circumstances the right of a Party of origin to determine whether to carry out an environmental impact assessment on the basis of its national law and practice is not prejudiced.
5. Upon receipt of a response from the affected Party indicating its desire to participate in the environmental impact assessment procedure, the Party of origin shall, if it has not already done so, provide to the affected Party:

(a) Relevant information regarding the environmental impact assessment procedure, including an indication of the time schedule for transmittal of comments; and
(b) Relevant information on the proposed activity and its possible significant adverse transboundary impact.

6. An affected Party shall, at the request of the Party of origin, provide the latter with reasonably obtainable information relating to the potentially affected environment under the jurisdiction of the affected Party, where such information is necessary for the preparation of the environmental impact assessment documentation. The information shall be furnished promptly and, as appropriate, through a joint body where one exists.
7. When a Party considers that it would be affected by a significant adverse transboundary impact of a proposed activity listed in Appendix I, and when no notification has taken place in accordance with paragraph 1 of this Article, the concerned Parties shall, at the request of the affected Party, exchange sufficient information for the purposes of holding discussions on whether there is likely to be a significant adverse transboundary impact. If those Parties agree that there is likely to be a significant adverse transboundary impact, the provisions of this Convention shall apply accordingly. If those Parties cannot agree whether there is likely to be a significant adverse transboundary impact, any such Party may submit that question to an inquiry commission in accordance with the provisions of Appendix IV to advise on the likelihood of significant adverse transboundary impact, unless they agree on another method of settling this question.
8. The concerned Parties shall ensure that the public of the affected Party in the areas likely to be affected be informed of, and be provided with possibilities for making comments or objections on, the proposed activity, and for the transmittal of these comments or objections to the competent authority of the Party of origin, either directly to this authority or, where appropriate, through the Party of origin.

Article 4

Preparation of the Environmental Impact Assessment Documentation

1. The environmental impact assessment documentation to be submitted to the competent authority of the Party of origin shall contain, as a minimum, the information described in Appendix II.
2. The Party of origin shall furnish the affected Party, as appropriate through a joint body where one exists, with the environmental impact assessment documentation. The concerned Parties shall arrange for distribution of the documentation to the authorities and the public of the affected Party in the

areas likely to be affected and for the submission of comments to the competent authority of the Party of origin, either directly to this authority or, where appropriate, through the Party of origin within a reasonable time before the final decision is taken on the proposed activity.

Article 5

Consultations on the Basis of the Environmental Impact Assessment Documentation

The Party of origin shall, after completion of the environmental impact assessment documentation, without undue delay enter into consultations with the affected Party concerning, inter alia, the potential transboundary impact of the proposed activity and measures to reduce or eliminate its impact. Consultations may relate to:

(a) Possible alternatives to the proposed activity, including the no-action alternative and possible measures to mitigate significant adverse transboundary impact and to monitor the effects of such measures at the expense of the Party of origin;
(b) Other forms of possible mutual assistance in reducing any significant adverse transboundary impact of the proposed activity; and
(c) Any other appropriate matters relating to the proposed activity.

The Parties shall agree, at the commencement of such consultations, on a reasonable time-frame for the duration of the consultation period. Any such consultations may be conducted through an appropriate joint body, where one exists.

Article 6

Final Decision

1. The Parties shall ensure that, in the final decision on the proposed activity, due account is taken of the outcome of the environmental impact assessment, including the environmental impact assessment documentation, as well as the comments thereon received pursuant to Article 3, paragraph 8 and Article 4, paragraph 2, and the outcome of the consultations as referred to in Article 5.
2. The Party of origin shall provide to the affected Party the final decision on the proposed activity along with the reasons and considerations on which it was based.
3. If additional information on the significant transboundary impact of a proposed activity, which was not available at the time a decision was made with respect to that activity and which could have materially affected the decision, becomes available to a concerned Party before work on that activity

commences, that Party shall immediately inform the other concerned Party or Parties. If one of the concerned Parties so requests, consultations shall be held as to whether the decision needs to be revised.

Article 7

Post-Project Analysis

1. The concerned Parties, at the request of any such Party, shall determine whether, and if so to what extent, a post-project analysis shall be carried out, taking into account the likely significant adverse transboundary impact of the activity for which an environmental impact assessment has been undertaken pursuant to this Convention. Any post-project analysis undertaken shall include, in particular, the surveillance of the activity and the determination of any adverse transboundary impact. Such surveillance and determination may be undertaken with a view to achieving the objectives listed in Appendix V.
2. When, as a result of post-project analysis, the Party of origin or the affected Party has reasonable grounds for concluding that there is a significant adverse transboundary impact or factors have been discovered which may result in such an impact, it shall immediately inform the other Party. The concerned Parties shall then consult on necessary measures to reduce or eliminate the impact.

Article 8

Bilateral and Multilateral Co-Operation
The Parties may continue existing or enter into new bilateral or multilateral agreements or other arrangements in order to implement their obligations under this Convention. Such agreements or other arrangements may be based on the elements listed in Appendix VI.

Article 9

Research Programmes
The Parties shall give special consideration to the setting up, or intensification of, specific research programmes aimed at:

(a) Improving existing qualitative and quantitative methods for assessing the impacts of proposed activities;

(b) Achieving a better understanding of cause-effect relationships and their role in integrated environmental management;
(c) Analysing and monitoring the efficient implementation of decisions on proposed activities with the intention of minimizing or preventing impacts;
(d) Developing methods to stimulate creative approaches in the search for environmentally sound alternatives to proposed activities, production and consumption patterns;
(e) Developing methodologies for the application of the principles of environmental impact assessment at the macro-economic level.

The results of the programmes listed above shall be exchanged by the Parties.

Article 10

Status of the Appendices
The Appendices attached to this Convention form an integral part of the Convention.

Article 11

Meeting of Parties

1. The Parties shall meet, so far as possible, in connection with the annual sessions of the Senior Advisers to ECE Governments on Environmental and Water Problems. The first meeting of the Parties shall be convened not later than one year after the date of the entry into force of this Convention. Thereafter, meetings of the Parties shall be held at such other times as may be deemed necessary by a meeting of the Parties, or at the written request of any Party, provided that, within six months of the request being communicated to them by the secretariat, it is supported by at least one third of the Parties.
2. The Parties shall keep under continuous review the implementation of this Convention, and, with this purpose in mind, shall:

   (a) Review the policies and methodological approaches to environmental impact assessment by the Parties with a view to further improving environmental impact assessment procedures in a transboundary context;
   (b) Exchange information regarding experience gained in concluding and implementing bilateral and multilateral agreements or other arrangements regarding the use of environmental impact assessment in a transboundary context to which one or more of the Parties are party;

(c) Seek, where appropriate, the services of competent international bodies and scientific committees in methodological and technical aspects pertinent to the achievement of the purposes of this Convention;
(d) At their first meeting, consider and by consensus adopt rules of procedure for their meetings;
(e) Consider and, where necessary, adopt proposals for amendments to this Convention;
(f) Consider and undertake any additional action that may be required for the achievement of the purposes of this Convention.

Article 12

Right to Vote

1. Each Party to this Convention shall have one vote.
2. Except as provided for in paragraph 1 of this Article, regional economic integration organizations, in matters within their competence, shall exercise their right to vote with a number of votes equal to the number of their member States which are Parties to this Convention. Such organizations shall not exercise their right to vote if their member States exercise theirs, and vice versa.

Article 13

Secretariat

The Executive Secretary of the Economic Commission for Europe shall carry out the following secretariat functions:

(a) The convening and preparing of meetings of the Parties;
(b) The transmission of reports and other information received in accordance with the provisions of this Convention to the Parties; and
(c) The performance of other functions as may be provided for in this Convention or as may be determined by the Parties.

Article 14

Amendments to the Convention

1. Any Party may propose amendments to this Convention.
2. Proposed amendments shall be submitted in writing to the secretariat, which shall communicate them to all Parties. The proposed amendments shall be discussed at the next meeting of the Parties, provided these proposals have

been circulated by the secretariat to the Parties at least ninety days in advance.

3. The Parties shall make every effort to reach agreement on any proposed amendment to this Convention by consensus. If all efforts at consensus have been exhausted, and no agreement reached, the amendment shall as a last resort be adopted by a three-fourths majority vote of the Parties present and voting at the meeting.
4. Amendments to this Convention adopted in accordance with paragraph 3 of this Article shall be submitted by the Depositary to all Parties for ratification, approval or acceptance. They shall enter into force for Parties having ratified, approved or accepted them on the ninetieth day after the receipt by the Depositary of notification of their ratification, approval or acceptance by at least three fourths of these Parties. Thereafter they shall enter into force for any other Party on the ninetieth day after that Party deposits its instrument of ratification, approval or acceptance of the amendments.
5. For the purpose of this Article, 'Parties present and voting' means Parties present and casting an affirmative or negative vote.
6. The voting procedure set forth in paragraph 3 of this Article is not intended to constitute a precedent for future agreements negotiated within the Economic Commission for Europe.

Article 15

Settlement of Disputes

1. If a dispute arises between two or more Parties about the interpretation or application of this Convention, they shall seek a solution by negotiation or by any other method of dispute settlement acceptable to the parties to the dispute.
2. When signing, ratifying, accepting, approving or acceding to this Convention, or at any time thereafter, a Party may declare in writing to the Depositary that for a dispute not resolved in accordance with paragraph 1 of this Article, it accepts one or both of the following means of dispute settlement as compulsory in relation to any Party accepting the same obligation:

   (a) Submission of the dispute to the International Court of Justice;
   (b) Arbitration in accordance with the procedure set out in Appendix VII.

3. If the parties to the dispute have accepted both means of dispute settlement referred to in paragraph 2 of this Article, the dispute may be submitted only to the International Court of Justice, unless the parties agree otherwise.

Article 16

Signature

This Convention shall be open for signature at Espoo (Finland) from 25 February to 1 March 1991 and thereafter at United Nations Headquarters in New York until 2 September 1991 by States members of the Economic Commission for Europe as well as States having consultative status with the Economic Commission for Europe pursuant to paragraph 8 of the Economic and Social Council resolution 36 (IV) of 28 March 1947, and by regional economic integration organizations constituted by sovereign States members of the Economic Commission for Europe to which their member States have transferred competence in respect of matters governed by this Convention, including the competence to enter into treaties in respect of these matters.

Article 17

Ratification, Acceptance, Approval and Accession

1. This Convention shall be subject to ratification, acceptance or approval by signatory States and regional economic integration organizations.
2. This Convention shall be open for accession as from 3 September 1991 by the States and organizations referred to in Article 16.
3. The instruments of ratification, acceptance, approval or accession shall be deposited with the Secretary-General of the United Nations, who shall perform the functions of Depositary.
4. Any organization referred to in Article 16 which becomes a Party to this Convention without any of its member States being a Party shall be bound by all the obligations under this Convention. In the case of such organizations, one or more of whose member States is a Party to this Convention, the organization and its member States shall decide on their respective responsibilities for the performance of their obligations under this Convention. In such cases, the organization and the member States shall not be entitled to exercise rights under this Convention concurrently.
5. In their instruments of ratification, acceptance, approval or accession, the regional economic integration organizations referred to in Article 16 shall declare the extent of their competence with respect to the matters governed by this Convention. These organizations shall also inform the Depositary of any relevant modification to the extent of their competence.

Article 18

Entry into Force

1. This Convention shall enter into force on the ninetieth day after the date of deposit of the sixteenth instrument of ratification, acceptance, approval or accession.
2. For the purposes of paragraph 1 of this Article, any instrument deposited by a regional economic integration organization shall not be counted as additional to those deposited by States members of such an organization.
3. For each State or organization referred to in Article 16 which ratifies, accepts or approves this Convention or accedes thereto after the deposit of the sixteenth instrument of ratification, acceptance, approval or accession, this Convention shall enter into force on the ninetieth day after the date of deposit by such State or organization of its instrument of ratification, acceptance, approval or accession.

Article 19

Withdrawal

At any time after four years from the date on which this Convention has come into force with respect to a Party, that Party may withdraw from this Convention by giving written notification to the Depositary. Any such withdrawal shall take effect on the ninetieth day after the date of its receipt by the Depositary. Any such withdrawal shall not affect the application of Articles 3 to 6 of this Convention to a proposed activity in respect of which a notification has been made pursuant to Article 3, paragraph 1, or a request has been made pursuant to Article 3, paragraph 7, before such withdrawal took effect.

Article 20

Authentic Texts

The original of this Convention, of which the English, French and Russian texts are equally authentic, shall be deposited with the Secretary-General of the United Nations.

IN WITNESS WHEREOF the undersigned, being duly authorized thereto, have signed this Convention.

DONE at Espoo (Finland), this twenty-fifth day of February one thousand nine hundred and ninety-one.

Appendices

Appendix I

List of Activities

1. Crude oil refineries (excluding undertakings manufacturing only lubricants from crude oil) and installations for the gasification and liquefaction of 500 tonnes or more of coal or bituminous shale per day.
2. Thermal power stations and other combustion installations with a heat output of 300 megawatts or more and nuclear power stations and other nuclear reactors (except research installations for the production and conversion of fissionable and fertile materials, whose maximum power does not exceed 1 kilowatt continuous thermal load).
3. Installations solely designed for the production or enrichment of nuclear fuels, for the reprocessing of irradiated nuclear fuels or for the storage, disposal and processing of radioactive waste.
4. Major installations for the initial smelting of cast-iron and steel and for the production of non-ferrous metals.
5. Installations for the extraction of asbestos and for the processing and transformation of asbestos and products containing asbestos: for asbestos-cement products, with an annual production of more than 20,000 tonnes finished product; for friction material, with an annual production of more than 50 tonnes finished product; and for other asbestos utilization of more than 200 tonnes per year.
6. Integrated chemical installations.
7. Construction of motorways, express roads*/and lines for long-distance railway traffic and of airports with a basic runway length of 2100 metres or more.
8. Large-diameter oil and gas pipelines.
9. Trading ports and also inland waterways and ports for inland-waterway traffic which permit the passage of vessels of over 1350 tonnes.
10. Waste-disposal installations for the incineration, chemical treatment or landfill of toxic and dangerous wastes.
11. Large dams and reservoirs.
12. Groundwater abstraction activities in cases where the annual volume of water to be abstracted amounts to 10 million cubic metres or more.
13. Pulp and paper manufacturing of 200 air-dried metric tonnes or more per day.
14. Major mining, on-site extraction and processing of metal ores or coal.
15. Offshore hydrocarbon production.
16. Major storage facilities for petroleum, petrochemical and chemical products.
17. Deforestation of large areas.

For the purposes of this Convention:

'Motorway' means a road specially designed and built for motor traffic, which does not serve properties bordering on it, and which:

(a) Is provided, except at special points or temporarily, with separate carriageways for the two directions of traffic, separated from each other by a dividing strip not intended for traffic or, exceptionally, by other means;
(b) Does not cross at level with any road, railway or tramway track, or footpath; and
(c) Is specially sign-posted as a motorway.

'Express road' means a road reserved for motor traffic accessible only from interchanges or controlled junctions and on which, in particular, stopping and parking are prohibited on the running carriageway(s).

Appendix II

Content of the Environmental Impact Assessment Documentation
Information to be included in the environmental impact assessment documentation shall, as a minimum, contain, in accordance with Article 4:

(a) A description of the proposed activity and its purpose;
(b) A description, where appropriate, of reasonable alternatives (for example, locational or technological) to the proposed activity and also the no-action alternative;
(c) A description of the environment likely to be significantly affected by the proposed activity and its alternatives;
(d) A description of the potential environmental impact of the proposed activity and its alternatives and an estimation of its significance;
(e) A description of mitigation measures to keep adverse environmental impact to a minimum;
(f) An explicit indication of predictive methods and underlying assumptions as well as the relevant environmental data used;
(g) An identification of gaps in knowledge and uncertainties encountered in compiling the required information;
(h) Where appropriate, an outline for monitoring and management programmes and any plans for post-project analysis; and
(i) A non-technical summary including a visual presentation as appropriate (maps, graphs, etc.).

Appendix III

General Criteria to Assist in the Determination of the Environmental Significance of Activities Not Listed in Appendix I

1. In considering proposed activities to which Article 2, paragraph 5, applies, the concerned Parties may consider whether the activity is likely to have a significant adverse transboundary impact in particular by virtue of one or more of the following criteria:

   (a) Size: proposed activities which are large for the type of the activity;
   (b) Location: proposed activities which are located in or close to an area of special environmental sensitivity or importance (such as wetlands designated under the Ramsar Convention, national parks, nature reserves, sites of special scientific interest, or sites of archaeological, cultural or historical importance); also, proposed activities in locations where the characteristics of proposed development would be likely to have significant effects on the population;
   (c) Effects: proposed activities with particularly complex and potentially adverse effects, including those giving rise to serious effects on humans or on valued species or organisms, those which threaten the existing or potential use of an affected area and those causing additional loading which cannot be sustained by the carrying capacity of the environment.

2. The concerned Parties shall consider for this purpose proposed activities which are located close to an international frontier as well as more remote proposed activities which could give rise to significant transboundary effects far removed from the site of development.

Appendix IV

Inquiry Procedure

1. The requesting Party or Parties shall notify the secretariat that it or they submit(s) the question of whether a proposed activity listed in Appendix I is likely to have a significant adverse transboundary impact to an inquiry commission established in accordance with the provisions of this Appendix. This notification shall state the subject-matter of the inquiry. The secretariat shall notify immediately all Parties to this Convention of this submission.
2. The inquiry commission shall consist of three members. Both the requesting party and the other party to the inquiry procedure shall appoint a scientific

or technical expert, and the two experts so appointed shall designate by common agreement the third expert, who shall be the president of the inquiry commission. The latter shall not be a national of one of the parties to the inquiry procedure, nor have his or her usual place of residence in the territory of one of these parties, nor be employed by any of them, nor have dealt with the matter in any other capacity.

3. If the president of the inquiry commission has not been designated within two months of the appointment of the second expert, the Executive Secretary of the Economic Commission for Europe shall, at the request of either party, designate the president within a further two-month period.
4. If one of the parties to the inquiry procedure does not appoint an expert within one month of its receipt of the notification by the secretariat, the other party may inform the Executive Secretary of the Economic Commission for Europe, who shall designate the president of the inquiry commission within a further two-month period. Upon designation, the president of the inquiry commission shall request the party which has not appointed an expert to do so within one month. After such a period, the president shall inform the Executive Secretary of the Economic Commission for Europe, who shall make this appointment within a further two-month period.
5. The inquiry commission shall adopt its own rules of procedure.
6. The inquiry commission may take all appropriate measures in order to carry out its functions.
7. The parties to the inquiry procedure shall facilitate the work of the inquiry commission and, in particular, using all means at their disposal, shall:

   (a) Provide it with all relevant documents, facilities and information; and
   (b) Enable it, where necessary, to call witnesses or experts and receive their evidence.

8. The parties and the experts shall protect the confidentiality of any information they receive in confidence during the work of the inquiry commission.
9. If one of the parties to the inquiry procedure does not appear before the inquiry commission or fails to present its case, the other party may request the inquiry commission to continue the proceedings and to complete its work. Absence of a party or failure of a party to present its case shall not constitute a bar to the continuation and completion of the work of the inquiry commission.
10. Unless the inquiry commission determines otherwise because of the particular circumstances of the matter, the expenses of the inquiry commission, including the remuneration of its members, shall be borne by the parties to the inquiry procedure in equal shares. The inquiry commission shall keep a record of all its expenses, and shall furnish a final statement thereof to the parties.
11. Any Party having an interest of a factual nature in the subject-matter of the inquiry procedure, and which may be affected by an opinion in the matter,

may intervene in the proceedings with the consent of the inquiry commission.

12. The decisions of the inquiry commission on matters of procedure shall be taken by majority vote of its members. The final opinion of the inquiry commission shall reflect the view of the majority of its members and shall include any dissenting view.
13. The inquiry commission shall present its final opinion within two months of the date on which it was established unless it finds it necessary to extend this time limit for a period which should not exceed two months.
14. The final opinion of the inquiry commission shall be based on accepted scientific principles. The final opinion shall be transmitted by the inquiry commission to the parties to the inquiry procedure and to the secretariat.

Appendix V

Post-Project Analysis
Objectives include:

(a) Monitoring compliance with the conditions as set out in the authorization or approval of the activity and the effectiveness of mitigation measures;
(b) Review of an impact for proper management and in order to cope with uncertainties;
(c) Verification of past predictions in order to transfer experience to future activities of the same type.

Appendix VI

Elements for Bilateral and Multilateral Co-Operation

1. Concerned Parties may set up, where appropriate, institutional arrangements or enlarge the mandate of existing institutional arrangements within the framework of bilateral and multilateral agreements in order to give full effect to this Convention.
2. Bilateral and multilateral agreements or other arrangements may include:
   (a) Any additional requirements for the implementation of this Convention, taking into account the specific conditions of the subregion concerned;
   (b) Institutional, administrative and other arrangements, to be made on a reciprocal and equivalent basis;
   (c) Harmonization of their policies and measures for the protection of the environment in order to attain the greatest possible similarity in standards and methods related to the implementation of environmental impact assessment;

(d) Developing, improving, and/or harmonizing methods for the identification, measurement, prediction and assessment of impacts, and for post-project analysis;
(e) Developing and/or improving methods and programmes for the collection, analysis, storage and timely dissemination of comparable data regarding environmental quality in order to provide input into environmental impact assessment;
(f) The establishment of threshold levels and more specified criteria for defining the significance of transboundary impacts related to the location, nature or size of proposed activities, for which environmental impact assessment in accordance with the provisions of this Convention shall be applied; and the establishment of critical loads of transboundary pollution;
(g) Undertaking, where appropriate, joint environmental impact assessment, development of joint monitoring programmes, intercalibration of monitoring devices and harmonization of methodologies with a view to rendering the data and information obtained compatible.

Appendix VII

Arbitration

1. The claimant Party or Parties shall notify the secretariat that the Parties have agreed to submit the dispute to arbitration pursuant to Article 15, paragraph 2, of this Convention. The notification shall state the subject-matter of arbitration and include, in particular, the Articles of this Convention, the interpretation or application of which are at issue. The secretariat shall forward the information received to all Parties to this Convention.
2. The arbitral tribunal shall consist of three members. Both the claimant Party or Parties and the other Party or Parties to the dispute shall appoint an arbitrator, and the two arbitrators so appointed shall designate by common agreement the third arbitrator, who shall be the president of the arbitral tribunal. The latter shall not be a national of one of the parties to the dispute, nor have his or her usual place of residence in the territory of one of these parties, nor be employed by any of them, nor have dealt with the case in any other capacity.
3. If the president of the arbitral tribunal has not been designated within two months of the appointment of the second arbitrator, the Executive Secretary of the Economic Commission for Europe shall, at the request of either party to the dispute, designate the president within a further two-month period.

4. If one of the parties to the dispute does not appoint an arbitrator within two months of the receipt of the request, the other party may inform the Executive Secretary of the Economic Commission for Europe, who shall designate the president of the arbitral tribunal within a further two-month period. Upon designation, the president of the arbitral tribunal shall request the party which has not appointed an arbitrator to do so within two months. After such a period, the president shall inform the Executive Secretary of the Economic Commission for Europe, who shall make this appointment within a further two-month period.
5. The arbitral tribunal shall render its decision in accordance with international law and in accordance with the provisions of this Convention.
6. Any arbitral tribunal constituted under the provisions set out herein shall draw up its own rules of procedure.
7. The decisions of the arbitral tribunal, both on procedure and on substance, shall be taken by majority vote of its members.
8. The tribunal may take all appropriate measures in order to establish the facts.
9. The parties to the dispute shall facilitate the work of the arbitral tribunal and, in particular, using all means at their disposal, shall:

   (a) Provide it with all relevant documents, facilities and information; and
   (b) Enable it, where necessary, to call witnesses or experts and receive their evidence.

10. The parties and the arbitrators shall protect the confidentiality of any information they receive in confidence during the proceedings of the arbitral tribunal.
11. The arbitral tribunal may, at the request of one of the parties, recommend interim measures of protection.
12. If one of the parties to the dispute does not appear before the arbitral tribunal or fails to defend its case, the other party may request the tribunal to continue the proceedings and to render its final decision. Absence of a party or failure of a party to defend its case shall not constitute a bar to the proceedings. Before rendering its final decision, the arbitral tribunal must satisfy itself that the claim is well founded in fact and law.
13. The arbitral tribunal may hear and determine counter-claims arising directly out of the subject-matter of the dispute.
14. Unless the arbitral tribunal determines otherwise because of the particular circumstances of the case, the expenses of the tribunal, including the remuneration of its members, shall be borne by the parties to the dispute in equal shares. The tribunal shall keep a record of all its expenses, and shall furnish a final statement thereof to the parties.

15. Any Party to this Convention having an interest of a legal nature in the subject-matter of the dispute, and which may be affected by a decision in the case, may intervene in the proceedings with the consent of the tribunal.
16. The arbitral tribunal shall render its award within five months of the date on which it is established unless it finds it necessary to extend the time limit for a period which should not exceed five months.
17. The award of the arbitral tribunal shall be accompanied by a statement of reasons. It shall be final and binding upon all parties to the dispute. The award will be transmitted by the arbitral tribunal to the parties to the dispute and to the secretariat. The secretariat will forward the information received to all Parties to this Convention.
18. Any dispute which may arise between the parties concerning the interpretation or execution of the award may be submitted by either party to the arbitral tribunal which made the award or, if the latter cannot be seized thereof, to another tribunal constituted for this purpose in the same manner as the first.

Appendix 2:

Protocol on Strategic Environmental Assessment to the Convention on Environmental Impact Assessment in a Transboundary Context, Kiev (Ukraine), 21 May 2003

The Parties to this Protocol

Recognizing the importance of integrating environmental, including health, considerations into the preparation and adoption of plans and programmes and, to the extent appropriate, policies and legislation,

Committing themselves to promo-ting sustainable development and therefore basing themselves on the conclusions of the United Nations Conference on Environment and Development (Rio de Janeiro, Brazil, 1992), in particular principles 4 and 10 of the Rio Declaration on Environment and Development and Agenda 21, as well as the outcome of the third Ministerial Conference on Environment and Health (London, 1999) and the World Summit on Sustainable Development (Johannesburg, South Africa, 2002),

Bearing in mind the Convention on Environmental Impact Assessment in a Transboundary Context, done at Espoo, Finland, on 25 February 1991, and decision II/9 of its Parties at Sofia on 26 and 27 February 2001, in which it was decided to prepare a legally binding protocol on strategic environmental assessment,

Recognizing that strategic environmental assessment should have an important role in the preparation and adoption of plans, programmes, and, to the extent appropriate, policies and legislation, and that the wider application of the principles of environmental impact assessment to plans, programmes, policies and legislation will further strengthen the systematic analysis of their significant environmental effects,

Acknowledging the Convention on Access to Information, Public Participation in Decision-making and Access to Justice in Environmental Matters, done at Aarhus, Denmark, on 25 June 1998, and taking note of the relevant paragraphs of the Lucca Declaration, adopted at the first meeting of its Parties,

Conscious, therefore, of the importance of providing for public participation in strategic environmental assessment,

Acknowledging the benefits to the health and well-being of present and future generations that will follow if the need to protect and improve people's health is taken into account as an integral part of strategic environmental assessment, and recognizing the work led by the World Health Organization in this respect,

Mindful of the need for and importance of enhancing international cooperation in assessing the transboundary environmental, including health, effects of proposed plans and programmes, and, to the extent appropriate, policies and legislation,

Have agreed as follows:

Article 1

Objective

The objective of this Protocol is to provide for a high level of protection of the environment, including health, by:

(a) Ensuring that environmental, including health, considera-tions are thoroughly taken into account in the development of plans and programmes;
(b) Contributing to the consideration of environmen-tal, including health, concerns in the preparation of policies and legislation;
(c) Establishing clear, transparent and effective procedures for strategic environmental assessment;
(d) Providing for public participation in strategic environmental assessment; and
(e) Integrating by these means environmental, including health, concerns into measures and instruments designed to further sustainable development.

Article 2

Definitions

For the purposes of this Protocol,

1. 'Convention' means the Convention on Environmental Impact Assessment in a Transboundary Context.
2. 'Party' means, unless the text indicates otherwise, a Contracting Party to this Protocol.
3. 'Party of origin' means a Party or Parties to this Protocol within whose jurisdiction the preparation of a plan or programme is envisaged.
4. 'Affected Party' means a Party or Parties to this Protocol likely to be affected by the transboundary environmental, including health, effects of a plan or programme.
5. 'Plans and programmes' means plans and programmes and any modifications to them that are:

   (a) Required by legislative, regulatory or administrative provisions; and
   (b) Subject to preparation and/or adoption by an authority or prepared by an authority for adoption, through a formal procedure, by a parliament or a government.

6. 'Strategic environmental assessment' means the evaluation of the likely environmental, including health, effects, which comprises the determination of the scope of an environmental report and its preparation, the carrying-out of public participation and consultations, and the taking into account of the environmental report and the results of the public participation and consultations in a plan or programme.
7. 'Environmental, including health, effect' means any effect on the environment, including human health, flora, fauna, biodiversity, soil, climate, air, water, landscape, natural sites, material assets, cultural heritage and the interaction among these factors.
8. 'The public' means one or more natural or legal persons and, in accordance with national legislation or practice, their associations, organizations or groups.

Article 3

General provisions

1. Each Party shall take the necessary legislative, regulatory and other appropriate measures to implement the provisions of this Protocol within a clear, transparent framework.
2. Each Party shall endeavour to ensure that officials and authorities assist and provide guidance to the public in matters covered by this Protocol.
3. Each Party shall provide for appropriate recognition of and support to associations, organizations or groups promoting environmental, including health, protection in the context of this Protocol.
4. The provisions of this Protocol shall not affect the right of a Party to maintain or introduce additional measures in relation to issues covered by this Protocol.
5. Each Party shall promote the objectives of this Protocol in relevant international decisionmaking processes and within the framework of relevant international organizations.
6. Each Party shall ensure that persons exercising their rights in conformity with the provisions of this Protocol shall not be penalized, persecuted or harassed in any way for their involvement. This provision shall not affect the powers of national courts to award reasonable costs in judicial proceedings.
7. Within the scope of the relevant provisions of this Protocol, the public shall be able to exercise its rights without discrimination as to citizenship, nationality or domicile and, in the case of a legal person, without discrimination as to where it has its registered seat or an effective centre of its activities.

Article 4

Field of application concerning plans and programmes

1. Each Party shall ensure that a strategic environmental assessment is carried out for plans and programmes referred to in paragraphs 2, 3 and 4

which are likely to have significant environmental, including health, effects.

2. A strategic environmental assessment shall be carried out for plans and programmes which are prepared for agriculture, forestry, fisheries, energy, industry including mining, transport, regional development, waste management, water management, telecommunications, tourism, town and country planning or land use, and which set the framework for future development consent for projects listed in annex I and any other project listed in annex II that requires an environmental impact assessment under national legislation.
3. For plans and programmes other than those subject to paragraph 2 which set the framework for future development consent of projects, a strategic environmental assessment shall be carried out where a Party so determines according to article 5, paragraph 1.
4. For plans and programmes referred to in paragraph 2 which determine the use of small areas at local level and for minor modifications to plans and programmes referred to in paragraph 2, a strategic environmental assessment shall be carried out only where a Party so determines according to article 5, paragraph 1.
5. The following plans and programmes are not subject to this Protocol:

   (a) Plans and programmes whose sole purpose is to serve national defence or civil emergencies;
   (b) Financial or budget plans and programmes.

Article 5

Screening

1. Each Party shall determine whether plans and programmes referred to in article 4, paragraphs 3 and 4, are likely to have significant environmental, including health, effects either through a case-by-case examination or by specifying types of plans and programmes or by combining both approaches. For this purpose each Party shall in all cases take into account the criteria set out in annex III.
2. Each Party shall ensure that the environmental and health authorities referred to in article 9, paragraph 1, are consulted when applying the procedures referred to in paragraph 1 above.
3. To the extent appropriate, each Party shall endeavour to provide opportunities for the participation of the public concerned in the screening of plans and programmes under this article.
4. Each Party shall ensure timely public availability of the conclusions pursuant to paragraph 1, including the reasons for not requiring a strategic environmental assessment, whether by public notices or by other appropriate means, such as electronic media.

Article 6

Scoping

1. Each Party shall establish arrange-ments for the determination of the relevant information to be included in the environmental report in accordance with article 7, paragraph 2.
2. Each Party shall ensure that the environmental and health authorities referred to in article 9, paragraph 1, are consulted when determining the relevant information to be included in the environmental report.
3. To the extent appropriate, each Party shall endeavour to provide opportunities for the participation of the public concerned when determining the relevant information to be included in the environmental report.

Article 7

Environmental report

1. For plans and programmes subject to strategic environmental assessment, each Party shall ensure that an environmental report is prepared.
2. The environmental report shall, in accordance with the determination under article 6, identify, describe and evaluate the likely significant environmental, including health, effects of implementing the plan or programme and its reasonable alternatives. The report shall contain such information specified in annex IV as may reasonably be required, taking into account:

    (a) Current knowledge and methods of assessment;
    (b) The contents and the level of detail of the plan or programme and its stage in the decision-making process;
    (c) The interests of the public; and
    (d) The information needs of the decision-making body.

3. Each Party shall ensure that environmental reports are of sufficient quality to meet the requirements of this Protocol.

Article 8

Public participation

1. Each Party shall ensure early, timely and effective opportunities for public participation, when all options are open, in the strategic environmental assessment of plans and programmes.
2. Each Party, using electronic media or other appropriate means, shall ensure the timely public availability of the draft plan or programme and the environmental report.

3. Each Party shall ensure that the public concerned, including relevant non-governmental organizations, is identified for the purposes of paragraphs 1 and 4.
4. Each Party shall ensure that the public referred to in paragraph 3 has the opportunity to express its opinion on the draft plan or programme and the environmental report within a reasonable time frame.
5. Each Party shall ensure that the detailed arrangements for informing the public and consulting the public concerned are determined and made publicly available. For this purpose, each Party shall take into account to the extent appropriate the elements listed in annex V.

Article 9

Consultation with environmental and health authorities

1. Each Party shall designate the authorities to be consulted which, by reason of their specific environmental or health responsibilities, are likely to be concerned by the environmental, including health, effects of the implementation of the plan or programme.
2. The draft plan or programme and the environmental report shall be made available to the authorities referred to in paragraph 1.
3. Each Party shall ensure that the authorities referred to in paragraph 1 are given, in an early, timely and effective manner, the opportunity to express their opinion on the draft plan or programme and the environmental report.
4. Each Party shall determine the detailed arrangements for informing and consulting the environmental and health authorities referred to in paragraph 1.

Article 10

Transboundary consultations

1. Where a Party of origin considers that the implementation of a plan or programme is likely to have significant transboundary environmental, including health, effects or where a Party likely to be significantly affected so requests, the Party of origin shall as early as possible before the adoption of the plan or programme notify the affected Party.
2. This notification shall contain, inter alia:

   (a) The draft plan or programme and the environmental report including information on its possible transboundary environmental, including health, effects; and
   (b) Information regarding the decision-making procedure, including an indication of a reasonable time schedule for the transmission of comments.

3. The affected Party shall, within the time specified in the notification, indicate to the Party of origin whether it wishes to enter into consultations before the adoption of the plan or programme and, if it so indicates, the Parties concerned shall enter into consultations concerning the likely transboundary environmental, including health, effects of implementing the plan or programme and the measures envisaged to prevent, reduce or mitigate adverse effects.
4. Where such consultations take place, the Parties concerned shall agree on detailed arrangements to ensure that the public concerned and the authorities referred to in article 9, paragraph 1, in the affected Party are informed and given an opportunity to forward their opinion on the draft plan or programme and the environmental report within a reasonable time frame.

Article 11

Decision

1. Each Party shall ensure that when a plan or programme is adopted due account is taken of:

   (a) The conclusions of the environmental report;
   (b) The measures to prevent, reduce or mitigate the adverse effects identified in the environmental report; and
   (c) The comments received in accordance with articles 8 to 10.

2. Each Party shall ensure that, when a plan or programme is adopted, the public, the authorities referred to in article 9, paragraph 1, and the Parties consulted according to article 10 are informed, and that the plan or programme is made available to them together with a statement summarizing how the environmental, including health, considerations have been integrated into it, how the comments received in accordance with articles 8 to 10 have been taken into account and the reasons for adopting it in the light of the reasonable alternatives considered.

Article 12

Monitoring

1. Each Party shall monitor the significant environmental, including health, effects of the implementation of the plans and programmes, adopted under article 11 in order, inter alia, to identify, at an early stage, unforeseen adverse effects and to be able to undertake appropriate remedial action.
2. The results of the monitoring undertaken shall be made available, in accordance with national legislation, to the authorities referred to in article 9, paragraph 1, and to the public.

Article 13

Policies and legislation

1. Each Party shall endeavour to ensure that environmental, including health, concerns are considered and integrated to the extent appropriate in the preparation of its proposals for policies and legislation that are likely to have significant effects on the environment, including health.
2. In applying paragraph 1, each Party shall consider the appropriate principles and elements of this Protocol.
3. Each Party shall determine, where appropriate, the practical arrangements for the consideration and integration of environmental, including health, concerns in accordance with paragraph 1, taking into account the need for transparency in decision-making.
4. Each Party shall report to the Meeting of the Parties to the Convention serving as the Meeting of the Parties to this Protocol on its application of this article.

Article 14

The Meeting of the Parties to the Convention serving as the meeting of the parties to the protocol

1. The Meeting of the Parties to the Convention shall serve as the Meeting of the Parties to this Protocol. The first meeting of the Parties to the Convention serving as the Meeting of the Parties to this Protocol shall be convened not later than one year after the date of entry into force of this Protocol, and in conjunction with a meeting of the Parties to the Convention, if a meeting of the latter is scheduled within that period. Subsequent meetings of the Parties to the Convention serving as the Meeting of the Parties to this Protocol shall be held in conjunction with meetings of the Parties to the Convention, unless otherwise decided by the Meeting of the Parties to the Convention serving as the Meeting of the Parties to this Protocol.
2. Parties to the Convention which are not Parties to this Protocol may participate as observers in the proceedings of any session of the Meeting of the Parties to the Convention serving as the Meeting of the Parties to this Protocol. When the Meeting of the Parties to the Convention serves as the Meeting of the Parties to this Protocol, decisions under this Protocol shall be taken only by the Parties to this Protocol.
3. When the Meeting of the Parties to the Convention serves as the Meeting of the Parties to this Protocol, any member of the Bureau of the Meeting of the Parties representing a Party to the Convention that is not, at that time, a Party to this Protocol shall be replaced by another member to be elected by and from amongst the Parties to this Protocol.

4. The Meeting of the Parties to the Convention serving as the Meeting of the Parties to this Protocol shall keep under regular review the implementation of this Protocol and, for this purpose, shall:

   (a) Review policies for and methodological approaches to strategic environmental assessment with a view to further improving the procedures provided for under this Protocol;
   (b) Exchange information regarding experience gained in strategic environmental assessment and in the implementation of this Protocol;
   (c) Seek, where appropriate, the services and cooperation of competent bodies having expertise pertinent to the achievement of the purposes of this Protocol;
   (d) Establish such subsidiary bodies as it considers necessary for the implementation of this Protocol;
   (e) Where necessary, consider and adopt proposals for amendments to this Protocol; and
   (f) Consider and undertake any additional action, including action to be carried out jointly under this Protocol and the Convention, that may be required for the achievement of the purposes of this Protocol.

5. The rules of procedure of the Meeting of the Parties to the Convention shall be applied mutatis mutandis under this Protocol, except as may otherwise be decided by consensus by the Meeting of the Parties serving as the Meeting of the Parties to this Protocol.
6. At its first meeting, the Meeting of the Parties to the Convention serving as the Meeting of the Parties to this Protocol shall consider and adopt the modalities for applying the procedure for the review of compliance with the Convention to this Protocol.
7. Each Party shall, at intervals to be determined by the Meeting of the Parties to the Convention serving as the Meeting of the Parties to this Protocol, report to the Meeting of the Parties to the Convention serving as the Meeting of the Parties to the Protocol on measures that it has taken to implement the Protocol.

Article 15

Relationship to other international agreements
The relevant provisions of this Protocol shall apply without prejudice to the UNECE Conventions on Environmental Impact Assessment in a Transboundary Context and on Access to Information, Public Participation in Decision-making and Access to Justice in Environmental Matters.

Article 16

Right to vote

1. Except as provided for in paragraph 2 below, each Party to this Protocol shall have one vote.

2. Regional economic integration organizations, in matters within their competence, shall exercise their right to vote with a number of votes equal to the number of their member States which are Parties to this Protocol. Such organizations shall not exercise their right to vote if their member States exercise theirs, and vice versa.

Article 17

Secretariat
The secretariat established by article 13 of the Convention shall serve as the secretariat of this Protocol and article 13, paragraphs (a) to (c), of the Convention on the functions of the secretariat shall apply mutatis mutandis to this Protocol.

Article 18

Annexes
The annexes to this Protocol shall constitute an integral part thereof.

Article 19

Amendments to the Protocol

1. Any Party may propose amendments to this Protocol.
2. Subject to paragraph 3, the procedure for proposing, adopting and the entry into force of amendments to the Convention laid down in paragraphs 2 to 5 of article 14 of the Convention shall apply, mutatis mutandis, to amendments to this Protocol.
3. For the purpose of this Protocol, the three fourths of the Parties required for an amendment to enter into force for Parties having ratified, approved or accepted it, shall be calculated on the basis of the number of Parties at the time of the adoption of the amendment.

Article 20

Settlement of disputes
The provisions on the settlement of disputes of article 15 of the Convention shall apply mutatis mutandis to this Protocol.

Article 21

Signature
This Protocol shall be open for signature at Kiev (Ukraine) from 21 to 23 May 2003 and thereafter at United Nations Headquarters in New York until 31 December 2003, by States members of the Economic Commission for Europe as well as States having consultative status with the Economic Commission for Europe pursuant to paragraphs 8 and 11 of Economic and Social Council resolution 36 (IV) of 28 March

1947, and by regional economic integration organizations constituted by sovereign States members of the Economic Commission for Europe to which their member States have transferred competence over matters governed by this Protocol, including the competence to enter into treaties in respect of these matters.

Article 22

Depositary
The Secretary-General of the United Nations shall act as the Depositary of this Protocol.

Article 23

Ratification, acceptance, approval and accession

1. This Protocol shall be subject to ratification, acceptance or approval by signatory States and regional economic integration organizations referred to in article 21.
2. This Protocol shall be open for accession as from 1 January 2004 by the States and regional economic integration organizations referred to in article 21.
3. Any other State, not referred to in paragraph 2 above, that is a Member of the United Nations may accede to the Protocol upon approval by the Meeting of the Parties to the Convention serving as the Meeting of the Parties to the Protocol.
4. Any regional economic integration organization referred to in article 21 which becomes a Party to this Protocol without any of its member States being a Party shall be bound by all the obligations under this Protocol. If one or more of such an organization's member States is a Party to this Protocol, the organization and its member States shall decide on their respective responsibilities for the performance of their obligations under this Protocol. In such cases, the organization and its member States shall not be entitled to exercise rights under this Protocol concurrently.
5. In their instruments of ratification, acceptance, approval or accession, the regional economic integration organizations referred to in article 21 shall declare the extent of their competence with respect to the matters governed by this Protocol. These organizations shall also inform the Depositary of any relevant modification to the extent of their competence.

Article 24

Entry into force

1. This Protocol shall enter into force on the ninetieth day after the date of deposit of the sixteenth instrument of ratification, acceptance, approval or accession.
2. For the purposes of paragraph 1 above, any instrument deposited by a regional economic integration organization referred to in article 21 shall not be counted as additional to those deposited by States members of such an organization.

3. For each State or regional economic integration organization referred to in article 21 which ratifies, accepts or approves this Protocol or accedes thereto after the deposit of the sixteenth instrument of ratification, acceptance, approval or accession, the Protocol shall enter into force on the ninetieth day after the date of deposit by such State or organization of its instrument of ratification, acceptance, approval or accession.
4. This Protocol shall apply to plans, programmes, policies and legislation for which the first formal preparatory act is subsequent to the date on which this Protocol enters into force. Where the Party under whose jurisdiction the preparation of a plan, programme, policy or legislation is envisaged is one for which paragraph 3 applies, this Protocol shall apply to plans, programmes, policies and legislation for which the first formal preparatory act is subsequent to the date on which this Protocol comes into force for that Party.

Article 25

Withdrawal

At any time after four years from the date on which this Protocol has come into force with respect to a Party, that Party may withdraw from the Protocol by giving written notification to the Depositary. Any such withdrawal shall take effect on the ninetieth day after the date of its receipt by the Depositary. Any such withdrawal shall not affect the application of articles 5 to 9, 11 and 13 with respect to a strategic environmental assessment under this Protocol which has already been started, or the application of article 10 with respect to a notification or request which has already been made, before such withdrawal takes effect.

Article 26

Authentic texts

The original of this Protocol, of which the English, French and Russian texts are equally authentic, shall be deposited with the Secretary-General of the United Nations.

IN WITNESS WHEREOF the undersigned, being duly authorized thereto, have signed this Protocol.

ANNEXES

Annex I

List of projects as referred to in article 4, paragraph 2

1. Crude oil refineries (excluding undertakings manufacturing only lubricants from crude oil) and installations for the gasification and liquefaction of 500 metric tons or more of coal or bituminous shale per day.

2. Thermal power stations and other combustion installations with a heat output of 300 megawatts or more and nuclear power stations and other nuclear reactors (except research installations for the production and conversion of fissionable and fertile materials, whose maximum power does not exceed 1 kilowatt continuous thermal load).
3. Installations solely designed for the production or enrichment of nuclear fuels, for the reprocessing of irradiated nuclear fuels or for the storage, disposal and processing of radioactive waste.
4. Major installations for the initial smelting of cast-iron and steel and for the production of non-ferrous metals.
5. Installations for the extraction of asbestos and for the processing and transformation of asbestos and products containing asbestos: for asbestos-cement products, with an annual production of more than 20,000 metric tons of finished product; for friction material, with an annual production of more than 50 metric tons of finished product; and for other asbestos utilization of more than 200 metric tons per year.
6. Integrated chemical installations.
7. Construction of motorways, express roads*/and lines for long-distance railway traffic and of airports**/with a basic runway length of 2100 metres or more.
8. Large-diameter oil and gas pipelines.
9. Trading ports and also inland waterways and ports for inland-waterway traffic which permit the passage of vessels of over 1350 metric tons.
10. Waste-disposal installations for the incineration, chemical treatment or landfill of toxic and dangerous wastes.
11. Large dams and reservoirs.
12. Groundwater abstraction activities in cases where the annual volume of water to be abstracted amounts to 10 million cubic metres or more.
13. Pulp and paper manufacturing of 200 air-dried metric tons or more per day.
14. Major mining, on-site extraction and processing of metal ores or coal.
15. Offshore hydrocarbon production.
16. Major storage facilities for petroleum, petrochemical and chemical products.
17. Deforestation of large areas.

- For the purposes of this Protocol:

  'Motorway' means a road specially designed and built for motor traffic, which does not serve properties bordering on it, and which:

(a) Is provided, except at special points or temporarily, with separate carriageways for the two directions of traffic, separated from each other by a dividing strip not intended for traffic or, exceptionally, by other means;
(b) Does not cross at level with any road, railway or tramway track, or footpath; and
(c) Is specially sign posted as a motorway.

'Express road' means a road reserved for motor traffic accessible only from interchanges or controlled junctions and on which, in particular, stopping and parking are prohibited on the running carriageway(s).

- For the purposes of this Protocol, 'airport' means an airport which complies with the definition in the 1944 Chicago Convention setting up the International Civil Aviation Organization (annex 14).

Annex II

Any other projects referred to in article 4, paragraph 2

1. Projects for the restructuring of rural land holdings.
2. Projects for the use of uncultivated land or semi-natural areas for intensive agricultural purposes.
3. Water management projects for agriculture, including irrigation and land drainage projects.
4. Intensive livestock installations (including poultry).
5. Initial afforestation and defores-tation for the purposes of conversion to another type of land use.
6. Intensive fish farming.
7. Nuclear power stations and other nuclear reactors*/including the dismantling or decommissioning of such power stations or reactors (except research installations for the production and conversion of fissionable and fertile materials whose maximum power does not exceed 1 kilowatt continuous thermal load), as far as not included in annex I.
8. Construction of overhead electrical power lines with a voltage of 220 kilovolts or more and a length of 15 kilometres or more and other projects for the transmission of electrical energy by overhead cables.
9. Industrial installations for the production of electricity, steam and hot water.
10. Industrial installations for carrying gas, steam and hot water.
11. Surface storage of fossil fuels and natural gas.
12. Underground storage of combus-tible gases.
13. Industrial briquetting of coal and lignite.
14. Installations for hydroelectric energy production.
15. Installations for the harnessing of wind power for energy production (wind farms).
16. Installations, as far as not included in annex I, designed:
    - For the production or enrichment of nuclear fuel;
    - For the processing of irradiated nuclear fuel;
    - For the final disposal of irradiated nuclear fuel;
    - Solely for the final disposal of radioactive waste;

- Solely for the storage (planned for more than 10 years) of irradiated nuclear fuels in a different site than the production site; or
- For the processing and storage of radioactive waste.

17. Quarries, open cast mining and peat extraction, as far as not included in annex I.
18. Underground mining, as far as not included in annex I.
19. Extraction of minerals by marine or fluvial dredging.
20. Deep drillings (in particular geothermal drilling, drilling for the storage of nuclear waste material, drilling for water supplies), with the exception of drillings for investigating the stability of the soil.
21. Surface industrial installations for the extraction of coal, petroleum, natural gas and ores, as well as bituminous shale.
22. Integrated works for the initial smelting of cast iron and steel, as far as not included in annex I.
23. Installations for the production of pig iron or steel (primary or secondary fusion) including continuous casting.
24. Installations for the processing of ferrous metals (hot-rolling mills, smitheries with hammers, application of protective fused metal coats).
25. Ferrous metal foundries.
26. Installations for the production of non-ferrous crude metals from ore, concentrates or secondary raw materials by metallurgical, chemical or electrolytic processes, as far as not included in annex I.
27. Installations for the smelting, including the alloyage, of non-ferrous metals excluding precious metals, including recovered products (refining, foundry casting, etc.), as far as not included in annex I.
28. Installations for surface treatment of metals and plastic materials using an electrolytic or chemical process.
29. Manufacture and assembly of motor vehicles and manufacture of motor-vehicle engines.
30. Shipyards.
31. Installations for the construction and repair of aircraft.
32. Manufacture of railway equipment.
33. Swaging by explosives.
34. Installations for the roasting and sintering of metallic ores.
35. Coke ovens (dry coal distillation).
36. Installations for the manufacture of cement.
37. Installations for the manufacture of glass including glass fibre.
38. Installations for smelting mineral substances including the production of mineral fibres.
39. Manufacture of ceramic products by burning, in particular roofing tiles, bricks, refractory bricks, tiles, stoneware or porcelain.
40. Installations for the production of chemicals or treatment of intermediate products, as far as not included in annex I.

41. Production of pesticides and pharmaceutical products, paint and varnishes, elastomers and peroxides.
42. Installations for the storage of petroleum, petrochemical, or chemical products, as far as not included in annex I.
43. Manufacture of vegetable and animal oils and fats.
44. Packing and canning of animal and vegetable products.
45. Manufacture of dairy products.
46. Brewing and malting.
47. Confectionery and syrup manufacture.
48. Installations for the slaughter of animals.
49. Industrial starch manufacturing installations.
50. Fish-meal and fish-oil factories.
51. Sugar factories.
52. Industrial plants for the production of pulp, paper and board, as far as not included in annex I.
53. Plants for the pre treatment or dyeing of fibres or textiles.
54. Plants for the tanning of hides and skins.
55. Cellulose-processing and production installations.
56. Manufacture and treatment of elastomer-based products.
57. Installations for the manufacture of artificial mineral fibres.
58. Installations for the recovery or destruction of explosive substances.
59. Installations for the production of asbestos and the manufacture of asbestos products, as far as not included in annex I.
60. Knackers' yards.
61. Test benches for engines, turbines or reactors.
62. Permanent racing and test tracks for motorized vehicles.
63. Pipelines for transport of gas or oil, as far as not included in annex I.
64. Pipelines for transport of chemicals with a diameter of more than 800 mmand a length of more than 40 km.
65. Construction of railways and intermodal trans-shipment facilities, and of intermodal terminals, as far as not included in annex I.
66. Construction of tramways, elevated and underground railways, suspended lines or similar lines of a particular type used exclusively or mainly for passenger transport.
67. Construction of roads, including realignment and/or widening of any existing road, as far as not included in annex I.
68. Construction of harbours and port installations, including fishing harbours, as far as not included in annex I.
69. Construction of inland waterways and ports for inland-waterway traffic, as far as not included in annex I.
70. Trading ports, piers for loading and unloading connected to land and outside ports, as far as not included in annex I.
71. Canalization and flood-relief works.

72. Construction of airports[**]/and airfields, as far as not included in annex I.
73. Waste-disposal installations (including landfill), as far as not included in annex I.
74. Installations for the incineration or chemical treatment of non-hazardous waste.
75. Storage of scrap iron, including scrap vehicles.
76. Sludge deposition sites.
77. Groundwater abstraction or artificial groundwater recharge, as far as not included in annex I.
78. Works for the transfer of water resources between river basins.
79. Waste-water treatment plants.
80. Dams and other installations designed for the holding-back or for the long-term or permanent storage of water, as far as not included in annex I.
81. Coastal work to combat erosion and maritime works capable of altering the coast through the construction, for example, of dykes, moles, jetties and other sea defence works, excluding the maintenance and reconstruction of such works.
82. Installations of long-distance aqueducts.
83. Ski runs, ski lifts and cable cars and associated developments.
84. Marinas.
85. Holiday villages and hotel complexes outside urban areas and associated developments.
86. Permanent campsites and caravan sites.
87. Theme parks.
88. Industrial estate development projects.
89. Urban development projects, including the construction of shopping centres and car parks.
90. Reclamation of land from the sea.

- For the purposes of this Protocol, nuclear power stations and other nuclear reactors cease to be such an installation when all nuclear fuel and other radioactively contaminated elements have been removed permanently from the installation site.
- For the purposes of this Protocol, 'airport' means an airport which complies with the definition in the 1944 Chicago Convention setting up the International Civil Aviation Organization (annex 14).

Annex III

Criteria for determining of the likely significant environmental, including health, effects referred to in article 5, paragraph 1

1. The relevance of the plan or programme to the integration of environmental, including health, considerations in particular with a view to promoting sustainable development.

2. The degree to which the plan or programme sets a framework for projects and other activities, either with regard to location, nature, size and operating conditions or by allocating resources.
3. The degree to which the plan or programme influences other plans and programmes including those in a hierarchy.
4. Environmental, including health, problems relevant to the plan or programme.
5. The nature of the environmental, including health, effects such as probability, duration, frequency, reversibility, magnitude and extent (such as geographical area or size of population likely to be affected).
6. The risks to the environment, including health.
7. The transboundary nature of effects.
8. The degree to which the plan or programme will affect valuable or vulnerable areas including lands-capes with a recognized national or international protection status.

Annex IV

Information referred to in article 7, paragraph 2

1. The contents and the main objectives of the plan or programme and its link with other plans or programmes.
2. The relevant aspects of the current state of the environment, including health, and the likely evolution thereof should the plan or programme not be implemented.
3. The characteristics of the environment, including health, in areas likely to be significantly affected.
4. The environmental, including health, problems which are relevant to the plan or programme.
5. The environmental, including health, objectives established at international, national and other levels which are relevant to the plan or programme, and the ways in which these objectives and other environmental, including health, considerations have been taken into account during its preparation.
6. The likely significant environmental, including health, effects*/as defined in article 2, paragraph 7.
7. Measures to prevent, reduce or mitigate any significant adverse effects on the environment, including health, which may result from the implementation of the plan or programme.
8. An outline of the reasons for selecting the alternatives dealt with and a description of how the assessment was undertaken including difficulties encountered in providing the information to be included such as technical deficiencies or lack of knowledge.
9. Measures envisaged for monitoring environmental, including health, effects of the implementation of the plan or programme.

10. The likely significant transboun-dary environmental, including health, effects.
11. A non-technical summary of the information provided.

- These effects should include secondary, cumulative, synergistic, short-, medium- and long-term, permanent and temporary, positive and negative effects.

Annex V

Information referred to in article 8, paragraph 5.

1. The proposed plan or programme and its nature.
2. The authority responsible for its adoption.
3. The envisaged procedure, including:

   (a) The commencement of the procedure;
   (b) The opportunities for the public to participate;
   (c) The time and venue of any envisaged public hearing;
   (d) The authority from which relevant information can be obtained and where the relevant information has been deposited for examination by the public;
   (e) The authority to which comments or questions can be submitted and the time schedule for the transmittal of comments or questions; and
   (f) What environmental, including health, information relevant to the proposed plan or programme is available.

4. Whether the plan or programme is likely to be subject to a transboundary assessment procedure.

Index